Palgrave Studies in Medieval and Early Modern Medicine

Series Editors

Jonathan Barry
Department of History
University of Exeter
Exeter, UK

Fabrizio Bigotti
Institute for the History of Medicine
Julius Maximilian University
Würzburg, Germany

CENTRE FOR THE STUDY OF MEDICINE AND THE BODY IN THE RENAISSANCE

The series focuses on the intellectual tradition of western medicine as related to the philosophies, institutions, practices, and technologies that developed throughout the medieval and early modern period (500-1800). Partnered with the Centre for the Study of Medicine and the Body in the Renaissance (CSMBR), it seeks to explore the range of interactions between various conceptualisations of the body, including their import for the arts (e.g. literature, painting, music, dance, and architecture) and the way different medical traditions overlapped and borrowed from each other. The series particularly welcomes contributions from young authors. The editors will consider proposals for single monographs, as well as edited collections and translations/editions of texts, either at a standard length (70-120,000 words) or as Palgrave Pivots (up to 50,000 words).

Justin Begley · Benjamin Goldberg

The Medical World of Margaret Cavendish

A Critical Edition

palgrave
macmillan

Justin Begley
Ludwig-Maximilians-Universität
München, Germany

Benjamin Goldberg
Department of Humanities
and Cultural Studies
University of South Florida
Tampa, FL, USA

ISSN 2524-7387 ISSN 2524-7395 (electronic)
Palgrave Studies in Medieval and Early Modern Medicine
ISBN 978-3-030-92926-8 ISBN 978-3-030-92927-5 (eBook)
https://doi.org/10.1007/978-3-030-92927-5

Cover design by eStudio Calamar

This Palgrave Macmillan imprint is published by the registered company Springer Nature Switzerland AG
The registered company address is: Gewerbestrasse 11, 6330 Cham, Switzerland

Acknowledgements

Prompted by a footnote in Hugh Trevor-Roper's *Europe's Physician* (2006), Justin Begley first encountered Pw V90 in 2014. At that time, he was working on a D.Phil. on Margaret Cavendish at the University of Oxford, and he has since taken regular trips to Nottingham to transcribe Pw V90 along with a number of other Cavendish-related manuscripts. Begley fortuitously met Benjamin Goldberg at the Folger Shakespeare Library in early 2017, and together they decided to embark upon the production of this edition. Both authors would like to thank the University of Nottingham Library as well as the British Library for agreeing to let them publish the material in this volume. They would also like to thank the librarians in Nottingham who have been exceptionally helpful in arranging trips to the archive and providing information about the manuscript and its history. Particular thanks go out to Jayne Amat, Hayley Cotterill, and Nicholas Blake. Because of Covid-19 induced travel restrictions, it was not possible for either author to visit Nottingham for most of 2020 and 2021. As a result, Mark Dorrington kindly arranged for Begley to meet Amat, Blake, or Rachael Orchard twice a week for over a month to finalise the transcription; for this, both authors are also extremely grateful.

In terms of institutional support, both authors have received generous funding over the last few years that has made substantial work on this volume possible. Begley would like to thank the University of Helsinki for funding a research trip to the University of Oxford in early 2019, at which point he also visited the University of Nottingham and discovered Harley MS 6491 in the British Library. Begley subsequently transcribed Harley 6491 as a Lisa Jardine Fellow of the Royal Society of London in 2020. He also worked on components of the introduction during a fellowship at the Institute for Research in the Humanities at the University of Bucharest in 2018, and completed the edition in 2021 with fellowships at the Herzog August Bibliothek and the Gotha Research Centre. Goldberg would like to thank the University of South Florida and especially the Department of Humanities and Cultural Studies for travel and research funding, as well as the US Fulbright Program, which provided an opportunity to work on this project with reduced teaching responsibilities. Goldberg's research was also supported by a Folger Shakespeare Library summer fellowship.

Over the past few years, Begley and Goldberg have presented material from Pw V90 at a number of conferences. Begley presented on components of the manuscript at several conferences in 2017 including at the Folger Research Seminar; *Humours, Mixtures, and Corpuscles* at the Fondazione Comel in Pisa; and *Scientiae* at the University of Padua. In 2018, Begley organised a panel at the *Renaissance Society of America Conference* in New Orleans at which he and Goldberg both presented on Cavendish and medicine, along with Lisa Sarasohn. Begley and Goldberg also delivered joint papers in 2019 at the *Second International Conference on Historical Medical Discourse* at the University of Helsinki and the *International Margaret Cavendish Society Conference* at the Norwegian University of Science and Technology in Trondheim. Further, Goldberg presented material relating to Pw V90 for the Studies in Science and Technology Colloquium at the University of Minnesota, Minneapolis, in 2018, as well as at the *Scientiae* 2019 conference in Belfast, and at an online conference on Margaret Cavendish's philosophy hosted by Trinity College Dublin in 2020.

Questions and comments from and discussions at these conferences have contributed immensely to this edition. In addition, the authors would like to single out Peter Elmer, William Poole, Patrick Wallis, Anne Stobart, Elaine Leong, Michelle DiMeo, Liza Blake, Dedo von Kerssenbrock-Krosigk, Joel Klein, and William Newman for being especially generous with their time. This edition also benefitted greatly from the feedback of two anonymous referees. Finally, it would not have been possible to bring this book to fruition without the help and guidance of the series editors, Jonathan Barry and Fabrizio Bigotti.

Contents

List of Figures

List of Tables

CHAPTER 1

Introduction

Among the papers in the University of Nottingham's "Portland-Welbeck" collection is a manuscript labelled "Pw V90". Behind this reference marker lies a fascinating "Booke wherein is Contain=ed Rare Minerall Receipts Collected at Paris from those Who hath had great Experience of them", which William and Margaret Cavendish, Marquess and Marchioness of Newcastle upon Tyne (and later Duke and Duchess of Newcastle), compiled between approximately 1647 and 1654. Totalling 196 small folio pages, this document contains an impressive range of medicinal recipes; a raft of letters from physicians and amateurs alike that counsel William, Margaret, and their kin on how best to maintain and improve their health; and various epistles that address broader physiological issues. The current volume provides the first transcription of Pw V90, which is not digitised and therefore can only be viewed in Nottingham, accompanied by a set of introductory chapters that situate the manuscript within its social, political, cultural, medical, and wider intellectual contexts.

As the above outline indicates, Pw V90 affords unique insights into the health, concerns, and ideas of William and Margaret, whose centrality to the social, medical, and intellectual life of mid-seventeenth-century England, and especially of the exiled Royalists during the Interregnum, means that the edition should have much to offer a wide array of scholars,

J. Begley and B. Goldberg, *The Medical World of Margaret Cavendish*, Palgrave Studies in Medieval and Early Modern Medicine,
https://doi.org/10.1007/978-3-030-92927-5_1

from historians of science, medicine, and politics to philosophers and literary critics. William, as is well known, was an acclaimed Royalist general who was stationed in Northern England during the early years of the British Civil Wars but fled to the Continent after his notorious defeat at the Battle of Marston Moor in July 1644. Having arrived on the Continent, he courted and soon married (in 1645) Margaret Lucas, who was one of Queen Henrietta Maria's ladies-in-waiting at the exiled English court in Paris. In the wake of the seminal publications of Lisa Sarasohn, Londa Schiebinger, and Sarah Hutton, scholars have gradually come to appreciate the historical significance of Margaret's literary, philosophical, and scientific output.[1] She has even recently been dubbed a "canonical" philosopher.[2] Yet the medical facets of Margaret's *oeuvre*, and the contexts in which they were initially formulated, are still in need of detailed analysis.[3] More generally, her intellectual concerns in the years leading up

[1] Lisa Sarasohn (1984), A Science Turned Upside Down: Feminism and the Natural Philosophy of Margaret Cavendish. *The Huntington Library Quarterly* 47.4: 289–307; Londa Schiebinger (1991), Margaret Cavendish, Duchess of Newcastle. *A History of Women Philosophers: Modern Women Philosophers, 1600–1900*, ed. Mary Ellen Waithe (Boston), 1–20; Sarah Hutton (1997), Anne Conway, Margaret Cavendish and Seventeenth-Century Scientific Thought. *Women, Science and Medicine 1500–1700*, eds. Lynette Hunter and Sarah Hutton (Thrupp), 218–234. Other important early discussions include Eileen O'Neill (1997), Disappearing Ink: Early Modern Women Philosophers and Their Fate in History. *Philosophy in a Feminist Voice: Critiques and Reconstructions*, ed. Janet A. Kourany (Princeton), 17–62; Susan James (1999), The Philosophical Innovations of Margaret Cavendish. *British Journal for the History of Philosophy* 7.2: 219–244; Stephen Clucas, ed. (2003), *A Princely Brave Woman: Essays on Margaret Cavendish, Duchess of Newcastle* (Aldershot). The most recent monographs on Cavendish's natural philosophy are Deborah Boyle (2017), *The Well-Ordered Universe: The Philosophy of Margaret Cavendish* (Oxford); David Cunning (2016), *Cavendish* (London); Brandie Siegfried and Lisa Sarasohn, eds. (2014), *God and Nature in the Thought of Margaret Cavendish* (Farnham); and Lisa Sarasohn (2010), *The Natural Philosophy of Margaret Cavendish: Reason and Fancy During the Scientific Revolution* (Baltimore).

[2] Alison Peterman (2018), Canonizing Cavendish. *HOPOS: The Journal of the International Society for the History of Philosophy of Science* 8.1: 191–197.

[3] But do see Boyle, *The Well-Ordered Universe*, 215–238; Lisa Walters (2014), *Margaret Cavendish: Gender, Science and Politics* (Cambridge), 138–194; Justin Begley (2016), *Margaret Cavendish, The Last Natural Philosopher* (Dissertation, University of Oxford), 90–154; Justin Begley (2017), "The minde is matter moved": Nehemiah Grew on Margaret Cavendish. *Intellectual History Review* 27.4: 493–514; Laura L. Knoppers (2019), Gender, Knowledge, and the Medical Marketplace: The Case of Margaret Cavendish. *The Routledge Research Companion to Women, Sex, and Gender in the Early British Colonial World*, eds. Kimberly A. Coles and Eve Keller (New York), 270–285; and

to her major publications of the early 1650s remain little understood, in part owing to the neglect of manuscript evidence. Paying careful attention to the material in Pw V90, the introductory chapters reconstruct the medical webs in which William and Margaret were enmeshed, and draw out some of the implications of these connections for understanding Margaret's thought. It is our hope that the transcription will stimulate further investigations into this important and fertile domain.

As we seek to show in our discussion of Pw V90, this document is a fascinating and distinctive hybrid of a recipe book and a letter collection that seamlessly blends Galenic and chymical cures, and that is more learned, discursive, and pan-European than many comparable documents that English individuals produced around the same time. These factors, along with the vast array of figures whom it features, make Pw V90 a powerful microcosm of mid-seventeenth-century courtly medicine. Of particular interest are the remedies and wide-ranging pieces of counsel from a plethora of British and French physicians and amateurs, including "Europe's physician" and the doctor to various English and French monarchs, Théodore de Mayerne; the explorer and self-styled Duke of Northumberland Robert Dudley; and the renowned pirate, chymist, and philosopher Kenelm Digby. Alongside the contributions of these names sit the recipes and advice of a spectrum of lesser-known practitioners, such as the wonderfully named English apothecary Calixt Rust and a Parisian apothecary known as "Mr Basincotes". The cast of characters also features medically attuned women of varied social standing, from an unnamed "Nun, att Antwerpe" to Elizabeth Throckmorton, Lady Raleigh, whose clandestine marriage to Sir Water Raleigh was viewed as a scandal in the court of Queen Elizabeth I.

Despite its diversity, the above and the biographies found in Appendix 8D point to the fact that Pw V90 is an unmistakably *Royalist* manuscript. For this reason, another objective of the introductory chapters is to complicate some of the bold assertions in Charles Webster's classic 1975 volume, *The Great Instauration*, about the relationship between medicine and political allegiance in mid-seventeenth-century England. According to Webster, who built upon the "Merton Thesis", it was not so much Anglican and Catholic Royalists as "puritans" who drove advances in

Benjamin Goldberg (2017), Epigenesis and the Rationality of Nature in William Harvey and Margaret Cavendish. *History and Philosophy of the Life Sciences* 39.8: 1–23.

seventeenth-century science and medicine.[4] Webster's ideas have had a lasting impact, with Roy Porter, for instance, contending that "Radical Puritans in the English Revolution (1642–60) wanted to transform the nation's institutions 'root and branch'" and especially "the existing medical Profession".[5] Various scholars have sought to nuance Webster's hypothesis by chalking up finer distinctions between religious and political communities. Andrew Wear, for instance, distinguishes between the views of several Protestant sects and associates chymistry with the more radical of the groups, while noting that, insofar as the primary aim of medicine was the utilitarian one of healing, medical pursuits necessarily extended across social, religious, and political lines.[6] Seeking to redraw confessional boundaries more fundamentally, Peter Elmer links Catholics and puritans as "outsiders" who had a central role in fostering mid-seventeenth-century medical progress, with an emphasis on how a figure who is central to this study, "the catholic Kenelm Digby", happily worked "alongside men of such radical persuasion as John French", the chymical physician and supporter of Oliver Cromwell.[7] Without going so far as to claim that Royalists made *the* key medical interventions of the period, or even that they formed a well bounded group, the introductory material in this edition demonstrates that numerous significant medical practitioners in England and beyond were inscribed within Royalist networks.[8] At the

[4] Along with Charles Webster (1975), *The Great Instauration: Science, Medicine, and Reform, 1626–1660* (London), the other classic study is Christopher Hill (1974), The Medical Profession and Its Radical Critics. *Change and Continuity in Seventeenth-Century England*, ed. Christopher Hill (London), 157–178. For the original formulation of what has become known as the Merton Thesis, see Robert K. Merton (1938), Science, Technology and Society in Seventeenth Century England. *Osiris* 4: 360–632.

[5] Roy Porter (1995), *Disease, Medicine, and Society in England, 1550–1860* (Cambridge), 5.

[6] Andrew Wear (1996), Religious Beliefs and Medicine in Early Modern England. *The Task of Healing: Medicine, Religion and Gender in England and the Netherlands 1450–1800*, eds. Hilary Marland and Margaret Pelling (Rotterdam), 145–171.

[7] Peter Elmer (1989), Medicine, Religion, and the Puritan Revolution. *The Medical Revolution of the Seventeenth Century*, eds. Andrew Wear and Roger French (Cambridge), 10–45 (43).

[8] In the background here are also issues surrounding the relations between Royalists, the English Universities, and medical theory and pedagogy. On these topics, see Mordechai Feingold (1984), *The Mathematician's Apprenticeship: Science, Universities and Society in England 1560–1640* (Cambridge). On the education of physicians in particular, see Ian Maclean (2002), *Logic, Signs and Nature in the Renaissance: The Case of*

same time, it suggests that political and religious allegiances can very often be disentangled: most of the individuals in Pw V90 had strong Royalist ties, but their religious loyalties spanned from Huguenot to Catholic. What united them were shared political allegiances in a time of crisis, as well as medical and intellectual projects, not least the collecting and sharing of medical recipes.[9] As we will see, Pw V90 does not engage as directly with politics as many printed recipe collections of the time, and is a testament to the fact that domains well beyond politics, with medical science being chief among them, underpinned Royalist networks.

Not only does this edition give a sense of the nature and composition of Royalist medical networks, but it also supplements, extends, and, in some cases, challenges the existing literature on several Royalist individuals beyond William and Margaret who are now widely recognised to have been central to the intellectual cultures of their day, especially Mayerne and Digby.[10] One letter, for example, in which Digby claims to have acquired his preferred universal cure, the "powder of vipers", from a skilled snake-catcher known as "Siginor Antonio Manfradi", provides tantalising evidence for his self-fashioning as a pirate-traveller. While Hugh Trevor-Roper's perceptive and far-reaching biography of Mayerne tracks this venerable physician's relationship to many of the highest-profile individuals in seventeenth-century political and intellectual life, his work only mentions Pw V90 in passing, and, accordingly, some fascinating aspects of Mayerne's practice are overlooked. This includes the fact that Rust was Mayerne's long-term apothecary and personal assistant, and that

Learned Medicine (Cambridge), especially 14–35 and 68–100 and Jonathan Barry (2019), Educating Physicians in Seventeenth-Century England. *Science in Context* 32: 137–154.

[9] Most analyses of Royalism and recipes have focused on a 1655 collection, *The Queens Closet Opened*, on which see Jayne Archer (2002), The Queens' Arcanum: Authority and Authorship in *The Queens Closet Opened* (1655). *Renaissance Journal* 1.6: 14–25; Laura L. Knoppers (2007), Opening the Queen's Closet: Henrietta Maria, Elizabeth Cromwell, and the Politics of Cookery. *Renaissance Quarterly* 60.2: 464–499; Edith Snook (2007), 'Soveraigne Receipts' and the Politics of Beauty in *The Queens Closet Opened. Early Modern Literary Studies* 15: 1–19; and Madeline Bassnett (2009), Restoring the Royal Household: Royalist Politics and the Commonwealth Recipe Book. *Early English Studies* 2: 1–32.

[10] For the significance and popular appeal of these thinkers, see Hugh Trevor-Roper (2006), *Europe's Physician: The Various Life of Sir Theodore de Mayerne* (New Haven); Danielle Dutton (2016), *Margaret the First: A Novel* (Melbourne); and Joe Moshenska (2016), *A Stain in the Blood: The Remarkable Voyage of Sir Kenelm Digby, Pirate and Poet, Courtier and Cook, King's Servant and Traitor's Son* (London).

Mayerne's chief remedy, at least in later life, was the *anima hepatis* or "salt of steel".[11]

As much as this edition offers fresh insights into the ideas and practices of major seventeenth-century figures, it also aims to raise the several little-known apothecaries and non-professionals who feature in it out of obscurity, and, in a wider historiographical sense, to offer counter-evidence to the still pervasive assumption that learned physicians and apothecaries were continually at odds from the 1630s onwards.[12] Dmitri Levitin and Jonathan Barry, among others, have convincingly shown that apothecaries of this period were more learned than was once supposed, and that the lines that divided them from physicians were by no means clear-cut.[13] Pw V90 substantiates this more nuanced picture of the relationship between medical practitioners of different social and institutional standings, not least by highlighting the close ties and productive interactions between Mayerne and Rust. More generally, rather than suggesting that dichotomies, clashes, and strict hierarchies spurred developments in seventeenth-century medicine, it foregrounds the tightly interwoven

[11] Trevor-Roper only mentions Pw V90 in passing at *Europe's Physician*, 374. The more technical work on Mayerne's thought, Brian Nance (2001), *Turquet de Mayerne as Baroque Physician* (Amsterdam), does not mention Pw V90 and mostly focuses on the more than twenty manuscripts of Mayerne's casebooks, "Ephemerides morborum" ("Diaries of Disease"). Mayerne's "'steel' in powdered form" is briefly mentioned in Elizabeth Furdell (2001), *The Royal Doctors, 1485–1714: Medical Personnel at the Tudor and Stuart Courts* (Rochester), 104; it is not discussed in Nance's work, despite the extensive discussion of medical theory (51, 52, 69, 71, 105, 129) and therapeutics (155–169, 194, 198). It would be useful in the future to compare the recipes and medicines in Pw V90 with "Ephemerides morborum". For a list of "Ephemerides", see Nance, *Turquet de Mayerne*, "Appendix One: A Guide to the Ephemerides morborum", 201–204.

[12] The seminal works here are Piyo M. Rattansi (1963), Paracelsus and the Puritan Revolution. *Ambix* 11.1: 24–32; Piyo M. Rattansi (1964), The Helmontian-Galenist Controversy in Restoration England. *Ambix* 12.1: 1–23; Charles Webster (1967), English Medical Reformers of the Puritan Revolution: A Background to the "Society of Chymical Physitians". *Ambix* 14: 16–41; and Christopher Hill (1972), *The World Turned Upside Down: Radical Ideas During the English Revolution* (London), 231–246. For a discussion of how Galenic and chymical medicine melded before the 1640s, see Allen Debus (1965), *The English Paracelsians* (London).

[13] See Dmitri Levitin (2015), "Made Up from Many Experimentall Notions": The Society of Apothecaries, Medical Humanism, and the Rhetoric of Experience in 1630s London. *Journal of the History of Medicine and Allied Sciences* 70.4: 549–587 and Barry, Educating Physicians in Seventeenth-Century England.

bonds that underpinned the medical marketplace.[14] While these ties are on show in the regular and personal correspondences in Pw V90 between individuals from different social backgrounds, they are also evident from instances where the social hierarchy is subverted on the basis of superior medical knowledge. Indeed, it was perhaps only within the medical domain that seventeenth-century individuals of relatively low social standing, such as apothecaries, could successfully defy the commands of the aristocrats whom they served. We see a transparent instance of such insubordination in Pw V90 when Rust, fearing that William was over administering a given remedy, informed him that "I Receaved yor Hon.^{rs} Commands for some Crocus=metal=lorum, which I have sent; But not in those quantities yor Lord.^{ip} directed, Because they are too great".[15]

As the fact that Rust was able to defy the social hierarchy by condemning the behaviour of a marquess suggests, medical knowledge had currency in the Cavendishes' world. But this should not be taken to mean that knowledge was free from social significance, and this edition shows that recipes were in fact often carefully devised to reflect (and improve) their inventor's social or intellectual status. Rust's cost-effective and overwhelmingly herbal treatments, for instance, differ markedly from those of a high-status figure such as the Duke of Northumberland, whose chymical cure-all was deliberately composed of many expensive and exotic ingredients that demonstrated his rank and worldliness.[16] By providing transcriptions of the manifold recipes and cures contained within Pw V90 and explanatory notes on them, this edition offers a tangible sense not only of how the Cavendishes selected judiciously from the resources at their disposal and deployed them in the pursuit of health and well-being, but also of how their affiliates used the gifting of recipes for personal advancement.

[14] On the issue of medical clashes, see Patrick Wallis (2007), Competition and Cooperation in the Early Modern Medical Economy. *Medicine and the Market in England and Its Colonies, c. 1450-c. 1850*, eds. Mark Jenner and Patrick Wallis (New York), 47–68 and Adrien Wilson (2007), Midwifery in the 'Medical Marketplace'. *Medicine and the Market in England and Its Colonies, c. 1450-c. 1850*, eds. Mark Jenner and Patrick Wallis (New York), 153–174.

[15] Transcription, 200.

[16] On oeconomy and ingredients in household medicine, see Simon Werrett (2019), *Thrifty Science: Making the Most of Materials in the History of Experiment* (Chicago).

With its emphasis on the interactions, practices, and expenditure of an aristocratic household that was fully *au fait* with (and contributed to) cutting-edge medical developments, the introductory material finally underscores the importance of situating even the most philosophically attuned early modern thinkers within their institutional and familial contexts. In particular, while scholars have long recognised the importance of natural philosophical principles to the medical writings of early modern physicians such as William Harvey and Francis Glisson, this volume, conversely, provides a medical backdrop to the ideas of Margaret, as a self-professed natural philosopher.[17] When Margaret has been read in relation to other thinkers of her day, she has typically been put in conversation with *novatores* or women writers, but this edition demonstrates that the development of her ideas profited greatly from a profusion of medical and domestic exchanges. In so doing, it indicates the outsized role that everyday health concerns played in the development of more abstract philosophical positions.

[17] See, for example, Roger French (1994), *William Harvey's Natural Philosophy* (Cambridge); Peter Anstey (2011), *John Locke and Natural Philosophy* (Oxford); and Peter Distelzweig, Benjamin Goldberg, and Evan R. Ragland, eds. (2016), *Early Modern Medicine and Natural Philosophy* (Dordrecht).

CHAPTER 2

Manuscript Composition and History

Pw V90 has not always been in its current location or present form, and a brief history of the document must accordingly precede an analysis of its contents. William Cavendish-Bentinck, the 7th Duke of Portland (1893–1977), bestowed the papers in his possession upon a number of libraries in England in 1949 and again in 1968. He bequeathed the personal and political correspondences of his family and most of their literary miscellany to the University of Nottingham; the Nottinghamshire Archives Office received papers concerning their estates in the county and those in Derbyshire and Northumberland; the Harley family manuscripts were passed onto the British Museum (and later to the British Library); the Hampshire Record Office obtained documents relating to the Duke's Hampshire estate; and the Bodleian Library was granted the Civil War papers of Reverend John Nalson.[1] This widely dispersed archive, which

[1] A great deal of the information concerning the history of this collection comes courtesy of librarians and archivists at The University of Nottingham's Manuscripts and Special Collections. Some of it can also be found at www.nottingham.ac.uk/manuscriptsandspecialcollections/collectionsindepth/family/portland/portlandofwelbeckabbey.aspx.

J. Begley and B. Goldberg, *The Medical World of Margaret Cavendish*, Palgrave Studies in Medieval and Early Modern Medicine,
https://doi.org/10.1007/978-3-030-92927-5_2

contains materials spanning from the sixteenth to the nineteenth century, is known as the Portland (Welbeck) Collection (Pw).[2]

It has already been noted that Pw V90 is part of the University of Nottingham branch of the archive, and their Manuscripts and Special Collections department reports that much of the material was "generated within the family, both of the Cavendish-Bentinck Dukes of Portland and their Cavendish, Holles and Harley forebears", though it is "difficult to be certain in all cases how the papers were collected and brought together".[3] The collection principally owes its current shape to the efforts of William J. A. C. J. Cavendish-Bentinck, 6th Duke of Portland (1857–1943), and his private librarian, Richard W. Goulding (1868–1929), who rebound numerous manuscripts as they reordered and extended the family library. Unfortunately, however, no rationale was offered for the reorganisation and the provenance records do not appear to have been preserved. Most of the documents (including those of William Cavendish) were accordingly allocated to the aforementioned libraries with minimal information about when they were composed, where they were formerly held, or whether they had been previously bound with other papers. Pw V90 was clearly rebound during the nineteenth century, yet the fact that the 76 blank folio pages that precede the final two entries were not removed strongly suggests that there was no tampering with the material. As we will see, it is fortunate that these empty pages went untouched insofar as they provide a major clue as to the genre of Pw V90: as a "starter collection". While it is uncertain what became of the manuscript from the late seventeenth until the late nineteenth century, it is notable that descendants often carefully preserved medical manuscripts such as Pw V90 as vital family records that contained potentially useful recipes, but also, more importantly, helped to establish and extend social networks and kinship bonds.

Marginally more is known about the construction of the manuscript than its provenance. A modern pencil marking on 1v (perhaps by

[2] Much of the contents of this collection are catalogued in William J. A. C. J. Cavendish-Bentinck (1891–1931), *The Manuscripts of His Grace the Duke of Portland, preserved at Welbeck Abbey* Vols. 1–10 (London).

[3] Pw V—Literary Manuscripts in the Portland (Welbeck) Collection, sixteenth–nineteenth centuries. University of Nottingham Manuscript and Special Collections, mss-cat.nottingham.ac.uk/DServe/Record.aspx?src=CalmView.Catalogid=PwVpos=1, Accessed August 2018.

Goulding) states that the "Title Page and many entries in the handwriting of Thomas Farr. Some pages in that of Wm. Cavendish Duke of Newcastle". A substantial number of the recipes and pieces of advice that follow—including the preponderance of the early items—are in scribal hands, one of which is, indeed, that of Farr, who was a longstanding member of William's household and assumed the role of steward to his son, Henry Cavendish, when he became the 2nd Duke of Newcastle in 1676.[4] But there are also two non-scribal hands in the manuscript. As the above note implies, we can confidently attribute Pw V90 ff. 106v–111v (Transcription, 210–216) to William Cavendish, with the title to a letter on 108v being tellingly labelled "A Letter frome Mr Rust an Apothecary att London aboute the Collick in *my* Stomack".[5] The letter proceeds to discuss cures for William's colic. We can also compare William's entries in Pw V90 with the hand found in some of his literary manuscripts in Nottingham and with the hand in British Library Harley 6491, where he signed his name at the end of the recipes that he had transcribed (see Appendix 8C).

While identifying the entries that William recorded in Pw V90 is a relatively straightforward matter, it is trickier to say with certainty whether or not the other non-scribal hand is Margaret's. We believe, however, that ff. 100r–104r (Transcription, 201–207) are probably in Margaret's hand, and that the same hand is responsible for many of the entries in the latter half of Harley 6491 (ff. 38r–51r; Appendix 8C, 280–290). In order to establish these points, we will compare the hands in these manuscripts with British Library Add. 70499, ff. 259r–297r, which contains Margaret's love letters to William from 1645, and, to a lesser extent, with Margaret's letters to Constantijn Huygens from 27 October 1658 and 11 January 1659 in the Royal Collections, Archive Constantijn Huygens, G1-9.1 (with the other letters to Huygens being in a scribal

[4] On Farr, see Lucy Worsley (2004), Building a Family: William Cavendish, First Duke of Newcastle, and the Construction of Bolsover and Nottingham Castles. *The Seventeenth Century* 19.2: 233–259 (247–248).

[5] Transcription, 210; emphasis added.

hand).[6] We will first describe some of the broad similarities and differences between the hands in these sources, and then examine particular letters and spelling patterns.

Overall, the *mise-en-page* is comparable in the relevant entries in Add. 70499, Pw V90, the Huygens letters, and Harley 6491, with a similar usage of space on the page and similar spacing between letters. Further, there is a stylistic affinity in terms of the "scratchiness" of the hand, most clearly in Add. 70499, but throughout all four manuscripts. Harley 6491 and Pw V90 are, upon close inspection, extremely similar, with the entries in Pw V90 appearing to be in a more careful and formal version of the hand in Harley. While different writing implements and reams of paper were of course used in all of these manuscripts, which naturally affects ink absorption, one noteworthy correspondence is that the entries in all of the documents contain ink blots and deeply inked letters. But perhaps the most important consideration here—and one that we believe is a crucial piece of evidence supporting the case that the relevant entries in Pw V90 and Harley 6491 are in Margaret's hand—is the paucity of punctuation, which is a notable feature of the letters that we know to be in her hand. Such a dearth of punctuation is unprecedented in the multi-handed Harley 6491 and Pw V90, and, more generally, mid-seventeenth-century manuscripts usually have at least *some* punctuation, however irregular.[7] In lieu of punctuation, gaps in the line (especially in Harley 6491 and the Huygens letters) are often used to mark breaks.

There are, however, also some notable dissimilarities between the entries in some of the documents in question. One point is that Add.

[6] See Liza Blake (2018), Pounced Corrections in Oxford Copies of Cavendish's *Philosophical and Physical Opinions*; or, Margaret Cavendish's Glitter Pen. *New College Notes* 10.6: 1–11. Some of what we say in our analysis of the hands in Pw V90 and Harley 6491, especially concerning the importance of context and genre, might also be used to challenge Blake's conclusion that most of the hand-written corrections in Margaret's books are not in her hand, but we do not have space to address the issue here. Cavendish's letters to Huygens can be accessed through the online collection of the Huygens's correspondences: http://resources.huygens.knaw.nl/briefwisselingconstantijnhuygens.

[7] Based on a detailed analysis of a long, multi-handed seventeenth-century recipe book, it has been argued that the problem with punctuation in such manuscripts is not usually its lack, but rather its diversity and irregularity: see Miriam Criado Peña (2020). Punctuation Practice in Early Modern English Scientific Writing: The Case of MS 3009 at the Wellcome Library, London. *Miscelánea: A Journal of English and American Studies* 61: 81–97. Also see Francisco Alonso-Almeida (2001), Punctuation Practice in a Late Medieval English Medical Remedybook. *Folia Linguistica Historica* 22.1–2: 207–232.

70499 is almost wholly devoid of majuscules, while there is at least the occasional majuscule in the relevant entries in Pw V90 and Harley 6491 (though fewer than in any other sections of these manuscripts). Much of the material in the Huygens letters and Add. 70499 also evinces a greater rightward italic than is the case in Pw V90 or Harley 6491. In addition, Add. 70499 is, by far, the messiest and most difficult to read, while the Huygens letters are the most stylistically formal. Pw V90 can indeed be described as a median between the formal Huygens letters and the informal Harley entries. In terms of letter formation, for example, Pw V90 tends to have connected, cursive letters, while Harley 6491 tends to have less formal, printed characters. Adding to this sense that the entries in Pw V90 were more carefully composed, a number of "curlicue" flourishes fill up space at the end of certain recipes; these do not appear in Add. 70499 or Harley 6491.

While acknowledging certain differences, we also want to raise two points that increase the probability that Margaret transcribed the relevant entries in Pw V90 and Harley 6491. The first point is that context matters. In this regard, there are significant temporal considerations: the earliest production was Add. 70499 (1645), followed by Harley 6491 (c.1645–1647), Pw V90 (c.1647–1654), and then the Huygens letters (1658–1659). Hands change over time, for all sorts of reasons, perhaps most importantly because of practise, and Add. 70499 is certainly the most unpractised hand. The generic context of these manuscripts also varies from Margaret's personal love letters to William in Add. 70499 to the medical and culinary recipes in Pw V90 and Harley 6491. Whereas the letters were only for William's eyes, diverse household members had to be able to decipher the hands in the medical manuscripts if the recipes were to prove useful. Indeed, a certain thematic similitude also binds the sections of Pw V90 and Harley 6491 in question: namely, the recipes are relatively simple, especially in comparison to some of the ostentatious chymical recipes found in both manuscripts. With that being said, Harley 6491 was (unlike Pw V90) written on the back of a medical treatise, and was not designed to be passed down as part of the family archive, which accounts for its less formal character. The second point is that natural variation must be kept in mind. Margaret's hand in Add. 70499, for instance, shows a wide range of differences throughout. Some of the letters are relatively tidy and well-spaced, while others are significantly more difficult to read, with numerous errors, scratched out words, and additions, as well as random spacing between letters, words, and lines. There is not only variation between these manuscripts, then, but also *within* documents in the same hand.

A close examination of Add. 70499, Pw V90, and Harley 6491 also reveals similarities and differences in the formation of letters and spelling. In all of the manuscripts in question, the descender of the minuscule "f" has a serif loop going down and left, then to the right. A more or less identical loop is also formed in the ascender of the "h", while the limb has a characteristic curved shape. The minuscule "b", too, has an almost identical looped stem, and a somewhat open lobe, though there is some variation in the degree of its closure between and within the manuscripts. The descender of the minuscule "p" has an otiose stroke in all of the manuscripts, though in the Huygens letters it tends only to move to the left, whereas in Add. 70499 and Pw V90 it loops back to the right, with Harley 6491 demonstrating both formations. The otiose stroke of the majuscule "P" is comparable in all of the manuscripts as is the looping lobe of the minuscule "k". Across all of the documents, the minuscule "s" and majuscule "S" are similar, with the majuscule evincing virtually identical loops, though there are a few variants of the minuscule found within (and between) each of the manuscripts. But there are differences between the manuscript entries too: the majuscule "A", for instance, tends to be less ornate in Harley 6491 than in Pw V90, where it is sometimes stylised with ascending upstrokes or loops, which is in keeping with the more formal nature of Pw V90. There are, however, some similarities between the "A" in Pw V90 and Add. 70499 (with the Huygens letters lacking instances of the majuscule "A"). The minuscule "d" shows some important dissimilarities, with Add. 70499, the Huygens letters, and Pw V90 all having examples of ascenders that curve to the left, which are not found in Harley 6491, though all of the manuscripts also have examples of the non-curved "d". In sum, despite some divergences in style (formal, italicised, etc.), the individual letters are similarly rendered in each of the manuscripts.

The evidence from spelling is more equivocal. As is well known, mid-seventeenth-century spelling was by no means standardised, especially in manuscripts, and all of the hands in question show remarkable variability in spelling, sometimes in the same sentence. A few examples will help to illustrate this. While many common words are spelt the same throughout all of the documents (e.g. "in" and "then"), in Pw V90 and Harley 6491 the word "be" is spelt as both "be" and "bee", though in Add. 70499 it is always rendered "be". Thorns (e.g. "y^{e}") are also only found in the

Pw V90 and Huygens letters. In general, the spelling in Add. 70499 has some of the most unexpected irregularities. One finds, for example, "satisfaction" spelt as "sattisfackshun", whereas in Pw V90 and Harley 6491 one tends to find "-tion" endings (e.g. "potion" and "observation"). Of course, generic differences make spelling comparisons tricky, with the recipe collections mostly consisting of medical terminology, while the vocabulary in Add. 70499 and the Huygens letters is more extensive. Some notable divergences in spelling include the word "half", which is spelt "Halfe" in Pw V90 and Harley and "half" in Add. 70499. A similar pattern exists with "week", which is rendered "Weeke" in Pw V90 and Harley 6491, but lacks the "e" in Add. 70499. "Again", however, is spelt consistently across the manuscripts as "again", as are "so" and "do" (rendered "soe" and "doe"). In the end, because there are both interesting correspondences and divergences, no firm conclusion can be made on the basis of spelling.

Finally, it is worth drawing attention to the general amateurishness of the sections that we believe to be in Cavendish's hand. At f. 101r in Pw V90, for instance, there is a rare moment of potential for substitution, where the scribe notes that, if one cannot find cumin, one can "take red sage[,] cammomille or sage or peney Royall or time".[8] Not only does this passage resonate with Margaret's love of lists, but it also underscores a willingness to adapt received recipes (which, as we will see, Mayerne deemed to be typical of her approach to medicine). Adjectives, such as the repeated use of "wynde" to modify "Collick" (on f. 102v), are also relatively rare elsewhere in the manuscript, but are common in Cavendish's wider output, such as her discussion of "The Windy Gyants" in *Poems, and Fancies*.[9] Based on the weight of evidence, including the lack of punctuation and the similarity in letter formations across the manuscripts, we lean towards the conclusion that the Harley 6491 and Pw 90 selections are in Margaret's hand, while recognising that nothing can be established with certainty based on palaeographical analysis alone.

Whether or not the hitherto unidentified non-scribal hands in Pw V90 and Harley 6491 belong to Margaret, both William and Margaret, as the heads of their household, would have almost certainly been responsible

[8] Transcription, 203.

[9] Transcription, 203 and Margaret Cavendish (1653), *Poems, and Fancies* (London), 155–157. This volume also interestingly includes a poem titled "What *Atomes* make the *wind Collick*" (15).

for selecting and organising the protracted physicians' reports and remedies tailored to them (and less frequently to Henry, William's second son from his marriage to Elizabeth Howard).[10] Indeed, while Isabelle Clairhout has taken Margaret's occasional claims about her lack of household employment literally—and has strangely treated them as indicative of a supposed opposition to "empiricism" in her writing—the available evidence suggests that Margaret's downplaying of her household duties should be taken with a pinch of salt and read partially as a marker of her Royalist pedigree.[11] Cavendish's central role in household duties is indicated, for example, by the fact that William Fuller, Bishop of Lincoln, recorded that Margaret was a harsh estate manager who was "severe in punishing those of the Forest in Nottinghamshire, taking away all the cattle that were not branded, as legally they ought to be".[12] Her sometimes overbearing tactics embroiled her in a 1671 controversy with three of her estate managers: John Booth, Francis Liddell, and Andrew Clayton. Annoyed by Margaret's punctilious attention to the estate books, which they feared might decrease their slices of the estate after William's death, the three men falsely claimed that she was having an affair with another estate manager, Francis Topp, the husband of Margaret's maid-in-waiting, Elizabeth Chaplain. The conspiracy was uncovered, and the perpetrators punished, but this unfortunate series of events corroborates Fuller's characterisation of Margaret as a demanding household manager whose daily commitments extended well beyond the production of an impressive range of literary progeny.[13]

As regards the notion that William and Margaret co-produced Pw V90, it is also notable that the manuscript is predominantly in English,

[10] Another case of a husband and wife collaborating on a medical manuscript is Mary and John Evelyn, British Library, Add. 78337. Some of this manuscript has been printed in Christopher Driver, ed. (1989), *John Evelyn, Cook: The Manuscript Receipt Book of John Evelyn* (London). Also see Michael Hunter (1989), *Science and the Shape of Orthodoxy: Intellectual Change in Late Seventeenth-Century Britain* (Woodbridge, Suffolk), 72–74.

[11] See Isabelle Clairhout (2014), Erring from Good Huswifry? The Author as Witness in Margaret Cavendish and Mary Trye. *Renaissance and Reformation* 37.2: 81–114.

[12] Lucy Worsley (2007), *Cavalier: A Tale of Chivalry, Passion, and Great Houses* (New York), 231, quotes from Francis H. Blackburne Daniell, ed. (1895), *Calendar of State Papers, Domestic Series, January to November, 1671* (London), 426, which contains a letter from William Fuller, Bishop of Lincoln, from Stamford, to Joseph Williamson (12 August 1671).

[13] See "A true narrative and Confession, of that horrid Consperacie, against her Grace Margarett Duchess of NewCastle acted at Welbecke, October the 31st last past, by Andrew Clayton, Francis Liddell, and John Booth, and now confessd by the said John Booth"

though there is a sprinkling of Latin in the form of key phrases, terms from medical botany, and a "receite for the papall or vaticall pills" from the churchman George Morley.[14] Plant names are also occasionally in French or Spanish.[15] Whereas William was a proficient French speaker who regularly conversed with the likes of René Descartes and Pierre Gassendi, Margaret's French appears to have been wanting.[16] Despite the fact that most of the letters are addressed to William, we can safely assume that she was central enough to the manuscript's composition for Frenchmen such as Mayerne to be asked to provide "counsell for your infirmity in English", which would have made it easier for Margaret to decipher the material and for the Cavendishes to delegate the composition of medicaments to the family servants.[17] This request is all the more striking because Mayerne usually compelled his correspondents (including kings and queens) to write to him in French, or, less frequently, Latin.[18] Further suggestive of Margaret's role in Pw V90 is, as we will see, the fact that a number of the specific topics in the manuscript also crop up in her later publications.[19] To give one example, she provided a (rather simple) receipt for over-boiled milk in Pw V90 titled "My Lady Marquesse of Newcastells Receipt to Cure a Fluxe or Dis=sentorye".[20] The concern here is amplified in her later dedication of an entire epistle in *Sociable*

(Nottingham University Library Pw1, ff. 315–317). For a fuller discussion of this controversy, see Katie Whitaker, *Mad Madge: Margaret Cavendish, Duchess of Newcastle, Royalist, Writer and Romantic* (London), 327–333.

14 Transcription, 207.

15 On medicine and the vernacular, see Paul Slack (1979), Mirrors of Health and Treasures of Poor Men: The Uses of the Vernacular Medical Literature of Tudor England. *Health, Medicine and Mortality in the Sixteenth Century*, ed. Charles Webster (Cambridge), 237–273 and Mary Fissell (1992), Readers, Texts, and Contexts: Vernacular Medical Works in Early Modern England. *The Popularization of Medicine, 1650-1850*, ed. Roy Porter (New York), 72–96.

16 Cavendish wrote in 1653 that she knew "no other *Language*" besides English, though we should perhaps take this with a pinch of salt since she proceeded to state that "Neither do I understand my owne *Native Language* very well" (Cavendish, *Poems, and Fancies*, A6r).

17 Transcription, 212.

18 Trevor-Roper, *Europe's Physician*, 6.

19 Along with the discussion below, see Begley, *Margaret Cavendish, The Last Natural Philosopher*, 90–117.

20 Transcription, 164.

Letters (1664) to the medicinal properties of milk, and particularly its effects on the stomach; in the same work, she also ruminated periodically on the causes and cures of fluxes.[21] While we will see in Chapter 4 that much of the scholarship on "housewifery" has gendered the creation and testing of household medicines as female, Elaine Leong has concluded from her comprehensive evaluation of archival sources that the collection, composition, and trial of early modern recipes "required a team of men and women armed with a range of skills and expertise", and depended more on the proclivities, abilities, and attitudes of its members than the normative literature tends to intimate.[22] Insofar as Pw V90 was produced collaboratively, it substantiates this picture.

In dating and attributing the letters in Pw V90, we are on firmer ground than when it comes to questions of composition, since the names of the epistolers, their locations, and the dates of most of the original letters are duly replicated. As the dated letters reveal, William and Margaret commenced Pw V90 two years after their marriage and augmented it for much of their time in exile. While Thomas Cademan, who was Henrietta Maria's physician in ordinary, sent the first of these dated epistles to William Cavendish on 29 August 1647, most of the letters are courtesy of Mayerne who (as this manuscript illustrates) was in routine contact with the Cavendishes from May 1648 until his death in 1654. Rust also sent a number of dated notes on Mayerne's behalf during these years. Given Mayerne's prominence in Pw V90, it is conceivable that his death had a bearing on the Cavendishes' decision to cease collecting recipes in this volume, though we will see that there may have been a more profound, personal reason. In any case, all of the dated material is arranged chronologically, and, since the manuscript was put together gradually, we can reasonably assume that many of the undated recipes and letters were received and then transcribed into Pw V90 during the months and years between the reception of the dated epistles.

[21] Margaret Cavendish (1664), *Sociable Letters* (London), 279, 322–324, and 448.

[22] Elaine Leong (2018), *Recipes and Everyday Knowledge: Medicine, Science, and the Household in Early Modern England* (Chicago), 45. Other wide-reaching analyses of similar material can be found in Elizabeth Tebeaux (1997), Women and Technical Writing, 1475–1700. *Women, Science and Medicine, 1500–1700*, eds. Lynette Hunter and Sarah Hutton (Thrupp), 29–62; Lynette Hunter (1999), English Cookery Books of the Period 1500–1700. *The Oxford Companion to Food*, eds. Alan Davidson, Soun Vannithone, and Tom Jaine (Oxford), 276–277; and Michelle DiMeo and Sara Pennell (2013), *Reading and Writing Recipe Books, 1550–1800* (Manchester).

With that being said, before composing Pw V90, the Cavendishes collected recipes in the back of manuscript treatises, and we surmise that some of these may have made their way into Pw V90. In particular, as Appendix 8C reveals, around the time of William and Margaret's marriage (in 1645), they gathered recipes in the back of "A short treatise of the nature of minerals" that "Doctor Davissone" (the renowned Scottish physician William Davisson) composed at William's behest "in Paris the 20 of July 1645". Harley 6491 features some of the same individuals (including Cademan) as Pw V90 and contains overlapping instructions, with both including, for example, "A receipt to fatt chickins". But perhaps the strongest evidence for the idea that some of the recipes in Pw V90 are clean copies of recipes previously collected elsewhere—with Pw V90 functioning, as we will see, as a sort of "starter collection"—is the existence of a recipe "For to make the Powder of Simp=athy" at the back of another manuscript, Harley 6940. The recipe in this manuscript—which is a 46-page duodecimo of "Chymicall Notions" that the physician Samuel Bispham bestowed on William—is virtually identical to the same recipe in Pw V90 (minus spelling differences).[23] While undated, it seems probable that Harley 6940 was composed in the mid-1640s, after William and Bispham met in Paris, and functioned much like Harley 6491: both are medical manuscript treatises dedicated to William where recipes were collected in a slapdash manner, and, in both cases, this likely occurred prior to the production of the more carefully constructed Pw V90. Many pages have unfortunately been torn out of the recipe section of Harley 6940, but perhaps some of the other recipes that were once in this manuscript were transferred to Pw V90 for their preservation. As all of this indicates, while the overall dating and chronology of Pw V90 are as sure as one might hope for a manuscript of this sort, the exact dating of the various recipes included is less certain.

Turning to the broader familial context should provide a deeper understanding of why Pw V90 was composed when it was and preserved in the ensuing years.[24] In this regard, its generic attributes are indicative. In

[23] See Harley 6940, ff. 54r-57v and Transcription, 172. For more on Harley 6940, see the entry on "Samuel Bispham" in Appendix 8D.III.

[24] Some of the most significant pieces in the substantial body of literature dealing with kinship are Tamara Hareven (1991), The History of the Family and the Complexity of Social Change. *The American Historical Review* 96.1: 95–124 and Naomi Tadmor (2010), Early Modern English Kinship in the Long Run: Reflections on Continuity and

particular, and as already suggested, it shares characteristics with an early modern genre known as a "starter collection", which Leong describes as consisting of

> a large section of copied text written in a uniform hand, occurring at the beginning of the manuscript. The notebook is then filled up with a substantial number of additional recipes in a different hand or hands. This combination of a group of recipes clearly transcribed from another collection and additional recipes gathered singly is one of the most common constructions. [...] Many starter collections are copied leaving blank pages for later additions.[25]

A starter collection, in other words, is a manuscript in which a critical mass of recipes has been transcribed from other documents, with the expectation that it will be updated and eventually passed down to family members (or, less frequently, friends) who will enlarge it with their own material. It should already be clear that Pw V90 conforms with the above characterisation in several ways: it begins with the uniform writing of an amanuensis, different hands contribute subsequent entries, many recipes are copied from letters and other sources, and blank pages pile up at the end. That said, most starter collections commence with an index of family (or community) supplied recipes, whereas the Cavendish manuscript begins, as noted, with a slew of epistles from learned physicians. While the inclusion of letters in a starter collection was by no means unprecedented, the sheer quantity in Pw V90 is abnormal for the genre.[26] Indeed, the Cavendishes consulted an unusually large number of physicians, much to the chagrin of Mayerne, who warned that William ought to "not hearken to soe many flying phisitians: vis all ye world y^t^ speaketh to him: of which every hath a receit: without ground or reason for ye most part".[27] While some of the material extracted from the letters has

Change. *Continuity and Change* 25.1: 15–48. For a focus on women in the context of kinship, see Susan Frye and Karen Robertson, eds. (1999), *Maids and Mistresses, Cousins and Queens: Women's Alliances in Early Modern England* (Oxford), and, for a discussion of a slightly later period, see Naomi Tadmor (1996), The Concept of the Household-Family in Eighteenth-Century England. *Past and Present* 151: 111–140.

[25] Leong, *Recipes and Everyday Knowledge*, 22.

[26] Leong, *Recipes and Everyday Knowledge*, 159.

[27] Transcription, 220. It is notable, however, that Mayerne himself often recommended the services of, and consulted, other physicians. Richard Baxter, for instance, consulted

wider relevance—as in cases such as "The Description of Salt of Steele" or "The virtues of the Antimoniall Cup"—most of the epistles that are transcribed in their totality address the specific concerns of then-living Cavendish family members, and would have been of little use to anyone other than the patients in question. Taken together, such dissimilarities speak against treating Pw V90 as a straightforward starter collection.

Pw V90 rather fuses the starter collection with the family medical manuscript, the latter of which was a document in a family archive that was not necessarily designed for augmentation. In the seventeenth century, family medical manuscripts and recipe collections of the English gentry took on a central role in family histories as "ledgers of social credits and debts, and maps of households' social networks" and they "were often preserved alongside other family records such as legal and financial documents and genealogies".[28] Although it would be a stretch to call Pw V90 an heirloom, it does provide important information on the Cavendish family, and especially the health of Margaret, William, and Henry. We learn, for instance, that Mayerne (using the standard humoural categories of sanguine, choleric, melancholic, and phlegmatic) classified both William and Margaret as "Melancholyk Hypo=condriack[s]", while Henry suffered afflictions of the "Spleene, & vapors ariseing to the head", with the spleen being the traditional site of melancholy.[29] It is also notable that Pw V90 was passed down through the Cavendish family line, and collected alongside legal, financial, and estate records.

The hybrid character of Pw V90 appears to stem from the specific circumstances of William and Margaret, and, most importantly, the fact that they never had children, despite their best efforts (as documented in the manuscript itself). Responding to William's request for advice on how he and Margaret might successfully conceive, Mayerne candidly remarked that "I know not, if in the estate she's in you ought Earnestly to desire" to have offspring, since

Mayerne about Anne Conway's headaches, and Mayerne, in turn, approached Walter Charleton (see Anne Conway (1992), *The Conway Letters: The Correspondence of Anne, Viscountess Conway, Henry More, and Their Friends, 1642–1684*, eds. Marjorie H. Nicolson and Sarah Hutton (Oxford), 20 and 29–31). Also see Elizabeth Connolly (2017), *A Dynamic Equilibrium: Doctors and Patients in Seventeenth-Century England* (Dissertation, University of Adelaide), 202–203.

28 Leong, *Recipes and Everyday Knowledge*, 11 and 134–135.

29 Transcription, 122 and 111.

> It is hard to get Children wth good Corage, when One is Melancholy, and after they are got and come into the World, they bring a great deale of Payne wth them, And after that very often one looses them, as I have try'd to my great greefe and am sory to have had them[;] Be in good health & then you may till yor grownd, otherwise it will be but tyme lost if you enter that race frowningly.[30]

The Cavendishes clung to the hope that they might start a family until at least 1649 when "Doctor Farrer"—or the physician Richard Farrar (see Appendix 8D.III)—dispatched a "Receipt for the Sterillitie in Women".[31] More generally, they regularly augmented their recipe book throughout the late 1640s, when William and Margaret lived together first in Faubourg St-Germain, which was a suburb of Paris, and then (from 1648) in Antwerp at the erstwhile home of the great (and recently deceased) Dutch artist Peter Paul Rubens.[32] Importantly, it is clear that they stopped collecting recipes in Pw V90 from late 1651 until early 1653. At this time, Margaret travelled from Antwerp back to England, where she petitioned, along with her brother-in-law, Charles Cavendish, for the rights to her husband's estates. The fact that no dated epistles were transcribed during her spell in England offers additional evidence that Margaret was a driving force behind Pw V90. When Margaret returned to Antwerp in 1653—where she remained with William until the Restoration—a flurry of additions were made to Pw V90. But, sometime in 1654, entries suddenly ground to a halt. It was in England that Margaret began to publish, and the trade-off between motherhood and writing is a theme that she touches upon at intervals in her early works. To give one instance, she noted in her 1653 *Poems, and Fancies* that, because she had "no *Children* to imploy my *Care*, and *Attendance* on", she instead attended to the economy of the mind.[33] While he was worried about the deleterious effect that writing had on Margaret's health, Mayerne appears to have regarded the Cavendishes' decision to cease trying for children as for the best, both because of William's old age and Margaret's constitution.

The turn from desiring biological offspring to embracing literary progeny may be the key to explaining why Pw V90 was abandoned. It

[30] Transcription, 127. See also the discussion in Smith, Claims to Orthodoxy, 22–24.

[31] Transcription, 182.

[32] For a much fuller account of Margaret's movements, see Whitaker, *Mad Madge*.

[33] Cavendish, *Poems, and Fancies*, A7r.

is most likely that, as was common at the time, William and Margaret originally intended to pass the manuscript down to their children as a storehouse of useful knowledge and a chapter of the family story. But, once they realised that they could not have a family of their own and that the manuscript would be left to the children from William's first marriage, they gradually neglected the manuscript and forsook it around 1654. This is despite the fact that, as piecemeal Restoration evidence indicates, William and Margaret were beset with physical afflictions, and no doubt continued to consult physicians, for the rest of their lives.[34]

[34] See, for instance, Nottingham University Library Pwl 31, f. 327.

CHAPTER 3

Content

This chapter outlines the content of Pw V90 and compares the quantity of recipes and advice that it contains with that of various English and European recipe collections. Before doing so, a few words must be said about how we have counted the recipes and ingredients. Calculating what counts as a recipe is a complicated matter because, in many cases, there is no simple way to prise apart an instance of physic from, say, a piece of advice. The difficulty is compounded by the fact that instructions are sometimes buried in protracted letters. In other instances, it is unclear whether a large, complex series of instructions and products should be regarded as a single recipe or as a collection of smaller, discrete ones.

Our counting is based on the following principles and procedures. Both authors made independent tallies after agreeing on the basic categories discussed below, and the reported numbers cover the ranges yielded by these independent reckonings. Yet, assuming that our sums are not off by an order of magnitude, nothing in the analysis hinges on the exact numbers. In our breakdown, then, treatments that involve multiple intermediate steps and products are handled as a single recipe unless a clear line of separation is drawn in the text. The course of therapy that Mayerne prescribed for William, for example, differentiates between "Opening Apozeme[s]" and "Purging Apozems" as well as between a

J. Begley and B. Goldberg, *The Medical World of Margaret Cavendish*, Palgrave Studies in Medieval and Early Modern Medicine,
https://doi.org/10.1007/978-3-030-92927-5_3

"Fomentation" and a "Broath". Since each of these is clearly differentiated and involves the use of different preparations and ingredients, they are counted as different recipes.

More significantly, we have endeavoured to distinguish, in keeping with the seventeenth-century understanding of the different aspects of therapeutics, between recipes, physic, and advice. Recipes were listed ingredients that, when combined through various procedures, yielded a specific medical product. Physic covered how medication should be taken (e.g. "glysterwise" or by an "electuary"); the proper dosage for a given patient (e.g. the differing quantities of the salt of steel that Mayerne prescribed for William, Margaret, and Henry); as well as recommendations for bleeding, purgation, sweats, dietetics, exercise, and other individualised pieces of counsel (e.g. Mayerne's reprimanding William for smoking tobacco). The final category, advice, included general discussions of disease and treatment (e.g. Mayerne's ruminations on William's "sower" stomach); considerations of the properties of medicines and other substances (e.g. Digby's reflections on vipers); as well as any vague expression of concern (e.g. Mayerne's reproach of Margaret for following his directions in a piecemeal manner).

Based on these principles, we have generated the following numbers (with those without a range indicating consensus). We take the basic agreement of our independent calculations as a sign that this procedure is accurate, while accepting that other modes of categorisation would have inevitably yielded different results. There are thus:

- 148–150 medical recipes.
- 4 veterinary recipes.
- 6–9 culinary recipes.

As is well known, cooking and pharmacy were not fully discrete in early modern medicine, especially in the context of Galenic theory.[1] Within this tradition, humoural imbalances were often thought to be the cause of disease, with an overabundance of black bile in the liver, for example, being said to block the yellow bile produced by the liver, resulting in an

[1] For a discussion of the gradual prising apart of food and medicine, see Elizabeth Spiller (2016), Recipes for Knowledge: Maker's Knowledge Traditions, Paracelsian Recipes, and the Invention of the Cookbook, 1600–1660. *Renaissance Food from Rabelais to Shakespeare*, ed. Joan Fitzpatrick (Farnham), 55–72.

overflowing of bile that caused pain and sensitivity (called jaundice). Just as it was held that certain climates and activities (such as exercise) could affect one's constitution, consumables (whether categorised as medicaments or food) were also thought to affect a person's humoural balance. Barley, for instance, was often included in a tisane for medical purposes, and almond butter, too, was thought to have therapeutic uses. With that caveat, the broadly culinary recipes are restricted to these:

- 3 almond butter recipes.
- 2 barley cream recipes.
- 1 recipe for white sugar candy.
- 1 way to boil and dress a carp.[2]
- 1 recipe for fattening chickens.
- 1 note on how to boil tea.

Finally, Pw V90 contains the following materials in addition to recipes:

- 141–143 pieces of physic.
- 86–88 pieces of advice.
- 1 table for weight conversions.
- 22 letters/excerpts from letters.

While it is relatively easy to characterise the smattering of culinary recipes, despite the sometimes blurry line between food and medicine, the large quantity of medical recipes, the great variation among them, and the complicated ways in which they interact with physic and advice all mean that they require a much fuller, comparative analysis.

A large-scale, quantitative analysis of early modern household manuscript recipes does not currently exist, but, in her pioneering study, Anne Stobart has given a breakdown of the recipes in twenty-two seventeenth-century English collections (eleven in print and manuscript respectively). She has calculated that they contain a total of 8209 recipes, a mere 14% of which are attributed. Of these, the likely gender of the author could only be inferred for 829 recipes, 44% of which are

[2] One might also categorise these last three—boiling a carp, boiling tea, and fattening chickens—as household management recipes rather than as culinary ones.

attributed to women and 56% to men. With respect to the means calculated in Stobart's sample, Pw V90 is highly anomalous: named individuals contribute approximately 66% of the recipes, and the vast majority of these (nearly 91%) are by men.[3] The principal reason for the disproportionate number of named, male contributors to Pw V90 is that the Cavendishes (unlike most examples in Stobart's study) were wealthy and well connected and thus in continual contact with learned British and French physicians, apothecaries, and natural philosophers.[4]

Another factor that contributes to the disproportionate number of attributed recipes in Pw V90 in comparison to Stobart's study is her inclusion of printed collections. Whereas the names associated with recipes tended to function as a gauge of their efficacy in the manuscript tradition, printed collections often obscured the origin of individual recipes, and rather gained their authority from the compiling author.[5] Indeed, even when she investigates manuscript sources, many of Stobart's late seventeenth-century cases, such as that of Margaret Boscawen (1627–1688), are of individuals who extracted many of their recipes from printed books.[6] By contrast, the Cavendishes were in a position to collect recipes directly from friends and acquaintances, as the next chapter shows in greater detail. Some of the recipes that they acquired later became famous, such as the Countess of Kent's powder, which is a rare female-attributed recipe in Pw V90 that was copied and recopied in the ensuing years, moving back and forth from manuscript to the printed page.[7] In a few instances, recipes in Pw V90 even became common items in apothecaries'

[3] Anne Stobart (2016), *Household Medicine in Seventeenth-Century England* (London), 31–32.

[4] While Pw V90 certainly diverges from the average, Leong discusses the historian and political philosopher James Tyrrell, who similarly excerpted recipes from Locke, Digby, and other elite natural philosophers (*Recipes and Everyday Knowledge*, 33–34). Stobart also discusses medical advice and recipes obtained from Locke by Edward Clarke (*Household Medicine*, 18–19).

[5] On this point, see Lynette Hunter (1997), Women and Domestic Medicine: Lady Experimenters, 1570–1620. *Women, Science and Medicine 1500–1700*, eds. Lynette Hunter and Sarah Hutton (Thrupp), 89–107 (89).

[6] See Stobart, *Household Medicine*, 37–39.

[7] Transcription, 188. On the Countess of Kent's Powder, also see Leong, *Recipes and Everyday Knowledge*, 147–148 and Mary Fissell (2007), The Marketplace of Print. *Medicine and the Market in England and its Colonies, c.1450–c.1850*, eds. Mark Jenner and Patrick Wallis (Basingstoke), 108–132 (115).

shops. Indeed, while an English recipe for "salt of steel" does not seem to have appeared in a published work until a few years after Pw V90 was produced, by the time that the celebrated Leiden-educated physician Gideon Harvey published his highly influential *The Family-Physician, and the House-Apothecary* (1676), which contains a comprehensive list of the prices of medical ingredients, Mayerne's salt of steel (or *anima hepatis*) had been turned into *sal chalybis* and was sold for 2 shillings.[8]

As a document that contains mostly named recipes by men, Pw V90 is in some ways closer to collections from the German-speaking world than to those in Stobart's English study. Discussing recipes in the *Codices Palatini Germanici*—or the German-language manuscripts that belonged to the library of the Palatine electors, and are now held at the University of Heidelberg—Alisha Rankin remarks that male contributors dwarfed female ones during the sixteenth and seventeenth centuries.[9] It is notable that in both the English and German contexts, however, noblewomen were often named as *compilers* of recipe collections.[10] The reasons why men were more likely to author recipes in the German context than the English one are obscure, and indeed the relatively small portion of surviving manuscript collections—along with the incomplete study of even these documents—makes it impossible to draw a complete European picture. But why women more frequently appear as compilers than creators of medical recipes can be at least partially explained in light of the advice in Renaissance texts on the oeconomic roles of women. In England, these texts usually took the form of practical, pedagogical tracts on "housewifery", such as John Partridge's continually reprinted 1571 *The Treasurie of Commodious Conceits, & Hidden Secrets*. Partridge declared in the preface that his manual should prove an indispensable

[8] Or, perhaps better, *should* be sold for this amount: see Gideon Harvey (1676), *The Family-Physician and the House-Apothecary* (London), 139. Note that "*chalybs*", in Latin, can refer to either iron or steel, but, in the chymical context, almost certainly refers to steel. The earliest published English recipe for the "salt of steel", which is similar to Mayerne's, that we have been able to find is in George Starkey (1660), *George Starkey's Pill Vindicated from the Unlearned Alchymist* (London), 5–7; also see Chapter 6B.

[9] Alisha M. Rankin (2013), *Panaceia's Daughters: Noblewomen as Healers in Early Modern Germany* (Chicago), 7. Also see Stobart, *Household Medicine*, 30. Note, too, that the preferences of the Palatine Elector guided the composition of codices, and thus represent various idiosyncratic, and even changing, ideas about the worthiness of the material to be included in the collections.

[10] Rankin, *Panaceia's Daughters*, 62–69.

resource for women, since "Amongst the rest of Phisicks helps, the huswifes help is chiefe".[11] Far from encouraging women to propagate medical advice in print, he accordingly emphasised that the housewife is not to be more than the chief ***help*** to the learned physician. Gervase Markham's *The English House-Wife* (first published in 1615) was another perennial favourite that stressed the need for ladies to possess a "phisicall kind of knowledge", though he was sure to clarify that mastery of the "Art of Physicke is farre beyond the capacity of the most skillful woman".[12] Across Europe, texts on "housewifery" containing similar comments about women and medicine also took the form of theologically-laden humanist tracts, with perhaps the most famous being Juan Luis Vives' 1524 *De institutione foeminae christinae*, which was quickly translated into several European vernaculars.[13] While much of this literature stressed the need for women to know how to prepare remedies, it crucially did not recommend that they practice and understand the art and theory of medicine (as it was taught in the universities).[14] These texts are of course normative insofar as they present what was deemed to be an idealised vision of female duties, but women were, in practice, often encouraged to collect, compile, and produce pre-existing recipes rather than invent their own, or refine ancient ones. This deeply engrained attitude towards the limits of female medical knowledge is on display in Pw V90 in the letters of the chymically inclined physician Richard Farrar, who notes that a certain decoction is so "facile y^{t} any woman may make" it, and remarks that he has written his direction for the treatment of the military man Thomas Glemham so clearly that "every Cooke mayde, much more any more skilfull cannot fayle in following" them.[15]

[11] John Partridge (1591), *The Treasurie of Commodious Conceits* (London), A2v.

[12] Gervase Markham (1631), *The English House-Wife* (London), 5. Hugh Plat expressed a similar sentiment when he wrote in his 1653 *Jewel House of Art and Nature* that he will seek to make ladies "as cunning as my self" while proceeding to note that his discourse if often "too Phylosophical for women" (London), 3.

[13] On Vives's recommendations for women, see Juan Luis Vives (2000), *The Education of a Christian Woman: A Sixteenth-Century Manual*, ed. Charles Fantazzi (Chicago).

[14] Rankin, *Panaceia's Daughters*, 10–12. The best general study of the Renaissance conception of women remains Ian Maclean (1990), *The Renaissance Notion of Woman: A Study in the Fortunes of Scholasticism and Medical Science in European Intellectual Life* (Cambridge).

[15] Transcription, 174 and 179.

Stobart's study not only examines the gender of authors of recipe collections, but also analyses the content of these manuscripts. Having surveyed 8209 entries in such collections, she concludes that

> 6,513 (79.3 per cent) recipes had primarily medicinal purpose, 1,232 (15 per cent) were culinary, 153 (1.9 per cent) were cosmetic, 125 (1.5 per cent) were household-related, 90 (1.1 per cent) were veterinary and 96 (1.2 per cent) were miscellaneous items of advice.[16]

Such a quantitative analysis only further underscores the idiosyncrasies of Pw V90. In the Cavendish manuscript, a mere 6.7% of the recipes are for foodstuff—far less than Stobart's average—and no cosmetic or household-related recipes are present (apart from perhaps the chicken fattening recipe). At the same time, Pw V90 contains a relatively large quantity of veterinary recipes (2.7%), and particularly cures for horses. This is fully explicable given that William was a renowned horseman, and, during his time at Rubenshuis, converted the artist's studio into a popular riding house where he taught his mode of horsemanship to visitors from across the Continent.[17] Based on the available evidence, Pw V90 is thus at the *extremis* of the averages of English collections, containing few culinary recipes, no cosmetic or household recipes, and a relatively large number of veterinarian and named or attributed recipes, especially by men. While it was noted that the attribution of recipes to men was more common in the German-speaking world, it is interesting that the occasional culinary recipe is intermingled with medical ones in Pw V90, which is a practice that was usually resisted in a German context but that was quite common in England.[18] As a document that an English couple composed in Paris and Antwerp, Pw V90 accordingly seems to fuse the features of British and continental recipe books, granting it a distinctively international character.

[16] Stobart, *Household Medicine*, 31.

[17] See Lucy Worsley and Tom Addyman (2002), Riding Houses and Horses: William Cavendish's Architecture for the Art of Horsemanship. *Architectural History* 45: 194–229 and Richard Nash (2017), William Cavendish: Riding School and Race Track. *Authority, Authorship and Aristocratic Identity in Seventeenth-Century England*, eds. Peter Edwards and Elspeth Graham (Leiden), 317–330.

[18] Rankin, *Panaceia's Daughters*, 71–72. On culinary recipes in England, see Wendy Wall (2015), *Recipes for Thought: Knowledge and Taste in the Early Modern English Kitchen (Philadelphia)*, especially 20–111.

A further peculiarity of Pw V90 is its proportion of epistles. While letters were indispensable to the exchange of medical knowledge and advice among literate individuals, they were rarely collected in the same volume as recipes, and, when this did occur, they were typically stripped of all detail and made procedural.[19] As Stobart remarks, the recipe collections that she surveys "rarely identified details of other aspects of healthcare such as prevention, diet and lifestyle", with the "general incidence of dietary advice" being "less than 2 per cent", while medical "practices, such as bloodletting, that were widespread and had longstanding use within household self-help were not mentioned".[20] By contrast, Pw V90 is packed full of medical advice and deliberations on courses of physic, including blood-letting. As already suggested, the Cavendishes probably decided to log these details because they conceived of their manuscript as much as a historical and familial record as a useful storehouse of medical advice. On a more practical level, the fact that the Cavendishes had scribes such as Farr under their charge also made the laborious process of transcribing letters less daunting. Pw V90, then, is not only a hybrid of models (the English and the continental ways of defining and compiling medical advice) but also of genres: the recipe and letter book.

Whereas most family recipe collections were put together haphazardly or constructed around specific ailments or ingredients, the Cavendish manuscript is a truly historical document in the sense that, as noted, it was organised chronologically. Despite this sequential arrangement, it manages to preserve a relatively clear structure, with headings on each page that register, in the case of recipes, one or more of the following: the individual with whom it was associated, its basic ingredients, and the diseases or symptoms that it was intended to treat. Virtually all of the entries mark the source of the recipe (e.g. "Sr Thomas Cadamans way to make Pomander") even if the proper name of the individual is sometimes absent (e.g. "an Nunn, att Antwerpe"). The descriptions can

[19] On "epistolary medicine", see Nancy Siraisi (2012), *Communities of Learned Experience: Epistolary Medicine in the Renaissance* (Baltimore) and Ian Maclean (2008), The Medical Republic of Letters Before the Thirty Years War. *Intellectual History Review* 18.1: 15–30. For the example of an elite, early modern Englishwoman, see Michelle DiMeo (2016), The Rhetoric of Medical Authority in Lady Katherine Ranelagh's Letters. *Women and Epistolary Agency in Early Modern Culture, 1450–1690*, eds. James Daybell and Andrew Gordon (London), 112–125.

[20] Stobart, *Household Medicine*, 49.

otherwise range from simple designations such as Cademan's "use of the Antimoniall Cup" to highly descriptive ones including "Sr Theador Mayernes Description of Remedy's, for the Lord Henry Caven=dishe, agaynst the Spleene, & vapors ariseing to the head, with trembling and shakeing fitts". In the case of epistles, the name of the patient is frequently recorded (e.g. "Sr Theodor Mayerne Advice For My Lady and my Self"). Selected segments of the manuscript are, however, organised around types of therapy such as vomits. Margaret's milk-based emetic and "Maryes Medicin for the Same", for instance, follow the common emetic "crocus metallorum".

Several other sections of Pw V90 are organised around a regimen of physic and therapy with a recipe for a particular medicine. In these cases, the material is sometimes, though by no means always, ordered in the following manner: the central recipe (e.g. salt of steel), the method of administration (e.g. glister, pills, syrup, etc.), further physic (e.g. purging or bleeding), and additional recipes required to treat the ailment (e.g. juleps). While it is difficult to generalise about the letters, they often follow a standard therapeutic logic. This means that they typically begin with a diagnosis and discussion of the disease that is afflicting one or more of the main patients—paying particular attention to their temperaments—before outlining a course of physic and providing a variety of relevant recipes. In most cases, the epistles include protracted instructions about preparation and the dosages needed for each patient; such information is also occasionally present in standalone recipes. The letters are likewise the main conduits through which the medical and physical advice scattered throughout the manuscript is relayed.

While the above modes of organisation are most common, a few sections orbit around an ingredient (e.g. crystals), while others do not provide a recipe at all but rather other medically relevant information (e.g. Rust's opinion on honey, though even this entry is immediately followed by Rust's way to make the essence of honey).[21] Interestingly, the last two entries in Pw V90, which are separated from the rest by 76 empty folio pages, are entirely different from all of the material that precedes them and were probably placed here for precisely this reason. The penultimate item is a brief guide to measuring weights, which distinguishes between three systems of measurement. The first is "Haber de Poyos", which

[21] Transcription, 191–193.

refers to the avoirdupois system, from the Anglo-Norman French "aber de peyse". This was the standard system of weights used in Britain for all goods except, according to the manuscript, "Gold, Silver, & Bred". The avoirdupois system was equivalent to other, common systems of weights in early modern Europe, according to which there are 16 ounces in a pound.[22] Then there is the "Troyo weight", which is the system of the Troy ounce, used for gold, silver, and bread, according to which there are 12 ounces to the pound.[23] The system of the Troy ounce, which supposedly originated from the French town of Troyes, was officially adopted in England in 1527 by an order of Henry VIII.[24] The third and last system is that of "Venice weight" used by "silke-men", which is similar to the avoirdupois system, apart from the fact that the Venice ounce seems to have been ¾ of the avoirdupois ounce. There were also regional differences in measurement within seventeenth-century Britain, with Scottish measures being larger than English ones, for instance, while Irish ones were somewhat smaller.[25] Given these regional discrepancies, and the fact that the Cavendishes were in contact with physicians from Scotland, France, and beyond, such a key was indispensable, especially if they were to create medicines using their own scales and weights.

The final entry, which we have categorised as culinary, is more peculiar: it is the aforementioned guide for "How to fatt Chickens". Despite being anomalous in Pw V90, this recipe is common in other English recipe books, with maximising animal resources being a constant concern for householders.[26] While it is not completely clear why this recipe was

[22] Many weighing systems originate from the Roman system of *libra,* which was divided into 12 *uncia,* the same division given to the Troy pound, used by jewellers and others. On the history of the evolution of these systems, see Ronald E. Zupko (1977), *British Weights & Measures: A History from Antiquity to the Seventeenth Century* (Madison) and, for Italy, Ronald E. Zupko (1981), *Italian Weights and Measures from the Middle Ages to the Nineteenth Century* (Boston).

[23] The table of weights is found on Transcription, 224.

[24] William Hallock and Herbert T. Wade (1906), *Outlines of the Evolution of Weights and Measures and the Metric System* (London), 34.

[25] Magdalena Bator and Marta Sylwanowicz (2017), Measures in Medieval English Recipes–Culinary Vs. Medical. *Studia Anglica Posnaniensia* 52.1: 28–52 (28). For background on the systems of measurement used by English and French apothecaries, see Ronald E. Zupko (1990), Medieval Apothecary Weights and Measures: The Principal Units of England and France. *Pharmacy in History* 32.2: 57–62.

[26] See Leong, *Recipes and Everyday Knowledge*, 46–70.

placed at the end of Pw V90, it is another telling piece of evidence for Margaret's active role in procuring recipes and running the household, for William has scribbled at the end of this recipe that "Lady Boswell"—or Lady Margaret Boswell—sent it to "my wife".[27]

[27] Transcription, 225.

CHAPTER 4

Context

Given what has been said about the Cavendishes' collaborative production of Pw V90, it should come as no surprise that the household is a primary context for understanding the nature of the manuscript. A wealthy early modern household such as the Cavendishes typically consisted of dozens of servants and family members. They were collectively responsible for maintaining multiple living quarters and stables, tending to gardens and grounds, preparing food, as well as grooming and feeding an assortment of animals.[1] This is not to mention the theme of the present work: the composition and administration of medicines, which was often referred to as "kitchen physick". Managing such a large household was invariably laborious, and at its centre was the need to tame nature into more useful shapes, while optimising human, plant, and animal resources. The ingredients needed to produce foodstuffs, medicines, glues, and cosmetics were a major factor in determining what kinds of plants should be grown

[1] On the early modern household, see Alix Cooper (2006), Homes and Households. *The Cambridge History of Science* Vol. 3, eds. Katherine Park and Lorraine Daston (Cambridge), 224–237. Also relevant in this context are the various "closets" so often referenced in collections of household knowledge, for which see Jennifer K. Stine (1996), *Opening Closets: The Discovery of Household Medicine in Early Modern England* (Dissertation: Stanford University) and Wall, *Recipes for Thought*, 228–229.

J. Begley and B. Goldberg, *The Medical World of Margaret Cavendish*, Palgrave Studies in Medieval and Early Modern Medicine,
https://doi.org/10.1007/978-3-030-92927-5_4

(and on what scale), as well as which animals should be kept.[2] Historians have dubbed the skills needed to complete or oversee the above activities successfully as "maker's" or "artisanal" knowledge.[3] The demand for such knowledge resulted in the production of various kinds of texts, including herbals, handbooks of mechanical arts, surgical manuals, and, most significantly for our purposes, recipe books.

The tradition of craft knowledge to which the above genres contributed harkened back to pseudo-Aristotle's *Secretum Secretorum*.[4] Two of the most popular English works in the "secrets" tradition were Partridge's aforementioned *Treasurie of Commodious Conceits* and Hugh Plat's *The Jewell House of Art and Nature* (1594). As Elizabeth Spiller has noted, these books were hodgepodges of inventions, experiments, and "instructions for distilling, cryptography, medicinal salves, dyeing, and cooking", all of which were linked to the extent that they were arts of the hand, and, equally important, instances of the knowledge that such arts could yield.[5] In contrast to the more speculative philosophical concerns that underlay, say, Digby's 1644 *Two Treatises* or Cavendish's 1666 *Philosophical Letters*, historians have often categorised these more practical texts as "household science" insofar as they were orientated towards the creation of specific products for particular situations within the domestic setting. At the same time, rooted in Aristotle's *Nicomachean Ethics*, this preoccupation with households and their members was central to the early modern domain of "oeconomics", which was considered a branch

[2] For a fuller discussion, see Leong, *Recipes and Everyday Knowledge*, 10–11.

[3] On maker's knowledge, see Antonio Pérez-Ramos (1988), *Francis Bacon's Idea of Science and the Maker's Knowledge Tradition* (Oxford); Pamela H. Smith (2004), *The Body of the Artisan: Art and Experience in the Scientific Revolution* (Chicago); Deborah Harkness (2007), *The Jewel House: Elizabethan London and the Scientific Revolution* (New Haven); and Spiller, Recipes for Knowledge. For how this tradition intersected with alchemy and astrology, see Anthony Grafton and William R. Newman (2001), *Secrets of Nature: Astrology and Alchemy in Early Modern Europe* (Cambridge, MA) and William R. Newman (2001), *Promethean Ambitions: Alchemy and the Quest to Perfect Nature* (Chicago).

[4] The key text in this regard is William Eamon (1994), *Science and the Secrets of Nature: Books of Secrets in Medieval and Early Modern Culture* (Princeton). With a focus on the contributions of women, see Meredith Ray (2015), *Daughters of Alchemy: Women and Scientific Culture in Early Modern Italy* (Cambridge, MA), 46–72.

[5] Spiller, Recipes for Knowledge, 61.

of moral philosophy alongside ethics (concerned with individuals) and politics (concerned with the state).[6]

Margaret was a strong proponent of practical or "useful" arts such as architecture and medicine, and she railed against a number of early Royal Society associates—especially "Microscopical or Dioptrical Writers" who "employ their time in Artificial Experiments"—because she regarded their claims to "usefulness" to be vacuous (with their microscopes merely serving as "pretty toys to employ idle time").[7] Although Pw V90 may seem far removed from Margaret's more reified philosophical reflections, it must be kept in mind that recipe collections played an integral role in informing householders not only about how to maintain and improve their health but also about the wider functions of the natural world. As such, Pw V90 features discussions on topics ranging from the character of flowers, roots, chemicals, and humoural constitutions to how ingredients combine to create mixtures or compounds.[8] Given that texts within the "maker's knowledge" tradition were aimed at, and increasingly written by, women as a means by which to instruct them in household and broader natural matters concurrently, it is likely that Margaret downplayed her household activities precisely because she wanted to distance herself and her writing from what she regarded as a feminine genre.[9] Indeed, she

[6] See Jill Kraye (1988), Moral Philosophy. *The Cambridge History of Renaissance Philosophy*, eds. Charles Schmitt, Quentin Skinner, Eckhard Kessler, and Jill Kraye (Cambridge), 303–386.

[7] Margaret Cavendish (2001), *Observations Upon Experimental Philosophy*, ed. Eileen O'Neill (Oxford), 105. For why Cavendish's challenges to certain leading members of the Royal Society should not be taken to indicate an opposition to experimentation or practical knowledge, see Begley, *Margaret Cavendish, The Last Natural Philosopher*, 254–257.

[8] Elaine Leong and Sara Pennell (2007), Recipe Collections and the Currency of Medical Knowledge in the Early Modern 'Medical Marketplace'. *Medicine and the Market in England and Its Colonies, c. 1450–c. 1850*, eds. Mark Jenner and Patrick Wallis (Basingstoke), 133–152.

[9] The historical literature concerning household science and its connection to experimental science is now legion, but essential starting points include Hunter, Women and Domestic Medicine; Frances Harris (1997), Living in the Neighbourhood of Science: Mary Evelyn, Margaret Cavendish and the Greshamites. *Women, Science and Medicine 1500–1700*, eds. Lynette Hunter and Sarah Hutton (Thrupp), 198–217; Elizabeth Spiller (2008), *Seventeenth-Century English Recipe Books: Cooking, Physic and Chirurgery in the Works of Elizabeth Talbot Grey and Aletheia Talbot Howard* (Aldershot); and Elizabeth Spiller (2008), *Seventeenth-Century English Recipe Books: Cooking, Physic and Chirurgery in the Works of W.M. and Queen Henrietta Maria, and of Mary Tillinghast* (Aldershot).

noted in *Sociable Letters* that "most commonly when any of our Sex doth Write, they Write some Devotions, or Romances, or Receits of Medicines, for Cookery or Confectioners" all of which "Express our Brief Wit in our Short Works".[10] Despite such derogatory remarks, we will see that Cavendish's very decision to publish was indebted to the proliferation of female-authored recipe books, just as her publications drew on knowledge contained within Pw V90. Such debts probably lay behind her impatience to clarify that her own natural philosophical works—which she hoped that scholars and philosophers both within and outside of the universities would engage with seriously—were of a different pedigree altogether to other women's "Receits of Medicines".

While it is beyond the scope of this introduction to offer a comprehensive survey of how Cavendish grappled with medical concepts, a brief overview of her medical principles and understanding of the Galenic and chymical traditions will prove instructive. Her first and most obvious form of engagement with medical ideas (at least in printed form) is the provision of recipes and medical advice, which is perhaps most pronounced in *The World's Olio* (1655), the preponderance of which was completed during the later 1640s, around the time that the bulk of Pw V90 was produced.[11] Even the "olio" of the title is interestingly rooted in the household, being derived from the name of a Spanish stew, *Olla podrida*, which was made of pork, beans, and a variety of other meats and vegetables. In the mid-1640s, "*Olla podrida*" was integrated into English as "oleo", "oglio", or "olio podrido", the last of which literally meant "rotten pot".[12] At the same time, "olio" became a general term for the

Relevant recent titles include Sara Pennell (2016), *The Birth of the English Kitchen, 1600–1850* (London) and Ann C. Christensen (2014), Words about Women's Work: The Case of Housewifery in Early Modern England. *Early Modern Studies Journal* 6: 1–28. Also see Samantha Snively (2015), 'The Quintessence of Wit': Domestic Labor, Science, and Margaret Cavendish's Kitchen Fancies. *Newberry Essays in Medieval and Early Modern Studies* 9: 21–28.

[10] Cavendish, *Sociable Letters*, 225–226.

[11] Margaret Cavendish (1655), *The World's Olio* (London), B1v.

[12] The recipe can already be found in the early seventeenth century in a collection by the cook of the Colegio Mayor de Oviedo in Salamanca (see Domingo Hernández de Maceras (1607), Cap LIII. Come se ha de hazer una olla podrida. *Libro del Arte de Cozina* (Salamanca), 55). On de Maceras, see Ken Albala (2010), Cooking as Research Methodology: Experiments in Renaissance Cuisine. *Renaissance Food from Rabelais to Shakespeare*, ed. Joan Fitzpatrick (Farnham), 73–88.

mixture of heterogeneous elements, though Cavendish appears to have been the first to use it to refer to an assemblage of essays.[13] Beyond the fact that household language afforded Cavendish a plentiful supply of metaphors, a whole section of *The World's Olio* is dedicated to medical topics, and includes essays such as "Of letting Blood", "Of Diet", "Diseases of the Spring, Summer, Autumn, and Winter", "Of Fruits", "Of Roots", and "Of Herbs".[14] In these pieces, Cavendish presents a heap of recipes and medical advice, and often lists of herbs, in a manner that is strikingly reminiscent of Pw V90. She notes, for instance, that "cooling and clensing" medicines include

> Julips made of Burrage-water, Plantain-water, with Sirrop of Suckery, and Sirrop of Burrage, and Bugloss, and the like; Also, Broths with Cooling Herbs, as Strawberry-leaves, Violet, Suckery, and the like. But no Hot Sirrops, nor no Sharp Herbs, as Sorrel, and the like; nor no Hot Herbs, as Thyme, Rosemary, Winter-Savery, Marjerum.[15]

All of these ingredients also appear in Pw V90, and Mayerne and Davisson in fact both provided similar recipes for cooling juleps that Margaret was to take when she "shall finde her selfe hott".[16]

The herbal ingredients in the above passage were central to Galenic medicine in general, but, as Cavendish admitted in the preface to *Philosophical and Physical Opinions* (1655), she particularly enjoyed reading "Gerrards Herball, which no question is a very rare book, and certainly discribes the tempers of herbs, fruits, and drugs very learnedly".[17] Here of course Cavendish is referring to the massively influential *Great Herball, or, Generall Historie of Plantes* (1597) by the English herbalist and physician John Gerard.[18] Despite the importance of new chymical medicines, and their presence throughout Pw V90, herbal medicine, and traditional

[13] "Olio" (*OED*, n. 2a and 2b).

[14] Cavendish, *The World's Olio*, 178–204.

[15] Cavendish, *The World's Olio*, 193.

[16] Transcription, 134 and 155.

[17] Margaret Cavendish (1655), *Philosophical and Physical Opinions* (London), B6r.

[18] For a discussion of this herbal, see Rebecca Laroche (2016), *Medical Authority and Englishwomen's Herbal Texts, 1550–1650* (London), 11–20 and Elaine Leong (2014), 'Herbals she peruseth': Reading Medicine in Early Modern England. *Renaissance Studies* 28.4: 556–578.

modes of therapy, remained popular with the Cavendishes. Such eclecticism was indeed relatively common in the Cavendishes' medical circles, with most medical practices still depending on the traditional Galenic categories of humoural medicine for diagnosis even as Paracelsian and post-Paracelsian chymistry became increasingly central to household pharmaceutical production and therapeutics.[19] John and Mary Evelyn, for instance, tellingly commented that the principal recreations of woman are "in the Distillatory, the knowledge of Plants and their Virtues, for the comfort of the poor Neighbours, and Use of the Family, which wholsome plain Dyet, and Kitching Physick preserved in perfect Health".[20] They thus allude to the centrality of both Galenic herbalism ("Plants and their Virtues") and chymical practices ("Distillatory") to household medicine.

In addition to revealing her intimate knowledge of herbal remedies, Cavendish's early works also underscore the store that she set by diet, which was one of the Galenic "non-naturals".[21] Indeed, she announced in *The World's Olio* that "nothing preserves Health more, and lengthens Life, than due and just proportion of Diet".[22] Not only does Cavendish's preoccupation with herbs and diet betray the influence of Gerard, but it also resonates with the counsel of Mayerne, who had recently commended William, for instance, for asking for "Advice concerning yor Diett And yo^{u} doe well for it is from thence you ought to hope yor Principall good".[23] More generally, the advice in *The World's Olio* frequently echoes that which Mayerne gave to Margaret in Pw V90. Mayerne often repeated that "her La^{p} must use some moderate Exercise", and Cavendish, in turn,

[19] The literature on Paracelsus is now enormous, but the classic study is Walter Pagel (1958), *Paracelsus: An Introduction to Philosophical Medicine in the Era of the Renaissance* (Basel). Also see Bruce Moran (2019), *Paracelsus: An Alchemical Life* (London).

[20] Mary and John Evelyn (1690), *Mundus muliebris: or, The Ladies Dressing-Room Unlock'd, and Her Toilette Spread in Burlesque* (London), A4v. On the connections between housewifery and chymistry, see Jayne Archer (2016), Women and Chymistry in Early Modern England: The Manuscript Receipt Book (c. 1616) of Sarah Wigges. *Gender and Scientific Discourse in Early Modern Culture*, ed. Kathleen P. Long (London), 191–216; Katherine Allen (2016), Hobby and Craft: Distilling Household Medicine in Eighteenth-Century England. *Early Modern Women* 11.1: 90–114; and Penny Bayer (2021), Women Alchemists and the Paracelsian Context in France and England, 1560–1616. *Early Modern Women* 15.2: 103–111.

[21] On non-naturals, see Maclean, *The Case of Learned Medicine*, 251–258.

[22] Cavendish, *The World's Olio*, 181.

[23] Transcription, 122.

recommended that "the Patient must not use any Violent Exercise, so as to heat the Body, but must use Moderate Exercises".[24] Likewise, William is advised that "of meates yo^{u} should eate, It is no great matter as long as yo^{u} are well" and "be temperate in ye quantity", while Margaret maintained that one could eat any meat so long as "they be Fine Meats" but she warns "not to eat too much at one time".[25] Building upon this early output, the other text in Cavendish's *oeuvre* that contains medical recipes is *Sociable Letters*, which consists of correspondences with an imagined female friend and gives various pieces of medical advice in an epistolary form that reflects many of the entries in Pw V90. Having heard, for instance, that one "Lady G. D. takes Cooling Julips in the Morning, and Cordials when she goes to Bed", Cavendish recommends that her imagined correspondent instead take "Hot things first, and Coolling after". Grounded in the traditional Galenic idea of primary qualities and the usage of opposites, she went on to provide therapeutic advice, reasoning that "Hot things after Cooling do rather make a Smothering Heat, than a Concocting, Digesting, or Expelling Heat, so as it only fills the Body full of Vapors".[26]

Cavendish published a companion piece to *Sociable Letters* titled *Philosophical Letters*, and she addressed many of the letters in this work, which engage with more theoretical medical questions, to the renowned chymical physician Jan Baptiste van Helmont, along with William Harvey and Gideon Harvey.[27] In an attempt to get to grips with Cavendish's views on the relative roles of Galenic and chymical medicine, what follows looks at some of the ways that she drew on Helmontian chymistry as she developed her natural philosophical principles, while Chapter 6 explores her critique of Helmontian optimism towards supposed "panaceas" or "universal cures". Cavendish was a friend of the physician Walter Charleton, who had translated some of van Helmont's writings in 1650.[28] In his *Ternary of Paradoxes,* in particular, Charleton translates van

[24] Transcription, 197 and Cavendish, *The World's Olio*, 195.

[25] Transcription, 139 and Cavendish, *The World's Olio*, 196.

[26] Cavendish, *Sociable Letters*, 290–291.

[27] See Goldberg, Epigenesis and the Rationality of Nature, and Begley, *Margaret Cavendish The Last Natural Philosopher*, 117–144 and 180–204.

[28] See Liam Semler (2011), The Magnetic Attraction of Margaret Cavendish and Walter Charleton. *Early Modern Englishwomen Testing Ideas*, eds. Jo Wallwork and Paul Salzman (Farnham), 55–72.

Helmont's description of how he arrived at the idea of an *archeus*, or the vital power that interpenetrates matter. According to van Helmont, in the words of Charleton, there must be a principle that is "immaterial, yet real, and operative, to which may be justly attributed the power of figuration or delineation, by a sigillary impression upon the *Archeus,* or Regent Spirit of the *Seed*".[29] Here Charleton highlights that van Helmont's divine *archeus* is most condensed in the "seed", understood as a spiritual essence that carries the life-plan of each natural body.[30] In formulating her own substance theory in her 1653 *Philosophicall Fancies*, Cavendish, in turn, styled her vital matter as "the *Spirits* that are in the *Seed*".[31] She conceived of individual creatures as dynamic beings consisting of material spirits that actively gather and disband as they grow and deteriorate, which means that even "*Death*, or *corruption*, is not an absence of life, but an *expulsive motion*".[32] Stripping the Helmontian *archeus* of its immateriality, Cavendish maintained that seeds are not above and beyond matter but are instead the critical mass of material spirits that grant each creature a core that withstands the vagaries of generation and corruption. Some scholars have argued that Cavendish straightforwardly challenged "the theories and practices of alchemy and medical chemistry", but, rather than rejecting chymical ideas *tout court*, she accepted and creatively reinterpreted many, while rejecting those that she found to be the most far-fetched.[33]

More generally, as Begley has shown elsewhere, Cavendish's notion that nature is fundamentally composed of "rational", "sensitive", and "inert" matter grew in part out of her adaption of the Galenic concept

[29] Walter Charleton and Jan Baptiste van Helmont (1650), *A Ternary of Paradoxes* (London), 129.

[30] See Walter Pagel (1982), *Joan Baptista van Helmont* (Cambridge) and Hiro Hirai (2005), *Le concept de semence dans les théories de la matière à la Renaissance: De Marsile Ficin à Pierre Gassendi* (Turnhout), 437–462.

[31] Cavendish (1653), *Philosophicall Fancies* (London), 22. For a fuller discussion, see Begley, *Margaret Cavendish, The Last Natural Philosopher*, 117–143.

[32] Cavendish, *Philosophicall Fancies*, 20.

[33] Sandy Feinstein (2021), Experience, Authority, and the Alchemy of Language: Margaret Cavendish and Marie Meurdrac Respond to the Art. *Early Modern Women* 15: 133–142 (134). See also Jacqueline Broad (2011), Cavendish, van Helmont, and the Mad Raging Womb. *The New Science and Women's Literary Discourse: Prefiguring Frankenstein*, ed. Judy A. Hayden (New York), 47–46.

of vital and animal spirits.[34] That is, before framing spirits as the fundamental components of nature in *Philosophicall Fancies*, Cavendish initially conceived of "Animal Spirits" in the terms derived from the Arabic-Latin medical interpretations of Galen as "the Radical Vapour in the Body, produced from the Natural Heat, and Radical Moysture"; an account that is in keeping with the description of the "naturall Vitriollique spirits" in Pw V90.[35] As this suggests, Cavendish not only offered medical advice much like that in Pw V90, but she also adapted Galenic and chymical medical concepts, some of which similarly found voice in the manuscript, in her more theoretical natural philosophical works.

While Cavendish was careful to clarify that her works fell within the scope of natural philosophy (as opposed to medicine) given their largely speculative character and focus on natural causes, she would have been aware that she was toeing a fine line in publishing on chymical and physiological topics. As noted in the previous chapter, women, including those such as Cavendish from aristocratic backgrounds, were generally encouraged to engage with medicine insofar as it helped them to compose remedies, but they were certainly not expected to express physiological views in print. Moving into the mid-seventeenth century, the role of women as medical contributors was, if anything, viewed with more suspicion than in previous decades, and, in this climate, even the female production and administration of cures became increasingly suspect. To give just one instance, James Primrose, the renowned physician and critic of William Harvey, remarked in his 1651 *Popular Errours of the People in the Matter of Physick* that medical knowledge "cannot be known but by a skilful Physician" and thus that

> women ought not to rashly and adventurously intermeddle with them. Again they usually take their remedies out of English books, or else make

[34] See Justin Begley (2021), Margaret Cavendish and Vegetable Life. *Vegetative Powers: The Roots of Life in Ancient, Medieval and Early Modern Natural Philosophy*, eds. Fabrizio Baldassarri and Andreas Blank (Dordrecht), 305–324 (especially 306–309) and Begley, *Margaret Cavendish, The Last Natural Philosopher*, 90–119.

[35] Cavendish, *The World's Olio*, 183 and Transcription, 123. On "radical moisture", see Gianna Pomata (2018), Innate Heat, Radical Moisture and Generation. *Reproduction: Antiquity to the Present Day*, eds. Nick Hopwood, Rebecca Flemming, and Lauren Kassell (Cambridge), 195–208.

> use of such as are communicated to them by others, and then they think they have rare remedies for all diseases.[36]

As Margaret Pelling has established, the attacks on women healers in mid-century England are best understood against the backdrop of the Royal College of Physicians' efforts to make good on its authority and to regulate the medical marketplace, which often involved dragging "irregular" healers before the College to face disciplinary measures.[37] Despite its best efforts, however, the College proved incapable of effectively regulating practices even in London, let alone remote corners of the country. While Pw V90 revolves around the letters of learned physicians and includes relatively few recipes by women, it provides interesting evidence that home-based medical care, and especially the production of remedies, remained a standard practice (even among high-profile Royalists) at the moment that the College was keenest to enforce regulations. Rather than slavishly adhering to the dictates of the College and its mouthpieces, the Cavendishes sought out, and sometimes paid a high price for, the best possible medical advice regardless of its origin.

Partly because Cavendish penned works that drew creatively on medical ideas during a period when women were not expected to publish on theoretical subjects, extant scholarship has tended to assume that she must have resisted medical (among other forms of) authority. The most obvious example of a study that frames her in this light is a chapter that Laura Knoppers has written on the relationship between Mayerne and Margaret.[38] According to Knoppers,

[36] James Primrose (1651), *Popular Errours of the People in the Matter of Physick* (London), 19.

[37] See especially Margaret Pelling and Frances White (2003), *Medical Conflicts in Early Modern London: Patronage, Physicians, and Irregular Practitioners, 1550–1640* (Oxford) and Deborah Harkness (2008), A View from the Streets: Women and Medical Work in Elizabethan London. *Bulletin of the History of Medicine* 82.1: 52–85. On the structure of medicine in early modernity, and, in particular, on how the tripartite division between physicians, surgeons, and apothecaries fails to capture the diversity and complexity of the medical marketplace, see Harold Cook (1986), *The Decline of the Old Medical Regime in Stuart London* (Ithaca); Margaret Pelling (1998), *The Common Lot: Sickness, Medical Occupations and the Urban Poor in Early Modern England* (London); and Fissell, The Marketplace of Print, 108–113.

[38] Knoppers, Gender, Knowledge, and the Medical Marketplace. See also the earlier and quite similar interpretation in Smith, Claims to Orthodoxy. What we say about Knoppers,

> Mayerne stresses his own knowledge of the individual patients as well as displays his book learning by translating the experience of the Cavendishes into medical terminology. Yet anxiety can be glimpsed as, in doing so, he occludes or critiques the role of the patient's own experience and knowledge, particularly when it comes to Margaret.[39]

Summoning sundry passages from Margaret's works—especially *Philosophical and Physical Opinions* and *Sociable Letters*—Knoppers juxtaposes Margaret's household-based conception of medicine and her subservient role as the patient with Mayerne's bookish, "scholastic" learning and domineering role as the physician. In practice, however, the dynamic between Margaret and Mayerne was much more complex, which we know in part because of the content of Pw V90, but also because of the broader philosophical and social commitments of the parties involved.[40]

First and foremost, Knoppers' argument that Mayerne ignores his patients' experiences with his usage of elite medical—and, according to her, male—terminology is a highly loaded way to describe these interactions.[41] It is especially difficult to make a solid case for Mayerne's "occlusion" of Margaret's experience given that the epistolary reports that William sent him about her health are now lost. But it can already be glimpsed in Pw V90 that, *pace* Knoppers, Mayerne (like most doctors of his day) considered the reported experiences that he was given along with his notes and recollections from prior examinations as he pursued a diagnosis. As he noted to William, for instance, there "is no neede of any Physician to describe a Sick=nesse if yo^u^ undertake to doe it as it appeares by your expresing so well the indisposition of my Lady, which is enough knowne to me, haveing had formerly thee Honour to Administer Phisick unto her La:^p^".[42] Mayerne thus carefully considered William's detailed

and in particular about Cavendish's relationship to medical authority, equally holds for Smith's conclusions.

[39] Knoppers, Gender, Knowledge, and the Medical Marketplace, 271.

[40] For a more sophisticated take on the power relations between physicians and patients, see Connolly, *A Dynamic Equilibrium*.

[41] For a more complete assessment of the status and treatment of female patients in early modern English medicine, see Wendy D. Churchill (2016), *Female Patients in Early Modern Britain: Gender, Diagnosis, and Treatment* (London). Note that Churchill discusses Mayerne a few times, but never as a sexist, though he is not her focus; see 3, 8, 15, 19, 21–25, 29, 33, 64, and 180.

[42] Transcription, 132.

description of Margaret's illness, which squared with his observations from earlier in-person visits. Indeed, thanks to Trevor-Roper's careful analysis of Mayerne's meticulous and voluminous notes, we know that Mayerne, while he could be demanding with patients, remained popular among them in large part owing to "the infinite care which he took in treating patients".[43] Regarding the issue of language and experience specifically, to the extent that medicine involved the pursuit of *scientiae*, Mayerne believed that it was significant to establish the causes of his patients' illnesses, as opposed to regarding their experiences as ends in themselves. Accordingly, when he wrote to William that "you ought to hope yor Principall good" comes from your diet, Mayerne proceeded to note that a healthy diet leads to the "deminuition of the Antecedent, & notable correction of ye conjunct causes".[44] Mayerne was not so much seeking to "translate experiences" as to use his knowledge of his patients' experiences to form a general picture of their physiologies and determine their illnesses, which, for a learned physician, required deploying theoretical terms.[45] When read in this light, Knoppers' claim that learned medicine occluded female experience in favour of male theory amounts to the tautology that learned medicine was *learned*: possessing theoretical knowledge and a technical lexicon was what principally distinguished university-trained physicians from other medical practitioners.

As we have seen, Mayerne identified both William and Margaret as melancholic hypochondriacs, which is a diagnosis that Knoppers also argues is gendered in a sexist way. On her account, Mayerne attributes "knowing" to William, but mere "opinion" to Margaret, and she maintains that Mayerne "makes much of his own authority and of the female patient's need to follow his advice".[46] Yet the passage that Knoppers quotes (from one of Mayerne's letters to William) does not contain an

[43] Trevor-Roper, *Europe's Physician*, 351.

[44] Transcription, 123.

[45] On authority and English physicians, see Harold Cook (1994), Good Advice and Little Medicine: The Professional Authority of Early Modern English Physicians. *Journal of British Studies* 33.1: 1–31. Mayerne, and indeed most physicians in Pw V90, clearly does not fit the "little medicine" aspect of Cook's formula, though Mayerne did seem to think that a life lived "Physically"—a term that refers to medical advice and prescription—is lived "Miserably" (Transcription, 135).

[46] Knoppers, Gender, Knowledge, and the Medical Marketplace, 272.

appeal to authority, but rather a reference to Mayerne's previous success with this cure:

> For my Lady she doth farre Exceede yo^{u} for matter of the Hypochondry. I have had hir in Cure of that disease heretofore wth good successe and the same way I led hir then, I would have hir goe againe the next Autumne, and so here after yearely, soundly Courtinge the Steele wch is onely Capable to doe hir good in the degree shee stands in neede of.[47]

The contrast that Knoppers draws between William and Margaret falls flat because, in a slightly earlier passage, Mayerne makes the same point with respect to the salt of steel solution working on William:

> If the yeare was not to farr spent or if yo^{u} had advis'd wth me sooner, after the Universall things I had Counsell'd yo^{u} the use of my Steele for many weekes together which yo^{u} have so well aproved of heretofore as well as yor Wife when she was under my hands, and I am Perswaded that both she and you must make that yor Chefest remedy for some yeares together.[48]

As the above passages show, there is no salient difference between either the courses of treatment that Mayerne recommended to Margaret and William or the degree to which he asserts his authority over the patient. If anything, Mayerne chastised William more than Margaret for not having "advis'd wth me sooner".

In distinguishing between "knowledge" and "opinion" in order to paint Mayerne as an authoritarian who discounted the experiences of an unruly female patient, Knoppers locates a string of phrases that supposedly relate to Margaret's recalcitrance. This includes another passage in a later letter in which Mayerne states:

> I believe that to Cure my Lady Marquesse your wife will be yet harder. Not so much for ye nature of the Disease, which is Rebellious, as for ye disposition of the Patient, who will not willingly submit to ye Councell of her Physichians, be they never so good & so skillful.[49]

[47] Transcription, 126.

[48] Transcription, 125.

[49] Transcription, 132.

On the basis of this passage, and others in Margaret's *oeuvre*, Knoppers argues that "Cavendish resists the impositions of her learned male physician", and instead creates and implements her own medical knowledge, "implicitly countering and critiquing male claims to medical (and scientific) authority".[50] Yet this, again, seems to be a misreading of the evidence, and, indeed, an interpretation that Cavendish would have herself cried out against. As Cavendish declared in *Philosophical Letters*, even if "it may be that some Physicians do err through their own ignorance and mistake, must we therefore condemn all the skill, and accuse all the Schools of Negligence, Cruelty, and Ignorance? God forbid: for it would be a great Injustice".[51] To shore up her argument, Knoppers interprets Margaret's extensive use of household metaphors and similes as a subversion of established medical knowledge. Yet metaphorical appeals to the still, boiling pot, and alcohol were entirely standard in early modern medicine, as well as featuring heavily in descriptions of the functions and activities of the parts of the body. Descartes, for one, analogised what the heart does to the blood with what boiling or fermentation does to wine (i.e. turns it into spirits via rarefication).[52] The ubiquity of household metaphors in natural philosophical texts is also unsurprising given that early modern kitchens and still-rooms often doubled as laboratories, and, more generally, it has already been noted that the domain of oeconomics cannot simply be gendered female, since men also contributed to household maintenance according to their interests and abilities.[53]

Whereas Knoppers regards Mayerne's chastisement of Cavendish for refusing to follow his advice as further evidence of his general impulse to discount her experiences, the issue relates less to gender than an age-old problem facing physicians: patient non-compliance. This issue tended to be especially acute when treating patients in the upper echelons of society.

[50] Knoppers, Gender, Knowledge, and the Medical Marketplace, 275.

[51] Margaret Cavendish (1664), *Philosophical Letters* (London), 353.

[52] René Descartes (1983), *Oeuvres de Descartes*, Vol. 1, eds. Charles Adam and Paul Tannery (Paris), 530–531.

[53] While Leong's *Recipes and Everyday Knowledge* focuses on medicine, a statistical study of the general role of women in the early modern English household can be found in Jane Whittle and Mark Hailwood (2020), The Gender Division of Labour in Early Modern England. *Economic History Review* 73.1: 3–32. Also see Lisa Smith (2006), The Relative Duties of a Man: Domestic Medicine in England and France, ca. 1685–1740. *Journal of Family History* 31.3: 237–256.

In a particularly flagrant example, Cademan noted in one of his early epistles that King Charles I refused to purge (i.e. to be made to vomit) despite this being a core aspect of the physic surrounding the course of steel that he had been prescribed. According to Cademan, yielding to the uncompliant Charles and allowing him to purge only once a day

> was to treate his Highnesse as a Prince for servants are ever better cured then Kings and a servant should have beene Purged .2. or .3. times before he had begun his Steele and once every sixe dayes & twice afterward.[54]

In light of this example, it would seem that Mayerne was not attempting to police gendered medical knowledge or control Margaret's behaviour because it was "characteristic of an uncooperative female, who will not submit to authority" but rather to encourage his patient, in this case a Marchioness, to follow unpleasant, arduous, and expensive courses of treatment.[55]

The most striking piece of evidence that Knoppers summons in support of Mayerne's supposed sexism is a passage from Pw V90 that reads: "It would doe her La:p good If she had some [issues] in her thyes, foure fingers above her kneese, But fayre Ladyes who have dayntie husbandes doe not willingly admit of yt subjection".[56] For clarification, the word "issues" in this passage refers to an incision made to drain an accumulation of fluid (see Appendix 8A, "Glossary of Terms"). Based on this passage, Knoppers proceeds to ask:

> Does Margaret's fairness or beauty grant her too much power over her husband, making him 'dayn-tie' (delicate, fastidious, perhaps even effeminate)? Mayerne needs to be cautious, but his point is clear, and again, tellingly, he uses language not of knowledge but of proper authority ('subjection').[57]

If Knoppers' reading is correct, then this passage undermines her argument that Mayerne respected William more than Margaret on the basis of gender. But Knoppers is probably misinterpreting the term "dayntie".

54 Transcription, 109–110.

55 Knoppers, Gender, Knowledge, and the Medical Marketplace, 272–273.

56 Transcription, 134.

57 Knoppers, Gender, Knowledge, and the Medical Marketplace, 273.

Although in early modern England it was sometimes used to describe something that is "fastidious" or "delicate", usually food (especially meat), when applied to people the term primarily signified "Valuable, fine, handsome; choice, excellent; pleasant, delightful".[58] Far from implying William's effeminacy, the use of the term "dainty" here indicates that William is an excellent and choice husband. The passage should thus be interpreted as meaning that a beautiful ("fayre") lady, who has a choice ("dayntie") husband, is probably unwilling to mar her beauty by subjecting herself to this incision, while her husband would likely refrain from forcing her to undergo the process not due to her "power over him" but because he was an affectionate husband. The term "subjection" could, as Knoppers points out, denote dominion and control, but the fact that Mayerne writes "y[t] subjection", rather than "his subjection", indicates that the reference is to a particular procedure rather than male authority.[59]

Knoppers' depiction of Margaret as an "uncooperative female" patient finally requires questioning on the basis of the latter's views on medical authority more broadly. In "An Epistle to All Learned Physicians" in *Philosophical and Physical Opinions*, Cavendish dedicated her medical reflections to those "Most reverend, and grave Fathers of health" and praised their "sage judgements, your prudent practises, your great experience".[60] In light of Knoppers' analysis, it might be tempting to read such remarks as ironic. But we have already seen that Margaret's own medical advice resonates with Mayerne's, and, beyond this, such sentiments are fully in keeping with other statements in her *oeuvre*. In *The World's Olio*, for instance, Cavendish informed readers that her next work was chiefly geared towards "Doctors" who "have been bred in the Famous Universities", for she was convinced that her recommendations would fall on deaf ears if they were picked up by "Empiricks, or Mountebanks, such as take the Name, and never studied the Science, whose Practice is rather to kill than to cure".[61] She insisted elsewhere that she was "absolutely of

[58] "Dainty" (*OED*, adj. 1). It might also imply being held in estimation, honour, favour; esteem, regard; affection, love (*OED*, n. 1).

[59] "Subjection" (*OED*, n. 1).

[60] Cavendish, *Philosophical and Physical Opinions*, 99.

[61] Cavendish, *The World's Olio*, 200. One of William's manuscripts contains a witty couplet titled "Of Docter Emprick alias Mountebank" that expresses a similar sentiment: "Your art is Unserteyne, what then is your Skill? / Tis ~~for~~ to Gett Moneye; for your

opinion, that the Practice of the Schools is the best and wisest Practice, as well in making Issues, letting blood, Purging by Siege or Vomits, as any other means used by them", and clarified that, in her estimation, the best housewives should closely follow the advice of learned physicians (even if she often avoided doing so in practice).[62] Praising Galenic medics as well as Paracelsian chymists once again in *Philosophical Letters*, Cavendish proclaimed that

> without making now any difference betwixt the *Galenists* and *Paracelsians*, and examining which are the best, (for I think them both excellent in their kinds, especially when joyned together) I will onely say this in general; that the Art of Physick has never flourish'd better then now, neither has any age had more skilful, learned, and experienced Physicians, then this present.[63]

Far from evincing a "clearly skeptical stance toward medical authority", as Hilda Smith has argued, Margaret prized the learning and expertise of university-trained physicians, and especially appreciated those such as Mayerne who coupled Galenic and chymical treatments.[64]

In a more nuanced study, Sandy Feinstein has drawn attention to a disagreement that Cavendish had with van Helmont and an unnamed Irish physician in the course of commenting on Cavendish's supposed opposition towards *chymical* medicine.[65] Feinstein correctly points out that Cavendish argued against van Helmont's advice to use wine in fevers, which was in part because she was following the advice of Davisson, a medical authority who, as noted, was central to Pw V90 as well as Harley 6491. Instead of wine, Cavendish recommended a tisane recipe that includes barley and lemons, and that is almost certainly the one that Davisson provides in Pw V90.[66] She thus called upon one authority (and her own experiments and experiences) to disagree with another. Given

Medicins kill" (Nottingham University Library, Pw V25, f. 142). For more on attitudes towards mountebanks at this time, see M.A. Katritzky (2007), *Women, Medicine and Theatre 1500–1750: Literary Mountebanks and Performing Quacks* (London).

[62] Cavendish, *Philosophical Letters*, 383.

[63] Cavendish, *Philosophical Letters*, 352.

[64] Smith, Claims to Orthodoxy, 25.

[65] Cavendish, *Philosophical Letters*, 385–386.

[66] The recipe seems to be Transcription, 155; also see Mayerne's recipe on Transcription, 124.

that Feinstein recognises this fact, it is strange that she proceeds to draw the conclusion that, in this context, Cavendish attempted to "devalue, and diminish" male medical authorities.[67] Feinstein further argues that Cavendish used rhetorical misdirection to praise van Helmont's opinion before undermining it with the statement that "I am utterly against it" (supposedly referring to the use of wine in fevers that he recommended). But Feinstein here conflates two of Cavendish's claims. The first of van Helmont's opinions, which Cavendish approved of, is that "*Drink*" in general "*ought not to be forbidden in Fevers*".[68] But Cavendish then circumscribed her agreement by arguing, in contrast to van Helmont, that one should not drink *wine* in fevers. This subtlety is important, since it further demonstrates that Cavendish did not reactively oppose van Helmont's ideas—or medical authority in general—but rather picked out and praised those medical positions and authorities that she found most convincing or valuable.

The studies of Knoppers, Smith, Feinstein, and others should alert us to the caution required when reading Pw V90 (and Cavendish's wider *oeuvre*) through the lens of gender, especially considering the lengths that Margaret went to in order to distinguish herself from suspect female authors and practitioners. They should also indicate the pitfalls of approaching medical "authority" as a fixed category that Cavendish spurned outright. On the contrary, we have seen that Cavendish generally appreciated the views of medical authorities and the institutions in which they were trained, even if she occasionally disagreed with certain ideas or ignored advice, and the fact that Mayerne was granted pride of place in Pw V90 (closely followed by Davisson and Cademan) further attests to the high esteem in which medical professionals were held within the Cavendish household. Mayerne was not only a member of the Royal College and physician in ordinary to Charles I and II but also the driving force behind the *Pharmacopoeia Londinensis*, the official set of medical recipes that apothecaries were permitted to create.[69] As already noted,

[67] Feinstein, Experience, Authority, and the Alchemy of Language, 142.

[68] Cavendish, *Philosophical Letters*, 385.

[69] On this document and its evolution, see Clare J. Fowler (2018), *Pharmacopoeia Londinensis 1618 and Its Descendants* (London).

the fact that many of the recipes and letters in Pw V90 were from high-profile learned medics is a striking feature that differentiates it from most of the recipe collections that Leong and Stobart have described.

Turning to the letters of apothecaries and other physicians in Pw V90 leaves no question as to why the seventeenth-century *literati* put their trust in Mayerne. Against the bland scribbles of local doctors, Mayerne expended significant energy composing all of his letters. At one moment, for example, he discussed how "ye passion of ye minde: acting fierc^{e}ly upon ye body and upon its humers doe contribute much to ye generating: of ye materiall causes".[70] William and Margaret would have undoubtedly appreciated Mayerne's willingness to spell out the causes of their diseases, and to take apart and explicate the effects of these causes in a manner that accorded with an Aristotelian-Galenic orientation. The other pillar of traditional seventeenth-century medicine was of course the Hippocratic corpus, and, in response to William's queries about new-fangled cures, Mayerne noted that "there is another balsum caled balsum de tolu: which is very good: but I remember that hipocrates saith: sape optima medecina est non uti medecina" ["often the best medicine is not to use medicine"].[71] Recognising that untested cures gain their authority merely from hearsay, Mayerne also confirmed that he held firm to the Hippocratic adage that "experience is my reason".[72] In addition to integrating ancient principles and authorities seamlessly into his epistles, Mayerne flaunted his conversance with the latest literary and philosophical developments preoccupying members of the Republic of Letters. In a letter of 1649, for instance, he informed William that his melancholia arose "as the Pas=sions of the Soule (whereof the tymes furnish yo^{u} wth subject Enough) doe move yor speritts, and Humours", which is probably a nod to Descartes' *Les passions de l'âme*, published earlier that year.[73] At least in Pw V90, learned physicians appear to have been healers and men of letters in equal measure, and their epistles accordingly functioned not only as storehouses of recipes but also as capsules of philosophical and ancient wisdom.

[70] Transcription, 213.

[71] Transcription, 221.

[72] Transcription, 124. "*Experimenta*" during the Middle Ages and into the Renaissance referred to therapies that had to be justified on empirical evidence; not just doctrine. The classic study is Peter Dear (2006), The Meanings of Experience. *The Cambridge History of Science* 3: 106–131.

[73] Transcription, 122.

As well as housing philosophical and methodological musings, Pw V90 is packed with recondite reflections on the modalities of therapy (e.g. via glister, oil, or salve), the medicinal properties of ingredients like antimony, and frank surgical and dietary advice. William and Margaret were respected as intelligent patients who desired to understand prescribed treatments and were (mostly) capable of doing so. Beyond their intellectual curiosity, the Cavendishes appear to have hoped that such explanations might help them to decide between the sometimes-competing counsel that they procured from practitioners. The fact that William solicited Mayerne's opinion on recipes or courses of treatment that other physicians or amateurs had recommended suggests that he was, indeed, fully aware that even the advice of learned physicians was not created equal. While William (like Margaret) was sometimes displeased with Mayerne's instructions, especially to avoid tobacco, the Cavendishes persistently requested his direction even after he pleaded with them to "have better opinion of ye phisitians amongst whome you live" in Paris.[74] Mayerne's recommendation that the Cavendishes seek out local physicians—despite the tidy sums that he received from them—accords with a general desire among learned physicians to base their treatments on the humours of individual patients as well as on how these interacted with environmental factors that were difficult to discern from afar. As Mayerne put it, "health is subject to daly mutabilities and there is noe security to deale with a body at a great distance".[75]

One of the reasons why Pw V90 is more erudite than many manuscript recipe collections described in the literature is precisely because it is *less* grounded in a local community surrounding an estate.[76] As a document that emerged from the combined interests of the Cavendish household, none of the entries in Pw V90 are marked as being passed down from a father or mother to a son or daughter, and it only contains a handful of recipes from kin and none from village "goodmen" or "goodwives". If its roots lay in any one context, it was, as noted, the elite and international setting of Henrietta Maria's court in which courtiers were actively

[74] Transcription, 215.

[75] Transcription, 220.

[76] On the significance of locales in the history of science and medicine, see Lynette Hunter (2005), Women and Science in the Sixteenth and Seventeenth Centuries—Different Social Practices, Different Textualities, and Different Kinds of Science. *Men, Women, and the Birthing of Modern Science*, ed. Judith P. Zinsser (DeKalb), 123–140.

encouraged to partake in experimental and philosophical activities, and to work on and exchange recipes and advice. Along with containing items from members of Henrietta Maria's circles, Pw V90 includes entries from slightly earlier courtiers such as Thomas Aylesbury (great-grandfather to Queen Anne and Queen Mary II), Sir Walter Raleigh, and his wife Elizabeth Raleigh. In such high society, recipes and recipe collections were valuable not only for their healing potential but also as records of obligations, debts, duties, and favours between family members, friends, acquaintances, and indeed servants within an intricate gift economy.[77] In keeping with the all-important role of testimony in the seventeenth century, the consensus was that a recipe with the support of a trustworthy or expert donor, and the attestations of patients for whom the course of treatment had worked, was of the highest quality.[78] So as to establish the *bona fides* of his salt of steel, for example, Mayerne reported that it had previously healed "his Highnesse" (Charles I).[79] More generally, the bombardment of William and Margaret with recipes, which Mayerne usually advised them against using, can be explained, at least in part, as the expression of a desire among the wealthy and learned to establish and extend financial, social, and familial bonds.

As regards the issue of financial bonds in particular, it is significant that William notoriously struggled to procure funds and balance his debts during the Civil Wars, which is a recurrent theme in Margaret's 1667 biography, *The Life of William Cavendishe*.[80] William's ability to borrow and spend thus depended on the size and quality of his networks as manifested, to a certain extent, through a non-monetary gift economy that included recipe-exchange.[81] We know that in reciprocation for Digby's

[77] See Leong, *Recipes and Everyday Knowledge*, 132–133; Stobart, *Household Medicine*, 30; and Rankin, *Panaceia's Daughters*, especially 93–203.

[78] For a fuller discussion of testimony, see Richard Serjeantson (1999), Testimony and Proof in Early-Modern England. *Studies in History and Philosophy of Science* 30.2: 195–236.

[79] Transcription, 109.

[80] Margaret describes, for instance, how William could charm his creditors, and make "such an impression in them, that they had all a deep sense of my Lord's Misfortunes: and instead of urging the payment of his Debts, promised him, That he should not want any thing" (Margaret Cavendish (1667), *Life of William Cavendishe* (London), 73).

[81] Leong, *Recipes and Everyday Knowledge*, 37–40.

recipes, William sent him a "Medicum Arcanum" against "Licentia veneraria" (or syphilis), which Digby then passed onto the physician Frederick Clod, son-in-law of the Prussian intelligencer, Samuel Hartlib.[82] A recipe collection could also serve as a ledger of social bonds and thus help one to acquire letters of credit or permission to delay payments, which was particularly important when abroad. When William moved to Antwerp, for example, he managed to obtain a loan from George Villiers, 2nd Duke of Buckingham, which he subsequently used to impress a number of Antwerp shopkeepers, and to convince them to keep lines of credit open to his household.[83] William of course relied on these various shops for ingredients that did not grow in the Rubenshuis gardens but that were needed to concoct the recipes in Pw V90. All of this is to say that recipes had a dual function in an early modern credit system: they helped to resolve issues surrounding monetary debts and to increase or prove one's status within a hierarchical courtly culture.[84]

Despite his financial debts, William was a prominent nobleman and cavalier whose networks encompassed powerful men and women across England, France, and the Low Countries, and who patronised poets and philosophers from Ben Jonson and William Davenant to Thomas Hobbes and René Descartes, not to mention numerous painters and architects. William was consequently a tempting target for anyone who sought to curry favour or establish a kinship bond with a well-connected aristocrat. One figure with ample reason to make inroads with the marquess (and later duke) was the aforementioned Robert Dudley, the illegitimate son of Robert Dudley, Earl of Leicester, and grandson of the previous Duke of Northumberland, John Dudley. Unable to claim his grandfather's title in England, Dudley left the country in 1605 and was eventually recognised as the Duke of Northumberland by Emperor Ferdinand II (see Appendix 8D). Cultivating a friendship with William would have been attractive to Dudley for several reasons. Firstly, by sending William recipes under his newfound title, Dudley espied an opportunity to shore up his legitimacy with a confidant of the monarchy. Secondly, Dudley never managed to restore his estates or titles in England, and he would have realised

[82] The Hartlib Papers, 29/4/21A.

[83] Worsley, *Cavalier*, 175.

[84] For the latter type of credit, see Mario Biagioli (2006), *Galileo's Instruments of Credit* (Chicago), especially 21–44.

that a Royalist of high repute such as William might make for a useful intercessor if, or when, the monarchy was restored. Lastly, Dudley took a keen interest in medicine and technological innovations—as is evident from his sundry ship-building as well as map and instrument-making pursuits—and making inroads with William, as a man with similar interests, might prove helpful when he was seeking funds for future enterprises or voyages.[85] Although it is impossible to say whether William reciprocated and sent Dudley recipes, establishing a friendship with a man who had the ear of the Grand Dukes would hardly have been unattractive, given that William was in dire financial straits.

In addition to reflecting social, financial, and kinship bonds, Pw V90 also sheds light on the mores of the courtly setting in which it was compiled, and particularly the role of aristocratic women in devising and disseminating recipes. In addition to many manuscripts, there were three popular printed recipe collections published in the first half of the 1650s by members of the court in exile: Elizabeth Talbot Grey's *A Choice Manual of Rare and Select Secrets* (1653), Queen Henrietta Maria's *The Queens Closet Opened* (1655), and the Countess of Arundel, Aletheia Talbot Howard's *Natura Exenterata* (1655). The first two were bestsellers well into the 1670s. In the seventeenth century, these volumes seem to have been widely regarded as the products of the authors with whom they were associated, as is indicated by a note in which Hartlib attributes *Natura Exenterata* solely to Talbot Howard.[86] Yet all of the above were published posthumously, and scholars have debated the degree to which they were collaborative volumes or even composed by men but associated with the named women. In the case of *A Choice Manual*, for example, the publisher, W. I. Gent (perhaps one William Jervis), intimated in the preface that the collection was made for Elizabeth's *use*, writing that "this small Manuall" was "once esteemed as a rich Cabinet of knowledge, by a person truely Honorable".[87] Until the mid-seventeenth century, printed recipe collections and housewifery manuals

[85] His collection of scientific instruments is now held at the Museo Galileo in Florence.

[86] Hartlib notes that he must send Robert Boyle "Natura exenterata written by *the* countesse of Arundel" in exchange for "Basilius Valentinus" (The Hartlib Papers, 26/72A-B).

[87] Elizabeth Talbot Grey (1653), *A Choice Manual of Rare and Select Secrets* (London), A3r. For why these works were not written by the women with whom they are associated, see Leong, *Recipes and Everyday Knowledge*, 150–154.

(despite their target audience) had been almost exclusively written by and associated with men.

Whatever the case may be, the above publications helped to clear the way for women to publish recipe books in the proceeding decades, most famously Hannah Woolley with her 1670 *The Queen-like Closet*. The explosion of publications, including many by English women, during the Interregnum is commonly imputed to a greater freedom to print at this tumultuous time and tends to be associated with radical religious sects. Yet none of the above recipe books, as some of the earliest female-attributed technical publications in Europe, were credited to radicals but rather to the Queen and elite women in her court.[88] In this sense, the political turmoil was probably only an indirect spur insofar as it convinced Royalists of the need to prove to the population the indispensability of the monarchy to the metaphorical and indeed literal health of English society. Henrietta Maria's collection in particular is (as Knoppers has convincingly argued) an intriguing piece of political rhetoric that aimed to persuade sceptical English Protestants that this foreign-born, Catholic Queen was, notwithstanding her faith and status, a hard-working Englishwoman and a good housewife.[89] Grey and Howard, for their part, were scions of the well-connected and politically powerful Talbot family, and Howard was in fact William Cavendish's cousin (being the daughter of his aunt, Mary Cavendish (1556–1632), and her husband, Gilbert Talbot (1552–1616)). Rather than being limited to a coterie of exiled Royalists, their recipes circulated far and wide even before their popular publications appeared, and two of their respective recipes—"The Countes of Kents Powder" and "Gascons Powder" from "ye Countesse of Arendell"—also feature in Pw V90. Grey's powder is the most famous and often-copied recipe from *A Choice Manual*, but the variation found in the Cavendish manuscript is far less specific than in the printed book, and was probably recorded in 1649, a few years prior to the published version. The recipe

[88] On the publication of women with an emphasis on its radical political and religious roots, see Patricia Crawford (2005), Women's Published Writings 1600–1700. *Women in English Society, 1500–1800*, ed. Mary Prior (London), 158–209; Fissell, The Marketplace of Print; Joad Raymond (2006), *Pamphlets and Pamphleteering in Early Modern Britain* (Cambridge), especially 161–201; and Donald F. McKenzie (2002), The London Book Trade in 1644. *Making Meaning: "Printers of the Mind" and Other Essays*, eds. Peter D. McDonald and Michael F. Suarez (Amherst), 126–143.

[89] Knoppers, Opening the Queen's Closet.

for Howard's powder was transcribed around the same time, and, as intimated, might even be considered, rather loosely, as a "family recipe". Both Grey and Howard were close friends of Henrietta Maria, with Grey heading into exile with the Queen as early as 1642. Margaret would have almost certainly met Grey when she joined the Queen's court in 1644.[90]

The first of the medical publications associated with a Royalist woman (i.e. Grey's *A Choice Manual*) appeared during the brief interval in which the Cavendishes ceased to work on their medical manuscript, when Margaret herself made the bold decision to publish. While she occasionally framed Queen Elizabeth I as a female literary model, Cavendish more proximately participated in a trend among Royalist women in her immediate courtly circle to associate their names with works that address medical topics.[91] As Cavendish was sure to underscore, however, her more philosophically orientated output was of a rather different character than the recipe books of her female contemporaries.

[90] Hunter, *Women and Domestic Medicine*, 90.

[91] On Cavendish's admiration for Elizabeth, see Brandie Siegfried (2007), *Bonum Theatrale*: The Matter of Elizabeth I in Francis Bacon's *Of Tribute* and Margaret Cavendish's *The Blazing World*. *Resurrecting Elizabeth I in Seventeenth-Century England*, eds. Elizabeth Hageman and Katherine Conway (Madison), 185–204.

CHAPTER 5

Ingredients

Early modern European *pharmacopoeiae* were consistently more diverse than their ancient and mediaeval precursors, dealing not just with the legacies of these medical traditions, but also with an enlarged set of chymical insights and ingredients that had only recently been introduced to Europe. Two collections of pharmaceutical knowledge were particularly important in England during the seventeenth century: the official pharmacopoeia of the Royal College of Physicians, the *Pharmacopoeia Londinensis* (1618), and the unofficial (and, in some ways, radical) English translation and commentary on that pharmacopoeia, *A Physicall Directory* (1649), by the apothecary, astrologer, and herbalist Nicholas Culpeper.[1]

The *Pharmacopoeia Londinensis*, which was first published in 1618 and replaced by a revised edition within a year of its publication, formed a crucial part of the Royal College of Physicians' attempt to regulate

[1] For an account of Culpeper's radical politics and life, see Benjamin Woolley (2004), *The Herbalist: Nicholas Culpeper and the Fight for Medical Freedom* (Toronto). On his medical and astrological ideas, see Olav Thulesius (1992), *Nicholas Culpeper: English Physician and Astrologer* (New York).

J. Begley and B. Goldberg, *The Medical World of Margaret Cavendish*, Palgrave Studies in Medieval and Early Modern Medicine,
https://doi.org/10.1007/978-3-030-92927-5_5

medical practice.[2] The College began to plan the ambitious document in around 1585. From the outset, it contained hundreds of recipes, with ingredients ranging from common herbs such as St. John's Wort and animal parts like fox's lungs to chymical preparations including oil of sulphur. Its creation, as noted, largely resulted from the efforts of Mayerne, who not only wrote the preface to the first *Pharmacopoeia*, but also supplied many of its recipes. The combination of Galenic herbalism and Paracelsian chymistry in the *Pharmacopoeia*, in particular, reflects Mayerne's own medical predilections.[3] While the *Pharmacopoeia* largely failed as a weapon in the College's battle to regulate medical practice (with censorship proving a marginally more effective tool), it was nonetheless remarkably successful in shifting ideas of pharmacy towards an ecumenical position between traditional herbalism and the less conventional mineral and chymical ingredients that dramatically increased in popularity in the wake of the widespread adoption of Paracelsian and Helmontian medical ideas.[4]

Even beyond the relative regulatory failure of the original *Pharmacopoeia*, Culpeper's translation of the text dealt a further blow to the College's attempts to control pharmacy. Not only did it provide the non-Latin reading public with the full range of the College's medical recipes, but it also supplied its readers with a witty and often critical commentary on most of the recipes and even upon some of the principles of physic that the institution championed. Indeed, as Spiller and others have pointed out, Culpeper addresses his work to one group that the College specifically targeted in their war on non-licenced physic: those "kind Gentlewomen [...] who freely bestow your pains, brains and cost, to your poor wounded and diseased neighbours".[5] In his posthumously published

[2] The revised edition appeared on 7 December 1618. See William Brockbank (1964), Sovereign Remedies: A Critical Depreciation of the 17th-Century London Pharmacopoeia. *Medical History* 8.1: 1–14.

[3] See Allen Debus (2002), *The Chemical Philosophy: Paracelsian Science and Medicine in the Sixteenth and Seventeenth Centuries* (New York), 182–191 and Fowler, *Pharmacopoeia Londinensis*.

[4] See Spiller, Recipes for Knowledge, 55–72.

[5] Nicholas Culpeper (1649), *A Physicall Directory, or, A Translation of the London Dispensatory Made by the Colledge of Physicians in London* (London), A4r. See Spiller, *Cooking, Physic and Chirurgery in the Works of Elizabeth Talbot Grey and Aletheia Talbot Howard*, xxx; Hunter, Women and Domestic Medicine; and Margaret Pelling (1997), Thoroughly Resented? Older Women and the Medical Role in Early Modern

School of Physick (1659), Culpeper denounced physicians as "Monopolizers" of knowledge, and linked the search for cures to the search for religious truth that had roiled mid-century England. As he exclaimed, the "truth is, throughout the whole World there are not such slaves to Doctors, as the poor English are; most of them profess themselves Protestants, but their Practices have been like those of the Papists, to hide the grounds of Physick from the vulgar".[6] Culpeper's work was significantly more popular than its Latin original not just because it was published in English but also because it was far more useful, with its extensive commentaries on the uses of the various preparations and cures, as well as astrological notes on the proper times for collecting herbs and applying medicines (which were not to be found in the Latin version).[7] But, even as Culpeper opened the floodgates to vernacular medical publishing, the English pharmaceutical texts that emerged in his wake were by no means beholden to his specific ideas, as in the case of the aforementioned works associated with women at Queen Henrietta Maria's court in exile.

In spite of the multifaceted impact of Culpeper's work, it did have a standardising effect insofar as it contained the vast majority of the ingredients utilised in seventeenth-century household recipe collections, including Pw V90 (though, as we will see, those in Pw V90 tend to be on the more expensive end). Derived from animals, minerals, and herbs, some of the approximately 250 ingredients in Pw V90 were taken as simples, whereas others (such as *crocus metallorum*) were wrought through complex chymical procedures. While many of the products in Pw V90 could have easily been sourced from local gardens or slaughterhouses, Ursula Härting has pointed out that Rubenshuis, in particular,

London. *Women, Science and Medicine 1500–1700*, eds. Lynette Hunter and Sarah Hutton (Thrupp), 63–88.

[6] Nicholas Culpeper (1659), *Culpeper's School of Physick* (London), A8v. See Andrew Wear (1992), The Popularization of Medicine in Early Modern England. *The Popularization of Medicine, 1650–1850*, ed. Roy Porter (New York), 17–41 (30–33) and Mark Jenner and Patrick Wallis (2007), Introduction. The Medical Marketplace. *Medicine and the Market in England and Its Colonies*, c. *1450–c. 1850*, eds. Mark Jenner and Patrick Wallis (London), 1–23.

[7] On the larger context of magical medicine in England, with some discussion of Culpeper, see Lauren Kassell (2007), Magic, Alchemy and the Medical Economy in Early Modern England: The Case of Robert Fludd's Magnetical Medicine, *Medicine and the Market in England and Its Colonies*, c. *1450–c. 1850*, eds. Mark Jenner and Patrick Wallis (London), 88–107.

had a magnificent garden, and the famous "Portrait of a Married Couple in the Park" by the celebrated Flemish painter Gonzales Coques tellingly depicts William and Margaret gesturing proudly towards the garden's gate.[8] Upon entrance, one allegedly encountered myriad fruit trees that afforded cherries, peaches, oranges, lemons, melons, and apples, nestled between roses, tulips, marigolds, violets, and chamomile, all of which were thought to have had medical utility.[9] But a spread of other ingredients in Pw V90 required importation from the Americas, China, and various regions of Africa, including fantastic items such as the "orientall-bezarstone", a stony, mineral-rich growth found in the digestive system of certain ruminants. In their study of seventeenth-century recipe collections, Leong and Pennell concluded that "Domestic medicine was reliant upon easily accessible ingredients rather than newly introduced exotics, such as Peruvian bark, or the metallic and mineral components advanced in Paracelsian and Helmontian pharmacopoeia".[10] Indeed, a debate raged for most of the seventeenth century about the value of indigenous versus exotic medicaments, with many authors arguing that imported ingredients and medicines like tamarind or balsam were "more helpful for enriching pharmacists, than for curing sick people".[11] But given the ubiquity of exotics within Pw V90, the manuscript is once again an exception.

Some quantitative information will allow for a more precise understanding of the ingredients in Pw V90, but a few comments are first required about the tables that follow. First of all, the various oil and spirit preparations (e.g. oil of vitriol and spirit of wine) are not included in the calculations of their base ingredients (i.e. vitriol or wine), though simpler preparations such as conserves, syrups, or waters (of, e.g., rose) are

[8] See Ursula Härting (2002), Lord William Cavendish und Duchess Margaret Cavendish im Rubensgarten in Antwerpen. *Jahrbuch der Berliner Museen* 44: 15–28.

[9] Nottingham University Library, Pw V/26, f. 127. More generally, on the grounds of Rubenshuis during their stay, see Ben van Beneden and Nora de Poorter (2006), *Royalist Refugees: William and Margaret Cavendish in the Rubens House, 1648–1660* (Antwerp) and James Fitzmaurice (2000), Margaret Cavendish in Antwerp: The Actual and the Imaginary. *In-Between* 9.1–2: 29–40.

[10] Leong and Pennell, Recipe Collections and the Currency of Medical Knowledge, 136.

[11] Jan van Beverwyck (1644), *Autarkeia Bataviae, sive introductio ad medicinam indigenam* (Leiden), 68. This passage and the larger context are discussed in Alix Cooper (2003), The Indigenous Versus the Exotic: Debating Natural Origins in Early Modern Europe. *Landscape Research* 28.1: 51–60.

included. Secondly, complex ingredients that required a host of components are also excluded. For instance, sugar is a common ingredient in many recipes, but we do not count the sugar used to prepare, say, *aqua mariae* in the count of sugar as used in the recipes. Finally, numerous ambiguities make a precise count difficult. In the case of "The Earle of Chesterfields Red Diaphorettick Powder", for example, use is made of both coral and white coral.[12] It is debatable whether these should be counted as one or two ingredients: we have opted for the latter. As all of this suggests, much like in the case of calculating letters and recipes, different criteria of demarcation would yield different counts. Nothing that follows hinges on the precise count, however, since what is important is the overall magnitude of ingredients and the similarities and differences to those found in other collections.

To begin with, Table 5.1 compares the basic (and more traditional) plant-based ingredients in Pw V90 (enumerated in order of frequency) with Stobart's list of the twenty most common herbs as determined by her study of early modern printed recipe books and manuscripts.[13]

Since we have seen that Pw V90 departs markedly from Stobart's findings on several fronts, it is all the more striking that these parallel lists of twenty herbs correspond so closely. With that being said, several of the most expensive ingredients (including cinnamon, clove, and saffron) appear with greater frequency in the Cavendish manuscript than in Stobart's average collection.[14] The magnitude of this difference is partly a function of the sample size: we are comparing Stobart's set of twenty-two collections to a single collection, and, because of the size of Stobart's sample, her range of ingredients is much larger, which means that the incidence of any one ingredient is lower. But the general ordering of particular ingredients is still interesting insofar as it underscores how a recipe collection such as Pw V90 can reflect the social status of its compilers.

[12] Transcription, 187.

[13] Stobart, *Household Medicine*, Table 3.2.

[14] Stobart, *Household Medicine*, Appendix 2, 184–187. For the price of spices until the 1580s in London, Cambridge, and Oxford, see James E. Thorold Rogers (1872), *A History of Agriculture and Prices in England, Vol. IV: 1401–1582* (Oxford), 523–543. Also see John Munro (1988), The Consumption of Spices and Their Costs in Late-Medieval and Early-Modern Europe: Luxuries or Necessities? https://www.economics.utoronto.ca/munro5/SPICES1.htm.

Table 5.1 Stobart's most common herbal ingredients and those same ingredients listed in order of instances found in Pw V90

Stobart's list	*Frequency (%)*	*Pw V90*	*Frequency (%)*
Rose	2.9	*Rose*	14
Rosemary	2	*Ale*	12.7
Sage	1.8	*Balm*	12
Liquorice	1.7	*Nutmeg*	9.3
Aniseed	1.6	*Rosemary*	8.7
Nutmeg	1.5	*Chamomile*	8.7
Cinnamon	1.5	*Cinnamon*	8
Wormwood	1.5	*Clove*	7.3
Fennel	1.4	*Sage*	6.7
Rue	1.4	*Saffron*	6.7
Ale	1.4	*Fennel*	6
Borage	1.3	*Borage*	4.7
Chamomile	1.2	*Plantain*	4
Betony	1.1	*Wormwood*	3.3
Plantain	1.1	*Liquorice*	3.3
Clove	1	*Aniseed*	3.3
Saffron	1	*Betony*	3.3
Raisin	1	*Rue*	2.7
Mint	1	*Mint*	2
Balm	0.9	*Raisin (incl. currants)*	2

In terms of prices, according to Stobart, who analysed the Clarke household accounts (from 1685 to 1702), cinnamon was about 100d per pound and nutmeg was 6d per pound (though its price had spiked to 18d by 1701). Since we do not have the relevant accounts of the Cavendish household, we have turned to the work of the aforementioned Gideon Harvey, who, in this case, provided similar prices in *The Family-Physician*. Gideon's charting of costs was part of a wider attempt to encourage householders to compose their own remedies according to his principles. Indeed, he went so far as to advertise *The Family-Physician* as showing "that in preparing medicines thus at your own houses, that it's not onely a far safer way, but you shall also save nineteen shillings in twenty". Because his work would, in essence, make the apothecary redundant, it was reviled by many, with a piece of anonymous doggerel titled *The Apothecaries Vindication* castigating Harvey thus: "A Scavenger his dung doth fling abroad, / Bespattering such as pass by in his Road. / Just so a Scribling Quack with his Goose-quill / Men in their good names strive to wound or kill".[15] Harvey's price list, however, not only helped many contemporaries, but also provides present-day historians with a useful guide. In his estimation, cinnamon should cost 8s per pound (which is not far from 96d) and nutmeg 6d per pound, which is identical to Stobart's pricing.[16]

Having compared the herbs in Stobart's study with those in Pw V90, it is possible to turn to the most common ingredients, herbal or otherwise, in the manuscript. In Table 5.2, we have sorted the ingredients according to their frequency, and have, where possible, included prices in pence (d) per pound. We assume the following conversions for weight, which, while common, were not universal: 1 lb = 12 oz = 96 drams = 288 scruples = 5760 grains.[17] For monetary conversions, we use the following: 1 pound (*£*) = 20 shillings (*s*), and 1 shilling = 12 pence (*d*).

Overall, the ingredients in Pw V90 were tremendously expensive, averaging a mean price of between 1281 and 1305 d/lb for the 19 ingredients with available price data. Even if we exclude ambergris—a waxy substance that originates in the intestines of sperm whales, and was so expensive that it significantly skews the result—the average ingredient is still 73 to 98 d/lb. To put this into perspective, the weekly income of the average labourer c.1640–1649 was approximately 126d, and that of a craftsman about 210d.[18] It is notable, however, that we lack prices for some of the most common ingredients in Pw V90. These are mostly herbal simples that were grown in Britain, northern France, or the Low Countries: betony, sage, rosemary, borage, and wormwood. While not a simple, another common ingredient for which we lack a price is ale. Since the above were commonly grown in medicinal/herbal gardens, collected by local herbwives and other household members, or, as with ale, available at most grocers and widely brewed in the household, there was little need for Harvey to list their prices. More oddly, Harvey also excluded the price for balm (or balsam), which was not grown at any scale in Britain (or indeed Europe) until the nineteenth century. Why balm is neglected is unclear, as are Harvey's general criteria for inclusion. In any case, except for balm,

[15] Anonymous (1676), *The Apothecaries Vindication: In Answer to an Abusive Book Entitled The Family Physician and House Apothecary* (London), unpaginated.

[16] Harvey, *The Family-Physician, and the House-Apothecary*, 114–139.

[17] We use 12 ounces per pound instead of 16 ounces, since this is a more common measurement in the historical literature.

[18] Jeremy Boulton (1996), Wage Labour in Seventeenth-Century London. *The Economic History Review* 49.2: 268–290 (279).

Table 5.2 Thirty of the most common ingredients in Pw V90, with prices from Harvey 1676

Ingredient	*Price (d/lb)*	*Instances in Pw V90*
Wine (incl. sack, muscadel)	N/A	55
Rose (incl. syrup and water of roses)	48	21
Violets	N/A	20
Ale (incl. posset)	N/A	19
Balm (incl. balm water and balsam)	N/A	18
Sugar	N/A	17
Nutmeg	6	14
Chamomile	16	13
Sulphur (incl. brimstone)	8	13
Rosemary	N/A	13
Cinnamon (incl. cassia)	96	12
Cloves (incl. gillyflower)	160	11
Rhubarb	168	10
Sage	N/A	10
Tartar (incl. cream of tartar)	4 to 8	10
Spirit of Wine (incl. aqua vitae)	N/A	10
Amber	48	9
Senna	48 to 64	9
Manna	24 to 64	9
Crocus metallorum	64	8
Saffron	288 to 648	8
Gold	N/A	7
St. John's Wort	N/A	7
Sanders (sandalwood, red and yellow)	12 to 32	7
Ambergris	23040	7
Hartshorn	8 to 18	7
Mace	160	7
Vitriol	N/A	6
Juniper Berries	4	6
Sal chalybis (salt of steel)	144	6

gold, and perhaps the spirit of salt, the ingredients for which we lack prices were probably inexpensive. While it is therefore likely that the above calculation overestimates the average price, the ingredients in Pw V90 were undoubtedly costly compared to those in other recipe collections and beyond what most families could have afforded. We do not know exactly how many of these cures the Cavendishes prepared, but internal evidence from Pw V90 indicates that, at the very least, they procured the ingredients required to make Mayerne's cures, and it is safe to say that even these purchases would have stretched the resources of the, by then financially unstable, Cavendishes.

Beyond questions of price, most of the plant-derived ingredients in Table 5.2 (including wine, ale, rose, balm, chamomile, rhubarb, juniper berries, and anis seeds) have their medical origins in Greek antiquity, though some are described in mediaeval medical recipes as stemming from further afield (like nutmeg from Indonesia). Most of the other ingredients in the top thirty are mixtures of newer ones, especially stemming from minerals and chemical sources such as tartar, salt of steel, spirit of wine, antinomy, and saltpetre. These ingredients became increasingly popular in a post-Paracelsian environment, when chymical medicine precipitated an explosion of interest in the medicinal value of minerals, animals, and stones. As the divergences between Tables 5.1 and 5.2 indicate, Pw V90 itself embodies this shift: it not only contains a profusion of traditional herbal cures, but it is also deeply chymical in character. In this regard, as already suggested, it is significant that Mayerne was famous for his controversial attempt to iron out the tensions between Galenic and chymical medicine, arguing as early as in his *Apologia* (1603) that chymical remedies did not violate the principles or methods of Galen (see Chapter 6A for a further discussion).[19] And yet he was a leading light of the Royal College of Physicians, and Pw V90 makes clear that Galenic theory and terminology (humoural balance, hot and cold temperaments, seasonality, etc.) were indispensable to his diagnostic practices.[20] More generally, there is an attempt throughout the manuscript to harmonise traditional simples (like the fungus agaric or turmeric powder) with chymical ingredients (like *crocus metallorum*, tartar, or brimstone).

[19] See Trevor-Roper, *Europe's Physician*, 343, 356–567, and 418–419 along with Didier Kahn (2007), *Alchimie et Paracelsisme en France à la fin de la Renaissance* (Geniva), 358–363.

[20] See, for example, Transcription, 132–135.

Table 5.3 Twenty-five of the most common chymical ingredients in Pw V90, with prices from Harvey 1676

Ingredient	*Price (d/lb)*	*Instances in Pw V90*
Sulphur (incl. balsam of sulphur)	8 to 96	13
Tartar (incl. cream of tartar)	4 to 8	10
Spirit of Wine (incl. aqua vitae)	N/A	10
Crocus metallorum	64	8
Gold	N/A	7
Sal chalybis (salt of steel)	144	6
Vitriol	N/A	6
Oil of Brimstone (oil of sulphur)	120	5
Mercury (incl. mercurius dulcis)	48 to 120	5
Spirit of Vitroil	N/A	4
Turpentine	N/A	4
Antimony	216	4
Sal nitrum (saltpetre)	10	4
Oil of Vitriol	64	3
Oil of Nutmeg	288	3
Salt of Wormwood	144	3
Crocus martis	32	2
Magisterium perlarum	720	2
Oil of Rosemary	N/A	2
Spirit of Sulphur	120	2
Oil of Tartar	144	2
Sal prunellae	240	2
Spirit of Salt	96	2
Sugar of Saturn	216	1
Oil of Cinnamon	4608	1

The most common chymical ingredients, along with their prices in pence per pound (d/lb), can be found in Table 5.3.

The average price of those chymical ingredients for which we have a price is between 384 and 392 d/lb. As this calculation and the above table indicate, ingredients from chymical cures were, on average, far more expensive than Galenic herbs: the mean price for all 422 ingredients (including chymical ones) found in Harvey's list is 184 to 188 d/lb, compared to 472 d/lb for the chymical ones. Indeed, 193 vegetable ingredients from the following of Harvey's categories—roots, barks, woods, leaves, flowers, fruits, seeds, simples grown from plants, and ingredients bought from grocers—have an average price of merely 37 to 38 d/lb. It is notable here that early proponents of chymical medicine tended to advertise it as medicine for the masses, with Paracelsus and van Helmont setting themselves up as renegades against supposedly elitist Galenists.[21] While he leaned heavily on Galenic principles in practice, the radical English physician and astrologer Lancelot Coelson expressed this sentiment in his 1656 *The Poor-Mans Physician* when he resisted the temptation to call upon "famous Physicians who have lived before us in severall ages, as *Dioscorides, Galen, Paulus Aetius, Egyneta*, and others: For it will easily appear of what credit this Noble Science of Physick was in times past, the insolency and pride of which Physicians was so high, that many of them disdained the fellowship of Kings, and expected homage as Gods".[22] Coelson accordingly diverged drastically from, say, Mayerne, who did appeal to Galen, Hippocrates, and Paul of Aegina. But, given the egalitarian rhetoric of chymical physicians, it is all the more ironic that courtiers were among the most enthusiastic proponents of chymical cures, and that such remedies could often be more financially burdensome than herbal ones.

Some of the most self-evidently expensive and difficult to obtain ingredients required for chymical cures in Pw V90 are various gemstones and the horns of rhinoceroses or "Unicornes" (the latter of which were likely narwhal teeth or horns).[23] Less obviously, there was also a huge price

[21] On the monetary point, van Helmont writes in the course of critiquing "the Schools", for instance, that "the sick are more worthy of pity, who have suffered such helpers, hired for much money unto the destuction of their life" (Jean Baptiste van Helmont (1664), *Works Containing his Most Excellent Philosophy, Physick, Chirurgery, Anatomy*, trans. John Chandler (London), 331).

[22] Lancelot Coelson (1656), *The Poor-Mans Physician* (London), A5r. Thomas O'Dowde also claims in his similarly named 1664 *The Poor Man's Physician, Or the True Art of Medicine* (London) that he will "administer freely and publickly to all that shall desire it, not expecting those persons or places, where other Physicians of the dull road [Galenists] would be afraid to show themselves" (89).

[23] On early modern unicorn horns, see Bernd Roling (2014), Der Wal als Schauobjekt: Thomas Bartholin (1616–1680), die dänische Nation und das Ende der Einhörner. *Zoology in Early Modern Culture: Intersections of Science, Theology, Philology, and Political and Religious Education*, eds. Karl A.E. Enenkel and Paul J. Smith (Leiden), 172–196. On the medical use of gems in Britain, see Tom Blaen (2012), *Medical Jewels, Magical Gems: Precious Stones in Early Modern Britain* (Crediton), and, for the Netherlands,

tag on Gascoigne's powder (a concoction of crabs' eyes and claws), the aforementioned bezoars, and assorted viper parts.[24] To warrant the price, many of these were upheld as the most powerful ingredients of all, with Primrose writing, for instance, that it "can scarce be said, whether the Bezaar stone, or to the Unicornes horn the common people attributes greater vertues, for those are thought to be the prime Antidotes of all".[25] In general, however, ingredients derived from animal parts were consistently among the most expensive ingredients used in medical recipes.[26] The following table lists the most common animal products and their costs. It is also notable that European medicine has a long history of the cannibalistic consumption of, say, *mumia*, umbilical cords, and, as found in Pw V90, "the Skull of a man not bury'd", which are also included here (see Table 5.4).[27]

As per Table 5.4, the average cost of these ingredients is a staggering 1731 to 1820 d/lb. But, at an astounding 23,040 d/lb, the frequency of ambergris in Pw V90 radically affects the average price. Once again, though it is impossible to determine the extent to which the Cavendishes actually bought or used these ingredients, they certainly did not neglect to include any of them merely on account of their rarity or expense. It is also noteworthy that Pw V90 calls for numerous viper parts (see Chapter 6B),

Marieke Hendriksen (2018), The Repudiation and Persistence of Lapidary Medicine in Eighteenth-Century Dutch Medicine and Pharmacy. *Gems in the Early Modern World: Materials, Knowledge & Global Trade, 1450-1800*, eds. Michael Bycroft and Sven Dupré (Basingstoke), 197–220.

[24] When commercialised in recipe books, these ingredients were often substituted for less expensive items (Stobart, *Household Medicine*, 73). In Pw V90, however, there are no real emendations, and so, for those of the recipes that were made, we can assume that the expensive ingredients were used.

[25] Primrose, *Popular Errours*, 359–360.

[26] See, for instance, Wendy Wall (2002), *Staging Domesticity: Household Work and English Identity in Early Modern Drama* (Cambridge), 189–220.

[27] Transcription, 119, 121, and 131. On human medical ingredients in early modernity, see Kenneth P. Himmelman (1997), The Medicinal Body: An Analysis of Medicinal Cannibalism in Europe, 1300–1700. *Dialectical Anthropology* 22.2: 183–203 and Richard Sugg (2006), 'Good Physic but Bad Food': Early Modern Attitudes to Medicinal Cannibalism and its Suppliers. *Social History of Medicine* 19.2: 225–240. For general discussions of cannibalism at this time, see Cătălin Avramescu (2009), *An Intellectual History of Cannibalism* (Princeton), especially 162–182 and Louise Noble (2011), *Medicinal Cannibalism in Early Modern English Literature and Culture* (New York).

Table 5.4 Twenty of the most common animal ingredients in Pw V90, with prices from Harvey 1676 (note that not all ingredients with one instance are included)

Ingredient	*Price (d/lb)*	*Instances in PwV90*
Hartshorn	8 to 18	7
Ambergris	23040	7
Castor	144 to 480	4
Ivory	4 to 12	4
Coral, white	40	4
Crabs' Eyes	64	4
Musk	5760	4
Pearls	432 to 1152	3
Bezoar	2304 to 2880	3
Cranium humanum (human skull)	96 to 132	3
Vipers (heads, skins, etc.)	N/A	3
Crabs' Claws	18	2
Wax	16 to 24	2
Coral, red	48	2
Cantharides, Spanish Flies	48	1
Mumia	64	1
Os e corde cervi (bone of a stag's heart)	216	1
Sea-Horse's Tooth	48	1
Spermacaeti	480	1
Ichthyocolla (isinglass)	64	1

and, while we do not know the exact price of vipers, the various medicaments derived from them, *Trochisci de Viperis*, ranged from 216 d/lb (for the English version) to 288 d/lb (for the Venetian one).[28] As this indicates, animal parts and the medicines derived from them did not come cheap.

Many of the priciest items in Pw V90 also had to be imported to France, the Low Countries, and England, such as the many gums and balsams from the Middle East and North Africa, the aforementioned ambergris and nutmeg from Indonesia and India, and rhinoceros' horn from Africa. The materials recommended in early modern recipe books at large spurred on and reflected the growth of the already global spice trade,

[28] Harvey, *The Family Physician,* 133.

and, perhaps even more than in England, such supplies were readily accessible in Antwerp, which, after all, served as a major seventeenth-century centre for the importation of goods that were then distributed across Europe.[29] Along with the elite status of the Cavendishes, this helps to explain why, once again, their collection is an outlier with its extreme quantity of such ingredients. In the century prior, physicians such as Timothy Bright penned works on *The Sufficiencie of English Medicines* (1580), which railed against the use of foreign herbs and spices and promoted, instead, English ingredients for English people. As Andrew Wear has argued, early modern English medics from William Harrison to Nicholas Culpeper "extolled home-grown remedies simply on a nationalistic dislike of anything foreign" and "drew upon this view of how people were related to their country of birth".[30] This discourse was partly grounded in the Galenic—indeed, Hippocratic—notion that environment had a significant bearing on individual temperaments, and it also drew on the religious supposition that God, in His perfect knowledge, furnished each land with the appropriate remedies for diseases that would afflict those people.[31] Using foreign drugs was, from this perspective, not only unnatural but even unchristian.

The basic humoural attitude that led to the rejection of foreign ingredients is indeed reflected in Pw V90 when Mayerne, for instance, notes that to cure colic

[29] For commerce in the Netherlands at this time, see Harold Cook (2007), *Matters of Exchange: Commerce, Medicine, and Science in the Dutch Golden Age* (New Haven), and, for Antwerp, see Susanna Burghartz, Lucas Burkart, Christine Göttler, eds. (2014), *Trading Values in Early Modern Antwerp* (Leiden).

[30] Andrew Wear (1992), Making Sense of Health and Environment in Early Modern England. *Medicine in Society: Historical Essays*, ed. Andrew Wear (Cambridge), 119–148 (127).

[31] For the classical context, see Hippocrates, *De aere, aquis et locis* and Galen, *De medendi methodo*, along with his understudied treatise *De compositione medicamentorum secundum locus.* For the religious undertones, see, for example, Nicholas Culpeper, *School of Physick* (London, 1659), 7. On the early modern context, see J. Worth Estes (1995), The European Reception of the First Drugs from the New World. *Pharmacy in History* 37.1: 3–23 and Alix Cooper (2007), *Inventing the Indigenous: Local Knowledge and Natural History in Early Modern Europe* (Cambridge).

> is not very easie particularly wheare y^e^ [^infirmity] cause is not ordinary, as in this [*sic*] countryes: in france and in germany: it is to common: y^e^ people particu=larly in france: having a great deale more choler then flegme.[32]

As Mayerne notes here, different environments affect the humoural composition of those living there, making different diseases more or less likely to afflict the population. But, despite this conviction that place, airs, and waters were deeply linked to health, he did not ultimately restrict consumption to local fares. More generally, as Patrick Wallis has demonstrated, there was a steady increase in the importation of exotic medical goods into England over the course of the seventeenth century, representing a willingness to expand the repertoire of ingredients even as traditional diagnostics continued to hold sway.[33] Other ingredients in Pw V90 were, however, standard items that stemmed from the *Pharmacopeia Londinensis*, and only required local supplies.[34] One example is "Oxymel Julianizans", which was a common mixture of vinegar, honey, fennel, chicory, asparagus, and a variety of other herbs, but in the Cavendish collection even standard herbs were often an integrated part of a course of treatment that had, at its centre, an expensive chymical ingredient or compound.[35]

As intimated in the case of Cademan's remarks on Charles I (discussed in Chapter 4), much of the discourse surrounding the origins of ingredients revolved around the status and corresponding constitution of the patient in question: working men and women were traditionally regarded as closer to the earth and thus wedded to a particular local, whereas expensive, rare, and exotic ingredients were more apt to suit the temperaments of wealthy and internationally orientated aristocrats. In this regard, lavish and imported ingredients also served a symbolic function insofar as they signalled the nobility of the recipe's creator. As the next chapter

[32] Transcription, 213. The reference is to Paul of Aegina, the seventh-century Byzantine Greek physician.

[33] Patrick Wallis (2012), Exotic Drugs and English Medicine: England's Drug Trade, c.1550–c.1800. *Social History of Medicine* 25.1: 20–46 and Patrick Wallis (2008), Consumption, Retailing, and Medicine in Early-Modern London. *Economic History Review* 61.1: 26–53.

[34] See Spiller, *Recipes for Knowledge*, 64–65.

[35] Nicholas Culpeper (1653), *Pharmacopeia Londinensis, or, The London Dispensatory* (London), 112. Also see Transcription, 112.

discusses in greater detail, courtiers such as Digby and Dudley—the latter of whom includes "orientall-bezarstone", "orientall bolus", and the "Powder of vipers" in the same cordial—were strong proponents of expensive ingredients from far-flung places.

CHAPTER 6

Notable Recipes

A. Théodore de Mayerne's Salt of Steel

In the extant historiography on Mayerne and mid-seventeenth-century medicine, the chief cure that he prescribed in his later years, the salt of steel, has been almost entirely overlooked. While the salt of steel is recognised as "Mayernes Invention" in Pw V90, similar iron-based cures had grown in popularity in the wake of Paracelsus's chymical output.[1] These were often called, after the alchemical fashion, *sal martis* or *crocus martis*: that is, the "salt of Mars" or the "saffron of Mars" (with Mars being the astrological and alchemical symbol for iron). Mayerne's "role model", Joseph Duchesne, recommended such cures often, especially for wounds and to diminish scars.[2] Similar cures can be found in the work of other chymical physicians, such as Johann Rudolf Glauber, who noted that

[1] Transcription, 102. Francis J. Hebbert (1950), Some Historical Aspects of Iron Therapy. *Glasgow Medical Journal* 31.11: 385–396 and Leonard J. Goldwater (1935), A Short History of Iron Therapy. *Annals of Medical History* 7.3: 261–267.

[2] On Mayerne and Duchesne, see Trevor-Roper, *Europe's Physician*, 21–23, 37–43, 54–61, and 80–100. For the recommendation of iron-based cures, see Joseph Duchesne (1590), *The Sclopetarie of J. Quercetanus, Phisition, Or His Book Containing the Cure of Wounds Received by Shot of Gunne Or Such Like Engines of Warre*, trans. John Hester (London), 34–35 and 39.

J. Begley and B. Goldberg, *The Medical World of Margaret Cavendish*, Palgrave Studies in Medieval and Early Modern Medicine,
https://doi.org/10.1007/978-3-030-92927-5_6

crocus martis was "A most excellent ingredient for Chirurgeons".[3] While it is notable that Mayerne's recipe calls for steel instead of pure iron, even this was not unknown in the period. Michael Sendivogius, for instance, described a recipe for *crocus martis* made from steel or iron plates, spirit of salt (hydrochloric acid), oil of vitriol (sulphuric acid), and some other ingredients.[4] The use of steel, as a "purer" sort of iron, however, became more common in the wake of the circulation of Mayerne's recipe, and was deployed by medics such as Thomas Sydenham in England and Louis Lemery in France; as we will see, at least in England, the salt of steel increasingly became associated with Mayerne's name.[5] William Temple, for example, testified to the popularity of Mayerne's salt when he relayed that the "infusion of powder of steel have had their turns", though he lamented that "none that I find have established their authority", while the Swiss physician Felix Plater noted that "Crocus Martis is commended by the Chymists" especially "in the Cure of Diabetes and Dystentery".[6]

From Pw V90, it becomes clear that Mayerne did not design his cure to treat a particular individual or disease—as was typical within the Galenic tradition—but rather conceived of it as a remedy for a vast array of illnesses. While the recipe itself is relatively humble when compared to many Galenic and indeed chymical compounds, Chapter 5 has already intimated that it was by no means easy to administer, which was a purported goal of chymical remedies, and in fact required one to ingest a vast swathe of herbal ingredients including beets, chamomile, dill, and linseed.[7] Not only was such vegetable matter to be boiled with syrup of violets and red sugar, but the efficacy of the compound was also thought to be contingent on it being taken at precisely four o'clock in the afternoon. The following day the patient was instructed to take a purge that

[3] Johann Rudolf Glauber (1689), *The Works of the Highly Experienced and Famous Chymist, John Rudolph Glauber*, trans. Christopher Packe (London), 328.

[4] Michael Sendivogius (1650), *New Light of Alchymie Taken Out of the Fountaine of Nature, and Manuall Experience: To which is Added a Treatise of Sulphur*, trans. John French (London), 39–40.

[5] Hebbert, Some Historical Aspects, 389.

[6] William Temple (1720), Of Health and Long Life. *The Works of William Temple* (London), 272–289 (209) and Felix Plater (1664), *Platerus Golden Practice of Physick* (London), 163 and 656. Not all of these products, however, were made identically, but instead share their main ingredients: sulphurated and calcinated iron or steel.

[7] On the goal of simplicity within the chymical tradition, see Andrew Wear (2000), *Knowledge and Practice in English Medicine, 1500–1680* (Cambridge), 85–100.

included three drams of "Creame of Tarter pulveri[^z]ed" and a dram each of coriander and aniseeds, boiled in "halfe a Pinte of Spring water".[8] Having been steamed all night, this potion was ingested with "twelve drops of Cinamon water" in the morning. The next day nine ounces of blood were drained from the "ye Basi[^li]cke vaine" of the right arm, and "Julebs" were taken the subsequent "Morning at seaven a Clocke".[9] As the above displays, having posited a universal cure that resonates with chymical medicine, Mayerne proceeded to emphasise individual herbal compounds and the temporal specificity of their ingestion, in line with the Galenic tradition.

In terms of the universality of Mayerne's salt of steel and its compatibility with Galenic medicine, Robert Boyle tellingly relayed in his 1663 *Considerations Touching the Usefulness of Experimental Naturall Philosophy* that the

> most Learned of the Methodists themselves have much valued and celebrated some peculiar Processes and Receipts, as here amongst us (to mention no others) the Famous Sr *Theodore Mayerne,* was wont almost in all Obstructions, Cachexies, and Hydropicall Distempers to magnifie and use that peculiar Salt of Steel of his, which he was pleas'd to call *Anima hepatis*.[10]

Taking Mayerne as his inspiration, Boyle proceeded to argue that cure-alls were by no means inconsistent with Galenic medicine, and, in particular, that even Galen's *De compositione medicamentorum* commended "divers of the Compositions of Auncienter Physitians, and especially magnifies *Andromachus* His Treacle". By invoking Andromachus the Elder—who was Nero's physician, and famously developed a treacle that was an alexipharmic against all poisons—Boyle endeavoured to establish an ancient (and, indeed, pre-Galenic) precedent for cure-alls. Scholars such as William R. Newman and Levitin have rightly pointed out that Daniel Sennert was an inspiration for Boyle's views on the ancient roots of chymical principles, but, as the above suggests, Mayerne's emphasis on the compatibility of Galenic and chymical medicine, as embodied in his

[8] Transcription, 104.

[9] Transcription, 104–105.

[10] Robert Boyle (1663), *Some Considerations Touching the Usefulness of Experimental Naturall Philosophy* (London), 401–402.

salt of steel, seems to have provided an alternative model for the sceptical chymist's understanding of the history of medicine.[11]

Mayerne's entries in Pw V90, one of which describes the salt of steel as an example of the "Universall things I have Counsell'd", only confirm Boyle's characterisation of the remedy.[12] But beyond the specificities of regimen and dosage, the universal applicability of Mayerne's cure was deemed to be heavily contingent on environmental factors. Mayerne repeated in successive epistles "that this is not now ye season to begin" taking the salt of steel, "since ye heat of Summer is neare at hand, dureing which other remedyes are to be used to moisten that evell black humour".[13] There is accordingly a tension between the cure's supposedly universal power and the limited window for commencing a therapeutic cycle based upon it. Whereas more radical Helmontians broke this link between patients and their environments by holding that anyone could take their "pure" remedies year-round, the complex cures of learned physicians were typically of seasonal applicability and could prove detrimental if administered at the wrong time or to a healthy individual. The ideal Galenic physician, then, was supposed to be perfectly attuned to the intricate relationship between a patient's body, mind, and the external world.

As we know from Margaret's criticisms of chymical medicine, she would have appreciated Mayerne's balanced approach (in contrast to the bolder claims of Digby or Dudley that are discussed below). Particularly relevant here is Margaret's response to van Helmont in *Philosophical Letters*, where she rhetorically asks:

> what would the skill of Physicians be, if one remedy should cure all diseases? Why should they take so much pains in studying the various causes, motions, and tempers of diseases, if one medicine had a general power over all? Nay, for what use should God have created such a number

[11] William R. Newman (2006), *Atoms and Alchemy: Chymistry and the Experimental Origins of the Scientific Revolution* (Chicago), 157–215 and Dmitri Levitin (2015), *Ancient Wisdom in the Age of the New Science: Histories of Philosophy in England, c. 1640–1700* (Cambridge), 271–278.

[12] Transcription, 125.

[13] Transcription, 132.

of different simples, Vegetables, and Minerals, if one could do all the business?[14]

To Cavendish's mind, the promotion of a universal cure that did not make use of herbal compounds or consider temporal specificity not only imperilled the noble profession of the physician, but bordered on the irreligious. Cavendish proceeded to suggest that Helmontian medicine in general would be much improved if its practitioners concerned themselves with "Scammony, Coloquintida, Manna, Cassia, Sena, Rhubarb, &c.", most of which are supplements to Mayerne's salt.[15]

Beyond questions of universality, Mayerne's very decision to concoct a "salt" betrays a distinctly chymical predilection. As is well known, Paracelsus supplemented the mercury and sulphur that was stressed in Hellenistic chymistry with salt (broadly conceived of as a solid, soluble, and inflammable substance with a crystalline structure) as the third part of his famous *tria prima*.[16] Whereas learned physicians tended to focus on how internal humours acquired the destructive qualities that caused diseases, Paracelsus imputed the dissipation of moisture that spawned emaciation and hectic fever to the abnormal ignition of sulphur.[17] In contrast to Aristotelian-Galenic opposites, the Paracelsian notion that like cures like made salts a common chymical prescription. Indeed, in a work posthumously published as *The Treatise on the Gout* (1676), Mayerne explicitly followed Paracelsus in expressing the primacy of salt, pronouncing that one should not exclusively "regard the division of the humours into four parts" but rather "consider that all the Excrements of the body do admit into themselves, and contain in their Composition, *Salt*".[18] Rather than repudiating humoural theory, however, Mayerne posited salt as a fifth essence that was not composed of the four Aristotelian elements. With this distinction, he could avoid shocking his more learned patients with abnormal diagnoses.

[14] Cavendish, *Philosophical Letters*, 390.

[15] Cavendish, *Philosophical Letters*, 378.

[16] On salt during the early modern period, see Anne Marie Roos (2007), *The Salt of the Earth: Natural Philosophy, Medicine, and Chymistry in England, 1650–1750* (Leiden), and, on the *tria prima*, see Pagel, *Paracelsus*, 83–103.

[17] See Walter Pagel (1984), *Smiling Spleen: Paracelsianism in Storm and Stress* (London), 30.

[18] Théodore de Mayerne (1676), *A Treatise of the Gout* (London), 5–6.

In addition to presenting salt as a fifth element, Mayerne explained why his salt of steel could heal the Cavendishes by appealing to the terms of humoural medical discourse. As Cademan noted when discussing Mayerne's salt of steel, it would "Purge and Extenuate thicke & Limous obstructions in the Spleene" that cause melancholy.[19] It is worth highlighting here that some confusion has surrounded Mayerne's diagnosis in historical accounts. While Katie Whitaker and others have treated "melancholy" as essentially the modern-day equivalent of depression, melancholy was not so much regarded as a psychological illness as a physiological one in the early modern period (though early modern physicians did recognise that one's physiology could affect one's psychology, or, as they would term it, passions).[20] No doubt with Mayerne's diagnosis in mind, Margaret penned two relevant essays in *The World's Olio*, one "Of Melancholy" and another "Of a dull or Melancholy Disposition proceeding from the Body, and the Melancholy proceeding from the Soul", with the latter's title directly reflecting Mayerne's comment that melancholy arose as "the Pas=sions of the Soule" moved spirits and humours. In these essays, Cavendish maintained that melancholy "proceeds from the ill-affected Body, rather than from an ill-affected Mind", and she went so far to assert that "Melancholy Men may be said to be Idle, or Musing, but not Sorrowfull or Sad".[21]

Despite being a physiological diagnosis, melancholy was associated with the intellectual and creative excellence of philosophers and poets (tragic or otherwise), and, indeed, Cavendish's own intellectual curiosity probably had a bearing on Mayerne's verdict about her humoural disposition.[22] Shortly after she published her first two works and approximately a year before Mayerne died, he notified William that "her Ladiships occupation in writing of bookes: with a sedentay life is absolutely bad for

[19] Transcription, 129.

[20] For an example of the link between actions, passions, and disease in the anatomical tradition, see William Harvey (1628), *De motu cordis,* Cap. XV (Frankfurt), 60. Indeed, it was well understood in the ancient world that one's body influenced one's affective state, and vice versa (see Aristotle, *De anima* and Galen, *Quod animi mores corporis temperamenta sequuntur*).

[21] Cavendish, *The World's Olio*, 118.

[22] See, for instance, Whitaker, *Mad Madge*, 29. On melancholy more generally, see Angus Gowland (2006), *The Worlds of Renaissance Melancholy: Robert Burton in Context* (Cambridge) and Jennifer Radden (2000), *The Nature of Melancholy: From Aristotle to Kristeva* (Oxford).

health", and he recommended his "steell liquore" as a means by which to counteract the ill-effects of her spending all day at her writing table.[23] Such a diagnosis appears to have stuck with Margaret, for she noted years later, in her 1666 *Observations Upon Experimental Philosophy*, that

> if much writing be a disease, then the best philosophers, both moral and natural, as also the best divines, lawyers, physicians, poets, historians, orators, mathematicians, chemists, and many more have been grievously sick: and Seneca, Pliny, Aristotle, Cicero, Tacitus, Plutarch, Euclid, Homer, Virgil, Ovid, St. Augustine, St. Ambrose, Scotus, Hippocrates, Galen, Paracelsus, and hundreds more, have been at death's door with the disease of writing.[24]

Cavendish pointed out, in other words, that, in putting writing before health, she was in the best of company. To Margaret's list of suffering authors might be added another of Mayerne's famous patients, the great scholar Isaac Casaubon, who was warned by Mayerne that continuous study had tipped his constitution from bilious (hot and dry) to melancholic (cold and dry).[25]

Mayerne's prescription of the salt of steel would have made particular sense as a means by which to counterbalance this form of "writer's" melancholy, because, within the Galenic tradition, the spleen was thought to act as a sponge that swells and diminishes as it receives humoural impurities and blocks black bile from flowing to the head.[26] Mayerne conceived of his steel as a suitable antidote precisely because he deemed salt deficiency to be the fundamental cause of black humours. He accordingly clung to the notion that the central concern of hygiene (or the prevention of disease) was to maintain the precarious balance of humours, and that humoural cures had proved effective throughout history in temporarily offsetting symptoms.[27] But in therapeutics only a chymical

[23] Transcription, 220.

[24] Cavendish, *Observations*, 7.

[25] Théodore de Mayerne (1703), *Opera Medica* (London), 144.

[26] See Pagel, *Smiling Spleen*, 134–142.

[27] On hygiene, see Heikki Mikkeli (1999), *Hygiene in the Early Modern Tradition* (Helsinki), especially 32–40 and Richard Palmer (1991), Health, Hygiene and Longevity in Medieval and Renaissance Europe. *History of Hygiene*, eds. Yosio Kawakita, Shizu Sakai, and Yasuo Otsuka (Tokyo), 75–98.

remedy would be able to treat the fundamental cause of an illness. By working with these precepts, Mayerne could defend his role as a learned physician who supplied regular advice and assistance to some of the most renowned figures of his day, while promoting a chymical cure that promised them more permanent relief (and him some much-needed respite). Mayerne's salt of steel was perhaps the chief practical instantiation of his lifelong mission, which began with his *Apologia*, to wed chymical and Galenic approaches to disease.

B. Kenelm Digby's Powders of Vipers and Sympathy

While Digby's "viper wine" is well known, his "powder of vipers" has gone entirely unnoticed.[28] From Digby's protracted letter to the Cavendishes, however, it can be inferred that he devised his powdered cure partly in response to rumours that his wine was responsible for the death of his wife, Venetia. Digby vehemently denied such accusations, but the renowned seventeenth-century gossip John Aubrey memorably turned accidental poisoning into the best case scenario when he claimed that Venetia "dyed in her bed suddenly [...] which her husband imputed to her drinking of viper-wine; but spitefull woemen would say 'twas a viper-husband who was jealous of her".[29] To distance his powder of vipers from his earlier viper cure and indeed his name, Digby suffused his account of it, in an epistle to the Cavendishes, with tales of cavaliers, concubines, and unnaturally old apothecaries. As we will see, there is also good reason to believe that Digby's shift from a viper wine to powder inspired his decision to transform the standard "weapon salve"—a liquid potion containing a patient's blood that was thought to cure a wound by anointing the weapon that inflicted it—into his celebrated powder of sympathy.

Rather than passing off the powder of vipers as his own invention, Digby allegedly procured the cure from one "Siginor Antonio Manfradi". While this Manfredi is an obscure figure, the same name features as a source in the 1646 *Del vero opobalsamo orientale discorso apologetico dell'illustrissimo* by the canon at the University of Rome and later physician

[28] The notoriety of Digby's viper wine is due in part to Hermione Eyre's fictional 2015 *Viper Wine* (London).

[29] John Aubrey (1898), *Brief Lives, Chiefly of Contemporaries, Set Down by John Aubrey, Between the Years 1669 and 1696*, Vol. 1, ed. Andrew Clark (London), 231.

to Pope Innocent X, Baldo Baldi.[30] This work, which was published in Rome around the time that Digby was stationed there, must have either been Digby's source, or he must have had direct contact with Manfradi, since Digby's recipe is similar to Baldi's both in its use of opobalsum and in the prominent place that it gives to viper flesh.[31] According to Digby, Manfradi's powder of vipers was so sought after (perhaps in the wake of Baldi's publication) that

> he hath not a grayne left wth in. 3. monthes after y^e^ make=ing of it Excepting such store as his Concubin (for he hath no wife) maketh him keepe [50r] for himselfe for by meanes of it, & of his Treacle She findeth him a Cavalier of 25. yeares old, though indeede he bee. 40.[32]

Along with indicating the demand for the powder of vipers and establishing that it had been self-tested, Digby added the detail of Manfradi's concubine to depict his alleged acquaintance as mysterious and foreign. Since concubinage was forbidden in seventeenth-century England and would have been widely regarded as morally reprehensible, this detail bolsters the image of Manfradi's virility and suggests that it was partly thanks to his viper powder that he was able to attract and retain a woman who was unfettered by marital bonds. As Digby was probably aware, this would have appealed to William in particular, who was thirty years his wife's senior.

Digby may have framed Manfradi as the conduit through which he procured the powder of vipers, but he did not regard him as its chief architect. According to Digby, the powder was rather the brainchild of an unnamed "Apo=thycary" who "exceedeth Infinitely all others; he is old & cannot live Longe, and had bene dead Longe since but for y^e^ Continuall use of his owne Treacle".[33] To square the continued existence of this hoary sage with the claim that his cure reversed ageing, Digby was careful to clarify that this man had only lived to an unnaturally old age because

[30] Alexander Chalmers (1812), *The General Biographical Dictionary*, Vol. 3 (London), 361.

[31] Baldo Baldi (1646), *Del vero opobalsamo orientale discorso apologetico dell'illustrissimo* (Rome), 99–100.

[32] Transcription, 152.

[33] Transcription, 152.

of the remedy that Digby was seeking to promote. As these descriptions suggest, Digby's characters were straight out of romance literature and indeed forged a path for the emergent types of chymical adepts: the young, virile cavalier and the elderly wise man.[34] Lawrence Principe has underscored the significance of romance literature to the development of Boyle's prose style, but it was far less common for such tales to suffuse recipes.[35] The payoff of this unconventional move was twofold for Digby. By lacing his recipes with far-fetched biographical details, he could further shape his own life as a romance—continuing the project that began in 1628 with his quasi-autobiographical "Loose Fantasies"—*and* use this posture to sell his wares.[36] In one fell swoop, Digby's eccentric tales lent authority to his person and validity to his cures. They also allowed him to pass his remedy off as exotic while shaping himself as its sole English spokesman.

Similar to Mayerne, Digby merged and mediated between learned and chymical principles as he channelled his experiences (and imaginings) to promote his viper powder. In this regard, it is significant that Digby's elderly apothecary procured the components of his remedy from "Egipt, Arabia, & India" where "ye whole Coledge of Phisicians there have subscribed ye truth of all his Ingredients".[37] Not only does Digby point to the Eastern roots of the cure, but, by appealing to a "Coledge of Phisicians" beyond English shores, he also foregrounds his awareness of the high esteem in which potential customers such as the Cavendishes would have held traditional medical institutions. Following the advice of these Eastern physicians, Digby further insisted that the powder of vipers is made "just ye same as Gallen Proscribed, wch they wch observe Religiously & are accurate in ye Choyse of their Ingredients doe make perfect Compositions I may say Miraculus ones".[38] Digby indeed gestured towards Galenising his remedy when he stated, for instance, that the apothecary will "Take Vipers & Cut of their heads & tayles about the

[34] On Digby's absorption in romance literature, see Vittorio Gabrieli (1968), *Loose Fantasies* (Rome), ix–xxxvii.

[35] See Lawrence Principe (1995), Virtuous Romance and Romantic Virtuoso: The Shaping of Robert Boyle's Literary Style. *Journal of the History of Ideas* 56.3: 377–397.

[36] The manuscript is preserved in the British Library, Harley 6758, and has been cleaned up and printed by Vittorio Gabrieli as *Loose Fantasies*.

[37] Transcription, 152.

[38] Transcription, 153.

length of three fingers, & take out ye guts & fatt soe y^{t} there remaine nothing but ye flesh".[39] This accords with the advice in the Galenic *De theriaca ad Pisonem* that one should not put the whole viper into a theriac, but rather cut "off the heads and tails and use the rest of the body in the mixture".[40] Digby, following Galen, gave this instruction because it was thought that the head of the snake was its most deadly part, and might prove lethal if ingested. Taking a daily dose of an antidote was indeed a common practice in ancient Rome, with Mithridates, King of Pontus, having attempted to immunise himself against all poisons by regularly ingesting a venomous compound well before Andromachus the Elder produced his theriac.[41] For his part, Digby insisted that "Treacle & Metridate are likewise Imediatly against Poyson; But they are more valued here for their resisting of Putrefaction, for their digesting of Crudities & na[^u]tious humors, for their strengthening all ye Internall Partes espetialy ye stomack & ye head".[42] Even while appealing to Galen, he accordingly endowed his viper powder with far greater healing properties than the ancients had imputed to similar treacles.

Dating back to the times of the Black Death, the "venice treacle"—colloquially called "the treacle"—had been deemed the best of all possible antidotes.[43] Fully aware of this, Digby opined that the "Powder of viper & Troches is much better at Rome then at venice, or Padua because ye mountaine vipers is beter then in Lowe & marish places as also ye heates being greater there it makes them the purer".[44] Grounded in an essentially Galenic presupposition that plants and animals at a higher altitude are of a hotter nature, the specific location from which Digby's preferred vipers were procured supported his larger claim that "The Powder of vipers" is "ten-times more Effectuall then the Troches of vipers; And

[39] Transcription, 154.

[40] See Robert Leigh (2015), *On Theriac to Piso, Attributed to Galen: A Critical Edition with Translation and Commentary* (Leiden).

[41] See Laurence Totelin (2004), Mithradates' Antidote—A Pharmacological Ghost. *Early Science and Medicine* 9.1: 1–19.

[42] Transcription, 153.

[43] See Christiane N. Fabbri (2007), Treating Medieval Plague: The Wonderful Virtue of Theriac. *Early Science and Medicine* 12.3: 247–283 and Bart K. Holland (2000), Treatments of the Bubonic Plague: Reports from Seventeenth Century British Epidemics. *Journal of the Royal Society of Medicine* 93.6: 322–324.

[44] Transcription, 150.

by y^e great Ex=perience of y^e wonderfull good it doth in Restore=ing Natures, And even in makeing y^e outwarde sem=blance of People returne to a kinde of youth=fulnesse".[45] As this suggests, Digby supposed that a subtle deviation from the standard recipe could transform the cure's effectiveness, allowing it to immunise not only against poison and preserve health and beauty but also, more remarkably, to reverse ageing. He thus claimed to possess a cure that surpassed even the enigmatic and elusive philosophers' stone, which only ever promised to prolong life.[46] While Digby drew judiciously on Galenic precepts because he realised that the commercial viability of his powder depended on it according with the medical advice that his potential buyers were accustomed to hearing, he was less constrained, as an amateur, by the principles of learned medicine than Mayerne, who only claimed therapeutic value for his cure-all.

A related recipe that is not attributed to Digby in Pw V90 but that gradually became associated with him is "the Powder of Simpathy". It has already been mentioned that, as with his transformation of viper wine into a powder, Digby transmuted the traditionally wet "weapon-salve" into a dry "powder of sympathy".[47] He went to lengths to justify this move in his *A Late Discourse … Touching the Cure of Wounds by the Powder of Sympathy* (1658), which was based on a lecture that he delivered at the University of Montpellier and went through over forty editions in the latter half of the seventeenth century, appearing in French, English, Latin, Dutch, and German. Like his powder of vipers, Digby did not claim to have invented the powder of sympathy but rather to have introduced "this way of curing" into "this Quarter of the World".[48] Taking advantage of the early modern culture of testimony discussed in Chapter 4, he avowed in his lecture that, using the powder of sympathy, he had rid one James Howell of a gangrenous wound under the watchful eyes of

[45] Transcription, 150.

[46] See Georgiana Hedesan (2016), *An Alchemical Quest for Universal Knowledge: The 'Christian Philosophy' of Jan Baptist van Helmont (1579–1644)* (London), 172–192.

[47] On Digby's powder of sympathy, see Ernest Gilman (1999), The Arts of Sympathy: Dr. Harvey, Sir Kenelm Digby, and the Arundel Circle. *Opening the Borders: Inclusivity in Early Modern Studies*, ed. Peter Herman (Newark), 265–297 and Seth Lobis (2011), Sir Kenelm Digby and the Powder of Sympathy. *Huntington Library Quarterly* 74.2: 243–620.

[48] Kenelm Digby (1658), *A Late Discourse … Touching the Cure of Wounds by the Powder of Sympathy* (London), 5–6.

"King James of England" and "his Sonne, the late King *Charles*, and the Duke of *Buckingham*". When King James demanded Digby's source for the remarkable remedy, he eventually revealed that it was from "a religious *Carmelite*, that came from the *Indies* and *Persia* to *Florence*", who entrusted the cure to Digby and made him its sole European heir.[49] Digby thus deployed similar narrative *topoi* for his powders of vipers and sympathy, and, having gauged what evidence would persuade a courtly audience, he accompanied his oriental and Italian authorities with regal (and notably deceased) English ones.[50]

With this anecdote, Digby implied that he obtained the powder of sympathy after 1620—when he departed on his grand European tour—and before the death of King James in 1625. But this was almost twenty years before the powder of sympathy was widely promoted in England. When it began to gain traction in the 1650s, with the writings of Nathaniel Highmore and Charleton, it was not actually associated with Digby at all, but was rather known as "Sir Gilbert Talbots Powder" (after a colonel in Charles I's army).[51] As regards the question of attribution, it is also notable that the powder is unconnected to Digby in Pw V90, and that, in his *Two Treatises*, Digby seemed indifferent towards the texture of the sympathetic cure, discussing "the weapons salue, or the sympathetike pouder" interchangeably.[52] It was only after he advertised the powder of vipers (in the late 1640s) that Digby came to emphasise the texture of his sympathetic cure. As all of this suggests, Digby's viper powder is an unacknowledged sibling, or even forerunner, to his sympathetic powder. When he finally came to promote exclusively the powder of sympathy, he perhaps hoped to be associated with powdered remedies in general, and thus validate his priority claims when it came to the powder of sympathy in particular.

[49] Digby, *A Late Discourse*, 11–12.

[50] For Digby's fabrication of authorities, see Elizabeth Hedrick (2008), Romancing the Salve: Sir Kenelm Digby and the Powder of Sympathy. *British Journal for the History of Science* 41.2: 161–185.

[51] Nathaniel Highmore (1651), *The History of Generation* (London), 113 and Charleton and van Helmont, *A Ternary of Paradoxes*, C3v–C4r. This Gilbert Talbot is not to be confused with the aforementioned 7th Earl of Shrewsbury who married William's aunt.

[52] Kenelm Digby (1644), *Two Treatises* (London), 164.

Digby's disquisition on the powder of sympathy in the *Late Discourse* elucidates, from a philosophical angle, why he came to see texture as so significant. Within the Aristotelian tradition of *minima naturalia*—according to standard interpretations of which matter could be divided no further than the smallest components of the four (observable) elements—phenomena that acted at a distance and that could not be explained with reference to the elements (such as the loadstone and iron) were deemed to be occult (though natural) and inexplicable because insensible.[53] Paracelsus—the reputed inventor of the weapon salve—and van Helmont following him challenged the notion that such sympathetic relations were inexplicable and preferred to attribute them vaguely to an invisible magnetic force.[54] Mediating between these discourses, Digby proposed that the supposedly occult qualities involved in action at a distance were best explained through recourse to his idiosyncratic Aristotelian atomism, which he first expounded in *Two Treatises*.[55] On Digby's account, it was more effective to apply his powder to the blood on a garter far removed from the body than directly to the wound because this would guarantee that only the most volatile parts of the powdered vitriol travelled to the person's wound to cure it, with the duller parts remaining on the cloth. He considered heat to be the property of narrow and rare bodies and therefore supposed that the subtle parts of his powder could latch onto cold air—based on the Aristotelian principle of elemental opposites—and move towards the warm blood. As Digby explained it, "small Atomes being cut, and loosened from their trunk, being composed of the four

[53] On the weapon salve during the seventeenth century, see Pagel, *Joan Baptista Van Helmont*, 9–13 and Hedesan, *An Alchemical Quest*, 25–34. For attempts to explain such occult qualities, see Ron Millen (2002), The Manifestation of Occult Qualities in the Scientific Revolution. *Religion, Science, and Worldview: Essays in Honour of Richard S. Westfall*, eds. Margaret J. Osler and Paul L. Farber (Cambridge), 185–216.

[54] For a fuller discussion of van Helmont's defence of the weapon salve as non-magical, see Carlos Z. Caminietzki (2001), Jesuits and Alchemy in the Early Seventeenth Century: Father Johannes Roberti and the Weapon-Salve Controversy. *Ambix* 48.2: 83–101 and Mark A. Waddell (2003), The Perversion of Nature: Johannes Baptista Van Helmont, the Society of Jesus, and the Magnetic Cure of Wounds. *Canadian Journal of History* 38.2: 180–197.

[55] Levitin, *Ancient Wisdom in the Age of the New Science*, 369–372.

Elements, (as all bodies are) the heat of the light doth stick, and incorporate it self with the most humid, viscuous, and gluing parts of the said Atoms, and brings them along with her".[56]

Digby appears to have begun to appeal to powders because of a desire to cast aside the baggage associated with his viper wine and establish the superiority of his cure-all over the traditional theriac. But it ended up playing a substantial role in his adaptation of Aristotelian atomism towards medical ends. When it came to promoting his sympathetic cure, Digby would have undoubtedly realised that it was far easier to conceive of corpuscles of powder fastened to blood travelling through the air than to imagine a heavy liquid involved in the same process. Because atoms were unobservable, but the atomic hypothesis was designed to provide a convincing account of sensible reality, such imaginative plausibility was vital. Just as Digby harnessed his imagination to conjure the background stories that allowed him to sell his cures, the same faculty allowed him to formulate a substance theory that could explain them. For all of his rhetorical flare, however, it seems that Digby's cure was not popular in the Cavendish household, for Margaret was of the opinion that "if those Persons that have been wounded, had been left to be cured onely by the Magnetick Medicine, I believe, numbers that are alive would have been dead, and numbers would die that are alive; insomuch, as none would escape, but by miracle".[57]

C. Robert Dudley's Cordial

Dudley's cordial—which Digby delivered to the Cavendishes on his behalf—is unquestionably the most complicated and perhaps the most interesting concoction in Pw V90. The entry begins with a list of ingredients needed for the recipe, and proceeds to spell out the processes by which many of them are made, providing recommendations for administration and dosages in each case, along with a statement of efficacy and a brief curative record. The cordial in Pw V90 is in fact a more complicated version of Dudley's famous powder, known as "Pulvis Warwicensis", which consisted of scammony, sulphur of antimony, and tartar. A

[56] Digby, *Late Discourse*, 22.

[57] Cavendish, *Philosophical Letters*, 354–355. Also see her remarks on "Sympathetical Powder" at 389.

professor at the University of Pisa, Marco Cornacchini, first printed this recipe in a 1620 work in praise of chymistry titled *Methodus qua omnes humani corporis affectiones ab humoribus copia*, which he dedicated to Dudley.[58] In the course of relaying how to craft his medicine, Dudley indeed alluded to this text, writing that William should burn sulphur "in a Chafing=disshe as yor Lords.^{ip} knoweth by Doctor Coracckeno his Booke written thereof".[59] While the nature and contours of Dudley's relationship with William remain opaque, this quotation suggests that he was well enough acquainted with the marquis to know his reading habits. The fact that the cordial in Pw V90 is more intricate than the one in Cornacchini's *Methodus* also suggests that the recipe that Dudley sent to the Cavendishes was the culmination of his medical (and indeed experimental) exploits.

Compared to other cures in Pw V90, Dudley's cordial is strikingly elaborate. It consists of manifold expensive ingredients—many of which were also prescribed as standalone cures—including the tinctures of gold and saffron, liquor of silver, bone in the stag's heart, sapphires, rubies, emeralds, contrayerva root (a plant found in the tropical regions of the Americas), as well as unicorn- and rhinoceros-horns. These are accompanied by the occasional anodyne ingredient, such as cream of tartar, milk of sulphur, and nutmeg. As a combination of many concoctions and powerful ingredients, Dudley's cordial is a "super cure" of sorts that was supposed to do everything from guarding against "all Poysons and the Plague" to stopping "the too abun=dant Courses of blood of women".[60] The existence of Dudley's recipe in Pw V90 is particularly intriguing because it underscores a tension between the egalitarian presentation of the medical manuscript as a whole—in which the recipes of nuns, physicians, and dukes sit side-by-side—and the social and financial hierarchies that undergirded the nature and circulation of particular recipes in the increasingly competitive seventeenth-century medical marketplace.

[58] See Marco Cornacchini (1619), *Methodus qua omnes humani corporis affectiones ab humoribus copia* (Florence). Not everyone was impressed by Dudley's powder, and Stefano Rodrigo de Castro, Professor of Medicine at the University of Pisa, in fact penned a satirical poem on the subject (see Giovanni Targioni Tozzetti (1780), *Notizie degli aggrandimenti delle scienze fisiche accaduti in Toscana nel corso di anni LX del secolo XVII*, Vol. 3 (Florence), 42).

[59] Transcription, 149.

[60] Transcription, 140 and 146.

Whereas Digby, as a struggling courtier, devised a comparatively simple powder of vipers, snakes are only one feature of Dudley's remedy. It would seem that recipes such as Dudley's were less items of practicable medical therapy than markers of social standing that circulated, in part, to impress patrons or friends. Although not quite as excessive, recipes such as Talbot Grey's powder might be placed in a similar category.[61] It appears to have been universally recognised that such cordials were overpriced, and, when discussing the distinctly less grand "*greater Cordial Pouder*" in the *Pharmacopoeia Londinensis*, Culpeper noted that these are "are too high for a poor man's purse" and that one should not consume above half a dram "more for the cost of them than any ill effects they work".[62] With that being said, the Cavendishes were assured that Dudley's cure had been "gi=ven with admirable successe" to many highborn individuals, not least "the great Duke of Tuscany", whose constitution Dudley implicitly linked to William's.

While we have seen that learned medics and apothecaries could work in harmony, courtiers were not always so amicable or obliging, with Dudley repeatedly elevating himself above even the learned physician. After giving directions for his tincture of gold, for instance, he remarked that "Even des=perate Feavors, when y^e^ Partie hath bene given over by y^e^ Physitians, I have Cured Soddainely".[63] Mayerne, however, would have been sceptical about such advice insofar as he preferred "Imperfect Mettalls" (such as iron) to the perfect ones (such as gold) that Northumberland recommends, and held, more generally, that "when it is in the Pursse, doe not trouble yor self[;] very often great reamidyes fayle where some light ones doe the feat".[64] Remarking on Dudley's cordial in particular, Mayerne noted that "in ye most Part of such things Parturiunt monthes" ["mountain labour"].[65] This is a reference to Aesop's fable "The Mountain of Labour" wherein a mountain, after much ado, gives birth to a mouse (an implicit critique of speech acts that promise much but deliver little). Perhaps because he mixed assorted ingredients impetuously, in a manner that most medics would have deemed dangerous, Dudley's recipes are

61 See Leong, *Recipes and Everyday Knowledge*, 167–168.

62 Culpeper, *Pharmacopeia Londinensis*, 126.

63 Transcription, 145.

64 Transcription, 128.

65 Transcription, 137–138.

also the only ones in the collection that repeatedly advertise their safety. Dudley's advice was, more generally, frequently at odds with Mayerne's, and, perhaps hoping to ingratiate himself with William, he even recommended smoking tobacco "in ye Pipe at least six times a day".[66] While this latter suggestion undoubtedly appealed to William, the huge price tags on the ingredients in Dudley's cordial and Mayerne's negativity towards it mean that it is unlikely that the Cavendishes ever actually put it to use.

Margaret's writing on universal cures lends credence to the supposition that the Cavendishes would have heeded Mayerne's advice and not wasted their money on Dudley's recipe. In a section of *Observations Upon Experimental Philosophy* titled "Of the Universal Medicine, and of Diseases", Cavendish wrote, regarding "universal medicine", that she does "not believe there is any, nor that all diseases are curable".[67] Even though she found chymical principles interesting, she was invariably sceptical about the more optimistic claims of chymical medicine. Attacking cordials in particular, she maintained in *Philosophical Letters* that

> it is well known to experienced Physicians, that Medicines prepared by the art of fire are more poysonous and dangerous then natural drugs; nay, I dare say, that many Chymical Medicines, which are thought to be Cordials, and have been given to Patients for that purpose, have proved more poysonous then any Purging Physick.[68]

While both of these passages were written over a decade after Pw V90 was abandoned, we might surmise that Cavendish's doubt about such universal cures and specific hostility to cordials was an outgrowth of her incredulity towards Dudley's incredible recipes. One might ask why the Cavendishes, if they were unlikely to have used Dudley's cordial, carefully transcribed all of the sub-recipes. As has already been suggested, however, such recipes were preserved as much for their historical value, as ledgers of networks and debts, as for their medical utility.

Along with signalling his high social standing, Dudley (like Digby) recommended exotic ingredients to shore up this image as a traveller, and, though he is less insistent than Digby on this point, he perhaps similarly hoped to turn a profit by sourcing some of the ingredients for

[66] Transcription, 148.

[67] Cavendish, *Observations*, 242.

[68] Cavendish, *Philosophical Letters*, 378.

the Cavendishes through his connections abroad. Dudley was also known for his acumen in engineering mechanical devices—most remarkably new ships and a silk weaving machine—and he consistently recommended the implementation of specific chymical contraptions. That is, far more than is the case with other recipes in Pw V90, Dudley emphasised instruments, their mechanisms, and the indispensability of fire as he explained how to create his cure. He also provides more detailed descriptions of the steps required to make his recipe than other contributors to the manuscript, which indicates that he likely went through the experimental process of making each of the ingredients himself (as Boyle would later do). He refers frequently, for example, to the colour (e.g. "a yellow orange Colour"), smell (e.g. "The first y^t sublimeth is naught & stinking"), and taste (e.g. "Cleare & wth out any tast") of the concocted ingredients, as well as the precise tools that should be used to make up the remedy (e.g. "mingle them well wth a Spatule of wood").

Beyond these verbal descriptions, mechanical contrivances and processes are also occasionally sketched in the margins. The most interesting of the represented instruments is a "duble filosophers egge", which was used to make a "tincture of gold". The "philosophers' egg" had many meanings, so it is worth dwelling on it for a moment. Most immediately, as the sketch in Pw V90 reveals (see Fig. 7.1), Dudley was referring to an egg-shaped glass vessel with a long neck used for heating substances over a sustained period of time. But a philosophers' egg could also refer to a hermetically sealed pear-shaped vase (also called "The Vase of Hermes") that functioned as an incubator.[69] The basic idea of incubation here is that the philosophers' egg could do the work of nature, but in a greatly reduced time—thus the "art" (i.e. chymistry) both imitates and improves upon natural methods, with all of the traditional implications of transformation, growth, and life. Beyond these references to specific technologies, a "philosophers' egg" was also an alternative formulation for the "philosophers' stone", which, as is well known, was an elusive substance that was deemed capable of turning base metals into gold, and was an obsession of Renaissance chymists. It was said that the "egg of the philosophers" consisted (like the universe) of four components: white,

[69] See *OED*, "philosopher", phrases, "egg of philosophers, n." For the "Philosophers' Egg" as the "Vase of Hermes", see John Read (1995), *From Alchemy to Chemistry* (New York), 30–37 and William R. Newman (1995), The Philosophers' Egg: Theory and Practice in the Alchemy of Roger Bacon. *Micrologus* 3: 75–101.

yolk, skin, and shell.[70] These were sometimes thought to typify not only the four elements but also the four metals of transmutation: copper, tin, lead, and iron. It is significant, in this regard, that Dudley claimed that his philosophers' egg could transform wine and gold into the superior "Tincture of Gold" that could, in turn, "worketh miracles in all ocasions".[71] Indeed, the way in which Dudley describes the intricate workings of the philosophers' egg on the wine and gold is reminiscent of the transubstantiation of blood in the miracle of the Eucharist, for he writes that the spirit of wine will "dissolve its Tincture, & drop downe with it like bloud into ye rest of the Spirit of wine".[72] Playing on his status as a notorious Catholic convert, this allusion was probably intentional.

It generally appears from Pw V90 that for seventeenth-century Catholic aristocrats such as Digby and Dudley—as distinguished individuals whose confessional persuasions and continental orientations complicated their standing within English high society—the creation and promotion of cures had an outsized but mostly untold role in the process of self-fashioning.

[70] See John M. Stillman (1960), *The Story of Alchemy and Early Chemistry* (New York), 170–171.

[71] Transcription, 144–145.

[72] Transcription, 144.

CHAPTER 7

The Transcription

A. Principles

Pw V90 was not produced for publication or indeed circulation beyond the immediate Cavendish family, and, because of this, the spelling and punctuation can be highly irregular. In producing this transcription, we have endeavoured to balance remaining faithful to the original with making the transcription readable. Following Michael Hunter's advice, we have not created a facsimile of the original, but have rather aimed "to produce an edition which does justice to the content of the manuscript, paying attention to its actual appearance but not fetishising this".[1] Having said this, not every reader will have the same conception of where the line between fidelity and fetishisation ought to be drawn. In the case of Pw V90, the situation is especially complicated, since this is a fragile manuscript that is becoming increasingly difficult to access. For this reason, we have often steered closer to fidelity to the original, in hopes of making our edition useful for competent readers of seventeenth-century manuscripts who cannot access Pw V90. In order to assist those readers who may not have much experience reading early modern manuscripts, or who have a limited understanding of sometimes rather esoteric

[1] Michael Hunter (2007), *Editing Early Modern Texts: An Introduction to Principles and Practice* (Basingstoke), 85.

J. Begley and B. Goldberg, *The Medical World of Margaret Cavendish*, Palgrave Studies in Medieval and Early Modern Medicine,
https://doi.org/10.1007/978-3-030-92927-5_7

seventeenth-century medical terminology, we have provided extensive, clarificatory notes and a Glossary of Terms (Appendix 8A). By taking this approach, we hope that our edition will be of use to the widest possible readership.

In terms of our transcription and annotation principles, it is first notable that this transcription has been produced in its entirety by scrutinising the original manuscript. We have reproduced the original's foliation, and noted the existence of blank leaves, while also using the common abbreviations of "r" and "v" for *recto* and *verso*. Although we have attempted to decipher every word in the original, the reading is not always perfectly clear, and, even when the reading is clear, the meaning is sometimes obscure. In these instances, we have recorded any doubts in the annotations, along with our best guesses about how to read the word or passage.

We have normalised i/j and u/v variants, with the latter being sometimes difficult to distinguish. These are not used consistently across the manuscript, and, indeed, are sometimes used differently by the same scribe in a single section. At the same time, double letters (namely, "n", "m", "l", "t" "e", and "o") have been spelt out when they are marked in the manuscript with a macron. These marks are difficult for publishers to replicate in the case of some consonants and spelling out the letters makes the manuscript more readable in most cases. We have retained the thorns used throughout (e.g. retaining "y^{t}" rather than spelling out "that"). For the sake of readability, when a "y" replaces an "i" as the first letter of a relatively common word (e.g. "yce", "yvory", "yron") we have made a blind change. All ligatures have been replicated.

Concerning "[*sic*]", we do not use it for what we would now consider to be grammatical or spelling errors, but only for issues (such as the accidental repetition of the same word or the clear confusion of the singular and plural) that could otherwise be interpreted as transcription errors. Not only have we retained most spelling irregularities, but we have also tried to capture the *mise-en-page* as far as possible. We have, for example, provided annotations indicating the location of the rare marginal notes or drawings scattered throughout the manuscript. With respect to insertions, deletions, and crossed-out words, we have sought to replicate these, and, where doing so has not been possible, we have indicated as much in the notes.

Occasionally we have interpolated words and punctuation in square brackets "[]" where they are not evident but where it might otherwise be difficult to interpret the sentence. We were reticent about doing this at first because of the way that it changes the look and sense of the text, but because Pw V90 includes many lists of ingredients and recipes that are often hard to distinguish from one another, we decided that making such interventions would be a courtesy to our readers. Further, where the punctuation and syntax are particularly irregular, we have provided the most plausible interpretation of the sentence, phrase, or clause in the notes. In rare instances where the second half of a pair of parentheses is missing, we have silently restored it. We have retained the capitalisations of the original text, and the various forward slashes ("/") found throughout are original to the manuscript. In Pw V90, line breaks are often used instead of commas (especially in lists), but we have added commas in brackets, since the line breaks cannot be replicated. Other relevant linebreaks (between, say, parts of a word) are indicated by an equal sign "=", since this symbol is most commonly used in the text itself. We use an asterisk ("*") for the ornate marks at the end of some notes, and a hyphen ("-") where similar marks are found in the original. Finally, there are a number of places where additions have been made to the text, and these are indicated with a square bracket and a caret (e.g. "[^the]").

Taken as a whole, the annotations serve two principal ends. The first, and most important, is to make sense of the manuscript; the second is to point readers to relevant scholarly literature. In terms of the first, the footnotes provide standard spellings for particularly aberrant spellings (e.g. noting that "cynamond" means "cinnamon", etc.). When a word is consistently spelt incorrectly, we give a gloss for the first instance but not every subsequent one. We also occasionally provide glosses for minor misspellings that could obscure the sense of the sentence (e.g. the use of "of" for "off"). More substantively, the footnotes often indicate the healing properties that products were thought to possess, frequently citing Culpeper's *Pharmacopoeia*.

For the definitions of the key terms that appear in Pw V90 and Harley 6491 (with their aberrant spellings), we direct readers to the Glossary. Where synonymous terms are used in the manuscripts (e.g. "Carabe" and "Yellow Amber"), we provide the definition in the Glossary next to the most commonly used term, with the less commonly used terms referring to this main definition. The Glossary mostly excludes derivatives of main ingredients (e.g. it includes "Tartar" and "Sulphur" but not "Oil of Tartar" or "Spirit of Sulphur"), unless they are significantly different from the main ingredient (e.g. it includes "Balsum of Sulpher" because it is very different from "Sulphur"). The Glossary heavily relies on the *Oxford English Dictionary* where entries are available, but definitions are sometimes modified to make them more precise or accurate (e.g. "complexion" is referred to as a term used in "the Galenic tradition" rather than in "the physiology and natural philosophy of the Middle Ages"). Even in the cases in which a term can be found in the *OED*, we hope that gathering terms and variant spellings in a single, easily accessible Glossary will save scholars time and energy as they come to terms with the common (and obscure) ingredients found in not only Pw V90 but also other early modern medical manuscripts and printed books.

B. Pw V90 Text

[1r is blank]

[1v]

Title Page and many entries in the handwriting of Thomas Farr. Some pages in that of Wm. Cavendish Duke of Newcastle.[2]

[Title Page, 2r]
A Booke wherein is Contain=ed Rare Minerall Receipts Collected at Paris from those Who hath had great Experience of them.

Here is also in this Booke some other Excelent Receipts wch are not Mineralls./

[2v is blank]

[3r]

The Description of Salt of Steele or Anima He=patis[3]; the which is Sr Theador Mayernes[4] Invention.

Take oyle of Brimstone made in the Campane or Bell, foure ounces[;] Spirit of wine rectified wthout flegme two oun=ces, Shaveings of Steele

[2] Written in pencil in the hand of an archivist.

[3] Early modern chymists often used the term "anima" to refer to an active principle of a medicine or medical ingredient that is separated from its origin through some chymical process (Robley Dunglison (1857), *Medical Lexicon: A Dictionary of Medical Science* (Philadelphia), 65).

[4] Théodore de Mayerne (1573–1655).

well clensed from any dust three-ounces: Let all these be put in [^a] glasse dish in a valted Cellar and Leave them togeather till it turns all to a kinde of whitish salt, which afterwards yo^{u} must Expose to the heat of the sun the space of three-weekes: All this while you must be carefull least[5] any dust mixe wth it and therefore yo^{u} should have it all covered wth Paper./

The Method how to administer this Remedy

Glyster.[6]/ Take Mallowes, violets, Pellitory of the wall,[7] Groundsell, Beetes; of Each one hand full: flowres of Cammomile, Melilot[,][8] Dill, [3v] of Each halfe a handfull; Seeds of Anis,[9] Fæinle[10] of Either two dragmes[11]: Linseed[,] fænu=greek seed of Either three: dragmes: Let all these be Boyled in a quart of whey, till it comes to a pinte, in wch yo^{u}

[5] Lest. Expressive of something to be prevented or guarded against (*OED*, conj. 1a).

[6] In her *Philosophical Letters*, Cavendish wrote that glysters/clysters "are a safe, speedy, easie and profitable medicine, and far more safe then Chymical Salts, Tartars, Spirits, or the like" (381). For a wider context, see Derek Doyle (2005), Per Rectum: A History of Enemata. *The Journal of the Royal College of Physicians of Edinburgh* 35.4: 367–370. In addition, glysters were a fashionable subject of art (see Laurinda S. Dixon (1993), Some Penetrating Insights: The Imagery of Enemas in Art. *Art Journal* 52.3: 28–35).

[7] According to Culpeper, pellitory helps "the stone and gravel in the Kidnies, difficulty of Urine, sore throats, pains in the ears" (Culpeper, *Pharmacopoeia*, 19). Unless otherwise noted, all references to the *Pharmacopoeia* are to the 1653 edition.

[8] Culpeper claimed that melilot "provokes urine, breaks the Stone, clenseth the Reins and Bladder, cutteth and clenseth the Lungs of tough Flegm; the juyce dropped into the eyes, cleers the sight, into the ears, mitigates pain and noise there; the head bathed with the juyce mixed with Vinegar, takes away the pains thereof: outwardly in Pultisses, it asswageth swellings in the privities" (Culpeper, *Pharmacopoeia*, 21).

[9] Aniseed.

[10] Fennel.

[11] Dragme (or "drachm") is the middle term for the apothecaries' system of measurement, which was the primary system used for medicines (and sometimes precious stones or metals). Each term is also associated with a symbol, and these are used throughout this manuscript: Pound (℔; 12 ounces); Ounce (℥; 8 drams); Dram/Drachm/Dragme (ʒ; 3 scruples); Scruple (℈; 20 grains); Grain (gr). This system of measurement dates back to ancient Rome but has its roots in the Greek measure of the *drachm*, which was the name of the standard silver coin in Greece, and, as a measure of weight, likely meant something like "a handful". The apothecaries' system was widely used in England until the Medical Act of 1858, when the *avoirdupois* system was adopted. The latter system had, however, been unofficially used to measure most things, except medicines and some precious stones and metals, in England since at least the thirteenth century (see Chapter 3, 33–34). The *avoirdupois* system goes in the opposite direction of the apothecaries' system: from drams, to ounces, to pounds, to stones, to quarters, to hundredweights, and finally tons.

must dissolve loohoc[12] of Cassia,[13] & Lenitive of Each halfe an ounce: Syrup of violets & red suger[14] of Each two ounces; These being mingled shall be given at foure a Clock in the afternoone; Glysterwise[15]/

The next day yo^{u} shall take the following Purge[16]/

Purge./ Take now Tamarinds[17] 6: dragmes; Creame of Tarter pulveri[^z]ed .3 dragmes; Coriander, Ani=seeds of each a dragme; Let all these be boyled wth halfe a Pinte of Spring water till it comes to 6: ounces; then straine the Liquor cleare in wch you shall infuse Leives of Sena halfe an ounce: Rubarbe 2: dragmes; Let these steepe all night upon hot Embers: in the morning after two walmes[18] straine it & add to it an ounce of Man=na of Calabre[19] [^&] twelve drops of Cinamon water Let this potion be taken in the Morneing early three houres before yor broth, Keep your Chamber that day./

[4r]

Bleeding./ The next day yo^{u} shall Let bloud in the right arme of y^{e} Basi[^li]cke[20] vaine to the quan=tity of nine ounces./

12 Lohoch.

13 Culpeper argued that cassia, a kind of cinnamon, "strengthens the stomach, helps digestion, causeth a sweet breath, resists poyson, provokes urin, and the terms, causeth speedy delivety [*sic*] to women in travel, helps Coughs and Defluxions of humors upon the Lungs, Dropsies, and difficulty of Urin" (Culpeper, *Pharmacopoeia*, 11).

14 Sugar.

15 I.e. via a glyster.

16 Purges, along with other bodily evacuations (like bleeding), were a core therapeutic tool of early modern physicians, in keeping with the ancient tradition of humoural medicine.

17 Tamarinds were thought to "purge choller, cool the blood, stay vomiting, help the yellow Jaundice, quench thrist, cool hot stomachs, and hot livers" (Culpeper, *Pharmacopoeia*, 27).

18 Walm. The bubbling and heaving of water, etc., in the process of boiling (*OED*, n. 3a). The meaning here is that after the mixture comes to a boil twice, one should proceed to the next step.

19 "Manna of Calabre", according to the lexicographer Randle Cotgrave, is "the best and most lasting Manna" (Randle Cotgrave (1611), *A Dictionarie of the French and English Tongues* (London), unpaginated).

20 Basilic.

The day following you shall take ye following Julebs[21] in the Morning at seaven a Clocke And in the afternoon at foure. Abstaineing from all Meate & Drinke for two houres after and so continue the use of them for sixe dayes./

Julebs./ Take pareings[22] of Pippins one hand full, ordinary Barley one ounce, Harts horne,[23] & Ivory shaveings, Conserve of Gillyflowers[24] of Each halfe an ounce[,] Citrin sanders[25] two dragmes, Let these boyle wth a quart of spring water till it comes to a Pint & a halfe of strained liquor, to wch you shall add a pint of Rennish[26] wine, This liquor yo^u shall straine through a paper in a glasse fonnell[27]: To the Cleare Liquor you shall add oyle of Brimston two dragmes, oyle of Tartar so much as shall tak away all the acrimony; then you shall add Syrups of violets, of Gillyflowers & of Pippins of each an ounce & a half; Aqua Mariæ[28] .20: drops, Let this quantity be for three dayes./

After the sixe dayes of Julebs yo^u shalbe [*sic*] purged againe as before./ These being finished you shall begin to use the Steele drinke wch is to be used as followeth./

[4v]

[21] Julep.

[22] Paring. A shaving or peeling (*OED*, n. I, 1).

[23] Culpeper held that (like "unicorn's horn") hartshorn "resists poyson and the pestilence, provokes urine, restores lost strength" (Culpeper, *Pharmacopoeia*, 31).

[24] According to Culpeper, gillyflower can "resist the pestilence, strengthen the heart, liver, and stomach, and provokes lust" (Culpeper, *Pharmacopoeia*, 26).

[25] Here "citrin" refers to a pale yellow or greenish-yellow. The reference is thus to "yellow sandalwood". This common ingredient was thought to "cool the heat of Feavers, strengthen the heart, and cause cheerfulness" (Culpeper, *Pharmacopoeia*, 24).

[26] Rhenish wine. Wine produced in the Rhine region (*OED*, adj. A, 1).

[27] Funnel.

[28] The recipe for "Aqua Mariae" includes ingredients such as "Sugar Candy", "Imperial water", "Ambergreese", and "Saffron". As a result, Culpeper warned that it is "very costly, as far beyond the reach of a poor mans purse, as of his brains" (Culpeper, *Pharmacopoeia*, 67).

Comune[29] Liquor} Take dryed femal fearne roots,[30] of dockrootes of each halfe a Pound[;] Shuckory,[31] foenile[,][32] Brus=cus rootes[33] of Each .4: ounces; Shaveings of Nephretick=wood, of Ivory of each two ounces[;] Citrin sandar one once, Harts tongue[,] Maiden=haire[,] Wall=Rue, white adianth; Ceterach[,][34] tama=riscus[35] Leives of each two handfull; flowers of St John=worth, Gillyflowres, Cows lipe[36]=flowers of each one handfull; pareings of Apples a pound & a halfe; All these being used cut small & dryed, & shalbe [*sic*] put in a great glasse wth 4: gallons of water of Pippins and set to infuse in Mary's Bath the space of seven dayes in a moderate heate, then you shall straine the Liquor through a paper in a fonell./

Steele Liquor}./ Take of this Liquor three gallons, & add to it an ounce & a halfe of salt of Steele or Anima=Hepatis, This Liquor is to be taken every day halfe a pint at a time:/

Purging Livor wth Steele./

Take the other gallon of Comune[37] Liquor, and add to it Sena 4: ounces, Rhubarbe. 2 ounces; Let these infuse in Mary's Bath fowre[38] dayes [5r] And straine it after, as the first adding to the cleare Liquor halfe an ounce

[29] Common.

[30] The division of ferns into "male" and "female" goes back at least to the sixteenth century (see e.g. Rembert Dodoens (1578), *A Niewe Herball, or Historie of Plantes*, trans. Henry Lyte (London), 400).

[31] According to Culpeper, succory "strengthens the Liver and Veins, it opens obstructions, stoppings of the Liver and Spleen" (Culpeper, *Pharmacopoeia*, 5). This herb, and, as we will see, many others in the manuscript, was meant to combat melancholy.

[32] Fennel.

[33] Bruscus was used to "provoke urine, break the stone, and help such as cannot piss freely" (Culpeper, *Pharmacopoeia*, 9).

[34] According to Culpeper, ceterach "wasts and consumes the spleen" and "is excellent good for melahcholly people, helps the strangury, provokes urine, and breaks the stone in the bladder" (Culpeper, *Pharmacopoeia*, 16).

[35] Tamarisk was deemed "profitable for the Rickets, and Burnings" (Culpeper, *Pharmacopoeia*, 12).

[36] Cowslip.

[37] Common.

[38] Four.

of salt of Steele or Anima Hepatis, this purging Liquor is to be taken halfe a pint at a time every third day./

Both these Liquors of Steele are to be taken earely, about sixe a Clocke in the Morning, The Steele Liquor wch is not purging, 2: dayes together. The purging Liquor the next day and so continue it in that order. You must rest for an houre after the takeing of it; and then exercise yo^{r}selfe moderately the space of two howers[39] after./

Every 10th day yo^{u} shall Purge wth ye Generall purge prescribed in the beginning & when yo^{u} finished yor whole Course of Steele yo^{u} must conclude with this Purge

The Manner how to take the Steele in Syrope./

In stead of takeing it in the Liquor's before prescribed, yo^{u} must melt a scruple of Salt of Steele for every dose, in a spoonefull of syrup of Apples or pippins./

[5v] Every third day yo^{u} shall take the same Salt, in a spoonefull of syrup ad Melancoliamto[40] to make it Purging./

You shall, in all other respects, observe the same order and directions, as in the other

M. Boucheret[41] AR.[42]

The virtues of the Antimoniall[43] Cup./

1. It keepeth the Body from Repletion of humours, the cause of all diseases./
2. It helpeth all the evill effects of the Stomach./

[39] Hours.

[40] While the spelling is off, this is most likely a reference to "syrup of meconium", which, according to Culpeper, "prevails against dry Coughs, Phtisicks, hot and sharp gnawing Rhewms, and provokes sleep", but he warns that it should only "be used with a great deal of caution" (Culpeper, *Pharmacopoeia*, 106).

[41] Mathew Boucheret (d. 1671/2).

[42] The "AR" here is a ligature, which may well stand for "apothecarius regius". Many thanks to William Newman for this suggestion, but we have not been able to determine its exact meaning and provenance.

[43] The effect of antimony on the stomach was thought to be analogous to its use in the purification of gold. For a much fuller account, see R. Ian McCallum (1999), *Antimony in Medical History: An Account of the Medical Uses of Antimony and Its Compounds Since Early Times to the Present* (Edinburgh).

3. It Cureth all Head=ache comeing from the Stomach./
4. It Cureth all Agues comeing from Repletion./
5. It helpeth the Vertigo or swimming of y^e head./
6. It helpes the Lethargie or forgetfulnesse./
7. It Cureth the greene Sicknesse.[44]/
8. It helpeth the falling Sicknesse./
9. It emptieth the Stomach of Viscous flegme, the Liver of choler; [^&] the spleene of Melan=cholly; & cleeres the Brest stopt with flegme & purgeth the head & Throat, It restoreth the [6r] Lost Appetite, and causeth Rest./
10. Lastly, by takeing it in the Spring & Fall, or any time betwixt;/
    It prevent[^et]h all surfets,[45] agues, goute, stone, sciatica's, dropsies, Meazells, Poxe, Itch, Scabbs and innumerable evills./
    It never looseth quantitie or qualitie./

The use of the Antimoniall Cup./

Take the Antimoniall Cup every night at Sixe a Clock when yo^u meane to take it the next Morneing and fill it with sack, white wine or Muscadell and so Let it infuse till seaven a Clock next morneing, & then drinke it, Observing the ordinary rules you do after the takeing of any Physique,[46] Drinkeing some cleere[47] posset drinke after every vomiting./

A Child may take this Medicin also wth this caution, that yo^u shall onely give him the halfe of the quantitie, And in case it should not work two houres after yo^u may give him the other halfe. If yo^u desire to prepare wine for foure Persons at once, you must put the Cup in a Cilver[48] Canne wth a pint of wine, & put the Canne in a Skillet full of water, seting the

[44] On this disease, see Helen King (2004), *The Disease* of *Virgins: Green Sickness, Chlorosis, and the Problems of Puberty* (New York).

[45] Surfeit. Superabundance, superfluity; an excessive quantity or supply of anything. Also, an illness attributed to excessive eating or drinking (or, occasionally, to extremes of temperature or some other environmental factor); an instance of this (*OED*, n. 2 and 4a).

[46] Physick (i.e. medicine).

[47] Clear. "Clere ale" was a kind of diluted ale. In the (*c.*1485) *A Litill Boke Necessarye [and] Behovefull A[g]enst the Pestilence* (London), one Bengt Knutsson gives the instructions for making this drink as such: "Take a quātite of tryacle of a pese &. ij. sponeful of clere wyn or rosewater or ale dissolve the triacle in the cuppe & drynke hit" (unpaginated).

[48] Silver.

skillet upon the [^hot] Embers [6v] And so let it infuse all Night, & in the morneing give to every one a Cup full of the said Infused wine./

After you[u] have used the Cup you[u] must rub it wth salt, & washe it with water & dry it wth Linen, and put [^it] againe into his Boxe./

A Letter from S[er] Thomas Cademan[49]
to the Lord Marquesse of Newcastell,[50]/
Agust the 29: 1647

My Lord

You[u] gave leisure as well to him that brought the Letter, as to him that is to answere it for I received it not till yesterday by the miste=rys[51] of the Cariole,[52] Till this operation now in hand be finished w[ch] I hope will be in .10. or .15. dayes at most I Cannot give yo[r] Lordshipe a just acount of his proceedings, In the meane time I send yo[r] Lordshipe what his Highnesse[53] has done in his Steele Course. First refusing to purge, he tooke every Morneing a Large draught almost as much as a Posticarys[54] pint w[ch] Containes .12. ounces, of fountaine water [7r] Wherein has beene Steeped all night .4. ounces of the simple Fylings[55] of Steele, w[ch] every needle maker sells, without any manner of preparation at all, And the same Steele is every day steeped in the same quantitie of water put fresh to it, till w[ch] the steele is kept in the same Cloath, wherein it was first tied up; after it, his Highnesse walked an houre till he was warme, And this was done for .20. dayes, without Intermision/

Except onely in the mide way he was purged w[th] .3. dragmes of Senna infused all night in warme broath, made soure wth Juoice of oranges and two ounces of syrup of roses solutive. And this was to treate his Highnesse

[49] Thomas Cademan (1590?–1653).

[50] William Cavendish (bap. 1593–1676).

[51] Ministry. A ministry, office, or some sort of service (*OED*, n. 2, 1a).

[52] Carriole. A small open carriage (*OED*, n. 1). While the first known English usage is from 1770, the French is “*carriole*”, from Old Provençal “*carriola*”, meaning a small two-wheeled carriage.

[53] King Charles I (1600–1649).

[54] Apothecaries’.

[55] Filings.

as a Prince for servants are ever better cured then Kings and a servant should have beene Purged .2. or .3. times before he had begun his Steele and once every sixe dayes & twice afterward The steele too should have beene prepared soe as to make the water tast worse & have done more good./

My Ladyes vomit is made thus. Halfe an ounce of Crocus=Mettallorum[56] powdered & washed thrice in Baulme[57]=water or Cardus=[7v] Benedictus[58]=water, when it is dry againe and made into most fine powder is to be steeped in a pottle of the strongest Sherry sack in a glasse bottle very well stopt, Leaveing roome for a pint or thereabouts to be put in as also y^{t} the glass may be well shaked thrice a day for the three first dayes, then before it setteles fill up the glass wth Malego=Sack.[59] After it hath stood .10. dayes and is very cleare it may be taken. One ounce is an ordinary dose for strong men, weake folkes, wimen & Children must goe lesse. Children of two yeares old will have enough of a Dragme wch is but the Eight part of an ounce, And generally those wch are easily moved Either to vomit or purge, must have a Lesser dose. My Lady Marquesse of Newcastell[60] is one of those that is [^so] hardly moved to cast, That hir Ladyshipe must have a double dose at least & will not be moved to any purpose without two ounces, rather more this infusion of Crocus=Metallorum will keepe alwaies good & therefore one may [word crossed out to illegibility] make a double or a treble quantity as they [8r] Please when they are in a place of repose. Yo^{r} Lordshipe is very happie to have no neede of any of these things, for as yor $Lo.^{ip}$ will give me leave to speake like a Physitian and not

56 Also known in English as "liver of antimony"; see, e.g. Nicolas Lémery and Walter Harris (1686), *A Course of Chemistry* (London), 217–222.

57 Balm.

58 According to Culpeper, carduus benedictus "helps swimming and giddiness in the head, deafness, strengthens the memory, helps griping pains in the belly, kills worms, provokes sweat, expels poyson, helps inflamation of the liver, is very good in pestilences, and the French-pocks; outwardly applied, it ripens Plague-sores, and helps hot swellings, the bitings of mad-dogs, and venemous beasts" (Culpeper, *Pharmacopoeia*, 16).

59 Malaga sack.

60 Margaret Cavendish [née Lucas] (1623?-1673).

like a Courtier, I belive the Joye yo^{u} have in my sweete Lady & happy tranquillity at home in your blessed family is the best Preservative of yor health wch I wish so Constant & so great as yor Lordshipe may never neede any thing else from./

My Lord/

Yo^{r} Lordshipes most
Humble servant,
Tho. Cademan/

[8v]

Sr Theador Mayernes Description of Remedy's, for the Lord Henry Caven=dishe,[61] agaynst the Spleene,[62] & vapors ariseing to the head, with trembling and shakeing fitts./

Glyster. Take Mallow[,] Marshmallow[,][63] Mercury[,] Pelletory=Leives of each a handfull; Centaury, St John=wort, worme=wood, Stoecas[64] of each a hand full; Seeds of Annis[,] Foe=nell, Cumin, of each: three dragmes; Bayes[65] & Junipery=berrys, of each two dragmes; Boyle all these in a quart of clere Posset ale till it comes to a Pint of strained liquor, In which you shall disolve Catholicum an ounce & a halfe, Honey of Mercury two ounces, Wine of Crocus Mettalorum foure ounces, Let this be given Glyster=wise at foure of the clock in the afternoone./

[61] Henry Cavendish (1630–1691).

[62] In humoural medicine, the spleen is the seat of black bile (in Greek “μέλαινα χολή” or in Latin “*melancholia*”). It was associated with the disease of melancholy (also see Chapter 3, 26–27).

[63] Marshmallow was thought to “ease pains, help bloody fluxes, the stone and gravel” (Culpeper, *Pharmacopoeia*, 3).

[64] Stœchas was deemed “a great counterpoyson” that opens “obstructions of the Liver and Spleen, clenseth the matrix and bladder, brings out corrupt humors, provokes urine” (Culpeper, *Pharmacopoeia*, 24).

[65] Bays were reputed to “expel wind, mitigate pains”, cure “cold infirmities of the womb, and dropsies” (Culpeper, *Pharmacopoeia*, 27).

Cordiall Juleb.} Take Black Cherrys, & Peony water, of each an ounce & a halfe; Antepileptic water[66] of Sangius[67] one ounce; Spirit of Castor[68] halfe an ounce, Syrup of Peony flowers an ounce & a halfe, oyle of vitrioll[69] ten drops, Take two spoone full of this when he is in his fitt./

[9r]

Opening Apozeme./
Take Oxymel Julianizans[70] an ounce & a halfe; Oxy=mel Scilitic[71] halfe an ounce; Black Cherrys & Peony water of each two ounces & a halfe; Let this be taken at two times in the Day before the following vomit./

Vomitt./
Take Infusion of Crocus Me[^t]allorum an ounce & a halfe; Oxymel Scilitic halfe an ounce; Hysop[72] wa=ter an ounce; Let this be taken in the Morneing, takeing after each vomit, sume warme Posset./

Cordiall Bolus./

66 Various waters made to treat epilepsy.

67 "Sanguis draconis" (or "Dragon's blood") apparently "cools and binds exceedingly" (Culpeper, *Pharmacopoeia*, 29).

68 According to Culpeper, castor "resists poyson, the bitings of venemous beasts, it provokes the terms, and brings forth both birth and after birth, it expels wind, easeth pains and aches, convulsions, sighings, lethargies, the smell of it allaies the fits of the mother" (Culpeper, *Pharmacopoeia*, 31).

69 Vitriol was deemed to "asswageth thirst, allayeth the violent heat in feavers and pestilences, and a few drops of it gives a pleasant grateful tast to any medicine" (Culpeper, *Pharmacopoeia*, 184).

70 Culpeper maintained that "*Oxymel Julianizans*" "cuts and divides humors that are tough and viscus" (Culpeper, *Pharmacopoeia*, 113).

71 The "Oximel of Squils" was deemed "good against the Falling-sickness, Megrim, Head-ach, Vertigo" and to help "women not well clensed after labor" (Culpeper, *Pharmacopoeia*, 113).

72 Hyssop was thought to combat "Coughs, shortness of Breath, Wheezing", while it also "kils worms in the body, amends the whol colour of the body, helps the Dropsie and Spleen, sore Throats, and nois in the Ears" (Culpeper, *Pharmacopoeia*, 19).

Take Conserve of Cowslips a dragme; Confection of Alkermis[,][73] Hiacynths of each :15: graines; Con=trayerva :25: graines; Let this be taken the night of his vomitt./

Balsam./
Take white Amber[74] prepared an ounce; Oyle of double=wall flowers[75] .3. ounces; Beefe marrow, Sperma=Cetj[76] of each two ouncs; Oyle of Bayes an ounce; Styrax[77] Liquida .5. dragmes; Oyle of nutmegs Expresion halfe an ounce; Chymicall oyle of Rosemary and [9v] Sage of each a Dragme; Of these make an oynt=ment to anoynt the back bone twice a day before a good fire./

Purging Potion./
Take Polipode halfe an ounce; Peony rootes two dragms; Leives of Betony,[78] flowers of Cowslips of each halfe a handfull; Tragea Comunis[79]; Cræme of tartar of each two dragmes; Sena six dragmes; Let this boile

[73] Culpeper deemed "Confectio Alkermes" to be a restorer of "consumptions, a resister of pestilences and poyson, a great relief to languishing nature" (Culpeper, *Pharmacopoeia*, 125).

[74] Ambergris. According to Culpeper, this "helps the fits of the mother" and "strengthens the brain and memory, the heart and vital Spirit, warms cold stomachs, and is an exceeding strengthener of nature to old people" (Culpeper, *Pharmacopoeia*, 49).

[75] "Double wallflower" is a thicker and larger variety of wallflower (see John Gerard (1633), *The Herball or Generall Historie of Plantes* (London), 457).

[76] Along with killing ulcers, Culpeper maintained that spermaceti "cleers the sight, provokes sweat", "helps bruises, and stretching of the nerves, and therefore is good for women newly delivered" (Culpeper, *Pharmacopoeia*, 32).

[77] Styrax (or "storax") was said to help "coughs and distillations upon the lungues, hoarsness, want of voice, hardness of the womb, but it is bad for headaches" (Culpeper, *Pharmacopoeia*, 29).

[78] Culpeper claimed that betony counteracts "the falling sickness, and all headaches coming of cold, clenseth the breast and lungs, opens stoppings of the Liver and Spleen, as the Rickets, &c. procures appetite, helps sour belchings, provokes urine, breaks the stone, mitigates the pains of the reins and bladder, helps Cramps and Convulsions, resists Poyson, helps the Gout, such as piss blood, madness and headach, kills worms, help bruises, and clenseth women after their labor" (Culpeper, *Pharmacopoeia*, 15).

[79] The "common tragea" was a mixture of various powders with sugar, which often included cinnamon, powdered nutmeg, flowering plants like carnations, and other herbal ingredients.

in a sufficient quantity of fume=tory[80] water to five ounces of strained Liquor, to wch yo^u^ shall add the infusion of two dragmes of Rhuberb made in Peony water, Syrup of Roses an ounce; Syrup of spina Cervina[81] six dragmes; Manna two dragmes; Cynomon[82] water ten drops, Let this potion be taken in the morneing three howres before his Broth./

Purging Apozems./
Take Sarsaparilla[83] an ounce & a halfe, Lignum=Rhodium & Boxe=wood of each an ounce; Polipody three ounces; Female ferne rootes two ounces & a halfe; Spanish scorzonera[84] rootes .12. dragmes; [10r] Leives of Peony; Hysop, Speedwell, Chamepythis[,][85] fumitory[,] Cetrach,[86] of each a handfull; flowers of Cowslips, Elder[87] & Hypericon,[88] of each .3.

[80] According to Culpeper, fumitory "helps such as are Itchy, and Scabbed, cleers the skin, opens stoppings of the Liver and Spleen, helps Rickets, Hypochondriak Melancholly, madness, frenzies, Quartan Agues, loosneth the belly, gently purgeth Melancholly, and addust choler" (Culpeper, *Pharmacopoeia*, 18).

[81] When spina cervina was used in the "Syrup of Purging Thorn", it was supposed to help "against the Dropsie", but Culpeper advised to "let it alone, as an upstart Medicine" (Culpeper, *Pharmacopoeia*, 111).

[82] Cinnamon.

[83] Culpeper noted that sarsaparillas are "helpful against pains in the head, and joynts, they provoke sweat, and are used familiarly in drying Diet drinks" (Culpeper, *Pharmacopoeia*, 9).

[84] According to Culpeper, scorzonera root "strengthens the vital spirits, resists poyson, helps passions and tremblings of the heart, faintness, sadness, and melancholly, opens stoppings of the liver and spleen, provokes the terms, easeth women of the sits of the mother, and helps swimmings in the head" (Culpeper, *Pharmacopoeia*, 10).

[85] Chamaepitys apparently "helps the Jaundice, Sciatica, stoppings of the liver, and spleen, provokes the Terms, clenseth the entrails, dissolves congeled blood, resists poyson, cures wounds and ulcers" (Culpeper, *Pharmacopoeia*, 16).

[86] Ceterach.

[87] Culpeper doubted whether elder root had any healing properties (Culpeper, *Pharmacopoeia*, 9).

[88] Hypericon was considered "as gallant a Wound-herb as any" (Culpeper, *Pharmacopoeia*, 19).

Pugills[89]; of Rosemary, Lavender, & Sage flowers, of each halfe a hand-full; Seeds of male Peony, Juniper Berrys, of each .3. dragmms; Seed of Annis and Coriander of each :2: dragmes;

Let thes Boyle in a sufficient quantity of water in wch the space of .4. houres you^u^ shall have infused thes perticulers following viz, Sena .2. ounces, Turbith[90] & Hermodactyls[91] of each 4 dragmes & a halfe[,] Rootes of true black Ellebory .3. dragmes; flowers of water=Lillys .3. Pugills, white tartar ten dragmes till it comes to fifteene ounces of strained Liquor in which you shall disolve syrup Magistralis ad Melancholiam[92] & of Damaske roses[93] of each an ounce & a halfe, This whole quantity is for thre doses, to be taken at three severall morne=ings, keepeing his Chamber that day./

[10v]

Purging Pills./
Take of Pilulæ cochiæ minores[94] half a dragme make three gilded Pills,[95] to be taken at .5. a Clock in the morneing, drinkeing a litle poset upon, thre howres before his broth./

Fomentation

[89] Pugil. A small handful, or a large pinch, of something (*OED*, n. 1).

[90] Turbith root was thought to "purgeth flegm" (Culpeper, *Pharmacopoeia*, 10).

[91] Like turbith, hermodactyl was said to "purge flegm" (Culpeper, *Pharmacopoeia*, 7).

[92] Probably a reference to "syrup of meconium".

[93] Damask roses. A species or variety of rose, supposed to have been originally brought from Damascus; originally the *Rosa gallica* variant *damascena* (*OED*, "damask", 2d).

[94] Pilulae cochiae minores were relatively common compositions that were mentioned as early as Jean Fernel's 1567 *Universa medicina, tribus et viginti libris absoluta* (Paris), 429 and 464.

[95] Pills covered with a thin coating of gold, especially with gold leaf (*OED*, "gilded", adj. 1). Also see Renzo Console (2013), Pharmaceutical Use of Gold from Antiquity to the Seventeenth Century. *A History of Geology and Medicine*, eds. Christopher Duffin, Richard T. J. Moody, and Christopher Gardner-Thorpe (London), 171–191 (171 and 189).

Take Polypody, marshmallow rootes, Inner barke of Elme & Ash tree, of each .4. ounces; Ilycampany[,][96] Aristolochy[97] round & longe, gentiana[98] white & black[,] Briony[99] rootes of each .2. ounces: Leives of Champythis[,] Chamedrys[,][100] Sweet Mandelyn; Agremony[,][101] Lavender, cotton, Southerwood,[102] Bayes, of each a handfull & a halfe; Rue[103] two handfull; Flowers of Camomile, Melilot, Broome,[104] Elder of each .3. Pugills; flowers of St Johnwort, of Centaury of each a hand-full; Seeds of Agnus castus[,] Scorbutgrasse,[105] Dill, Annis of each one ounce; Lime and foenugreekseed of each an ounce & a halfe; Rinds of Lymonds[106] & oranges of each .2. ounces & a half; Lignum=rhodium[,] Barke of Sassafras ten dragmes;

[96] According to Culpeper, elecampane "resists poyson, helps old Coughs and shortness of breath, helps Ruptures and provokes lust" (Culpeper, *Pharmacopoeia*, 6).

[97] Aristolochia was thought to be useful for problems related to childbirth, and was divided into two types: "long" and "round" (e.g. Dodoens, *A Niewe Herball*, 314). On the toxicity of birthwort, and the uses of Aristolochia in ancient medicine, see John Scarborough (2011), Ancient Medicinal Use of Aristolochia: Birthwort's Tradition and Toxicity. *Pharmacy in History* 53.1: 3–21.

[98] Gentiana was upheld as "a notable counterpoyson" that also "clenseth the body of raw humors" (Culpeper, *Pharmacopoeia*, 7).

[99] Bryony was a purgative that was also thought to be "very good for dropsies", for "the fits of the Mother", and to "take away Freckles, Sunburning and Morphew from the face" (Culpeper, *Pharmacopoeia*, 4).

[100] Culpeper held that chamaedrys cuts through "tough humors, opens stoppings of the liver and spleen, helps coughs and shortness of breath, strangury and stopping of urine and provokes the terms" (Culpeper, *Pharmacopoeia*, 17).

[101] Culpeper maintained that, "outwardly applied", agrimony "helps old sores, ulcers &c. Inwardly it helps the Jaundice and the spleen" (Culpeper, *Pharmacopoeia*, 13).

[102] According to Culpeper, southernwood not only "resists poyson, kils worms, provokes lust" but also "dissolves cold swellings, and helps the bitings of venemous beasts, makes hair grow" (Culpeper, *Pharmacopoeia*, 13).

[103] Rue was deemed to remedy "difficulty of breathing, and inflamations" (Culpeper, *Pharmacopoeia*, 23).

[104] Broom was meant to "open the Stomach, break the Stone in the Reins and Bladder, help the green sickness" (Culpeper, *Pharmacopoeia*, 18).

[105] Scorbutgrass.

[106] Lemons.

Let these be boyled in two parts of Smithes water[107] and one of strong wine vinegar; adding to it at the end of the boyling vineger of scills[108] halfe a Pint [11r] The Liquor strained shall serve to make fomentations upon the place of the Spleene wth Spunges sowne be=twixt two woollen Clothes[109] every morneing the space of an houre at a time continuing this for 9: dayes./

Half an houre after this fomentation he shall take a draught of this ensueing broth in wch draught you shall dissolve .ʒii. of Creame of tarter & .ʒi. of creame of tarter of Steele./

Broath./
Take Spanish Scorzonira rootes one ounce & a half; Cichory, Dentde-lion's[110] rootes of each an ounce; shaveings of Ivory, Harteshorne & of Sea=horse-tooth[111] of each thre dragmes; Red Grocers Corinths, Capers well unsalted, of each a Spoone full;

Let these be put in an old Cocks Belly to be boiled in an Earthin Pott wth a neck of veale in three quarts of water adding in y^{e} midle of the boiling Leives of Borage, Buglosse,[112] Cherveil,[113] great Cherveil Balme, Mayden haire & Hartes=tongue of each one handfull; Flowers of Marygold, violets, Gillyflowers of each 2 Pugills; tops of S^{t}. Johnwort .3 Pugills; [11v] Pippins cut in slises .3; two flowers of Mace./

107 Smith's water. The trough of water used by a blacksmith to cool hot iron (see Culpeper, *Pharmacopoeia*, 174).

108 Squills.

109 Cloths.

110 Dandylion.

111 The reference is unclear, though Francis Bacon's 1626 *Sylva Sylvarum* mentions "rings of sea-horse teeth" ((1862), *The Works of Francis Bacon*, Vol. 5, eds. James Spedding, Robert Ellis, and Douglas D. Heath (Boston), 149). Peter L. Simmonds' dictionary gives this definition: "Sea-horse tooth, a name given to the teeth of the walrus, and of the hippopotamus, which yield ivory" ((1858), *Dictionary of Trade Products* (London), 334).

112 Bugloss apparently "cheers the heart, helps drooping spirits" (Culpeper, *Pharmacopoeia*, 4).

113 Chervils were said to "provoke urine, they stir up lust and desire of copulation, comfort the heart and are good for old people, help pleurisies and pricking in the sides" (Culpeper, *Pharmacopoeia*, 16).

Let all these boile till all come to .24. ounces of Broath, to be for .3. morneings to be [*sic*] taken as above./

Every other day an houre [^before this] broth he shall take the ensueing remedy./

Purging Ptysan[114]/
Take Sena well cleansed & cut smale 3 dragmes, Cloves thrde[115]; Let these infuse all night in water of Apples, made acide wth Juyce of Oranges, the next morneing straine the Liquor to be taken as before./

Purging Pills./
Let him be Purged again w^th^ .25. graines of Pills as before./

Cordiall Bolus./
Take Mythridate[116] two Scrupells, Confection of Al=chermes .15. graines[,] East Bezoar[117] .7: graines, Gascoigne Powder[118] .10: graines, mingle these in a bolus to be taken in going to bed, the day of his Purge./

Spleeniteque[119] Plaister./

[114] Tisane.

[115] Most likely cloves threaded together.

[116] There were many ways to prepare mithridate, but the College called for several expensive ingredients including saffron, frankincense, and roses (Culpeper, *Pharmacopoeia*, 131).

[117] Bezoars were deemed "good in feavers, pestilences, and consumptions" (Culpeper, *Pharmacopoeia*, 33).

[118] On "Gascoigne's Powder", see John K. Crellin (2019), Gascoigne's Powder: A British Prescription and Home Medicine, 1600s to early 1900s. *Pharmaceutical Historian* 49.1: 1–15.

[119] Splenetic.

Take Gumm Amoniack[120] dissolved in Canary wine .2. ounces[,] Gum Takamaka,[121] Caramna[122] of each an ounce, Labdanum[123] [12r] Mollified[124] in black Bal[^s]am[125] .3. ounces; Cloves pulver=ized halfe an ounce;

Let these in a hot Morter be made into a Masse to make a Plaister to Apply under the short ribbs; of the left side; & renew it every 10th day./

Electuary for steele./
Take Spanish Scorzanery rootes Preserved .2. ounces; Succory & Dentdilion[126] rootes preserved of each an ounce[,] Rindes of oringes & Citrons dryed Preserved of each halfe an ounce; Earthwormes prepared .5. dragmes; powder of fumitory, Chamedrys & of Chelendine rootes of each .3. dragmes; Crabbs eyes[127] prepared, Magisteres of Pearle[,] of Coralls[128] of each .2. dragmes, shaveings of Mans skull unburied & of

[120] Gum ammoniac was considered a great remedy "for infirmities of the spleen" and "all pains in the joynts" (Culpeper, *Pharmacopoeia*, 51).

[121] When it came to tacamahaca gum, Culpeper asserted that "whatsoever ach or swelling proceeds of wind or cold raw humors, I know no better Plaister coming from Beyond Sea" (Culpeper, *Pharmacopoeia*, 50).

[122] Caranna was thought to counteract "aches, and swellings", especially "the toothach" (Culpeper, *Pharmacopoeia*, 50).

[123] Laudanum was considered to "mitigate violent pains" and "stop the fumes that trouble the brain in feavers" (Culpeper, *Pharmacopoeia*, 143).

[124] Mollify. To make soft or supple; to make tender; to reduce the hardness of (*OED*, v. 1a).

[125] Used, in general, as a synonym for "balm". Also conceived of by Paracelsus as a healthful preservative essence, of an oily and softly penetrative nature, that exists in all organic bodies (*OED*, n. 4). Thomas Browne used the term "balsum" in the latter sense in 1643 when he scoffed, for example, that they who "found themselves on the radicall balsome or vitall sulphur of the parts, determine not why *Abel* liv'd not so long as *Adam*" (*Religio Medici* (London), 96).

[126] Dandylion.

[127] Crab's eyes were thought to break "the stone, and open stoppings of the bowels" (Culpeper, *Pharmacopoeia*, 31).

[128] According to Culpeper, "Pearls, are a wonderfull strengthner to the heart" and "both they and the red Corral preserve the body in health, and resist feavers" (Culpeper,

Harts Skull of each two dragmes, & a half; white Amber prepared .5. scruples Mingle all these wth Syrup of Citrons rind and make an Electuary.

Take of this Electuary an ounce & a half; Anima Hepatis .18. graines divide this into .3. doses to be taken at three severall times in the morneing, drinking instantly Upon it a litle Mace Ale or thin broth, walking for .2. houres Moderatly./

[12v]

Stomach Pills./
Every other day half an howre before Supper, Let him take Pills of Mastick[129] & stomachick, of each .15. graines./

Every third day yo^{u} shall add to each three doses two graines of anima Hepatis, till it comes to .20: graines of anima Hepatis for each doses [*sic*]./

Digestive Powder
Take annis, foenile & Coriander seeds prepared with Juyce of quinces of each halfe an ounce, rinds of oringes & Citrons dryed preseserved of each .5. dragmes[;] Majoerem tops dryed preserved .2. dragmes; balme and Rosemary tops dryed preserved of each a dragme and a half, Red roses .4. scruples; red-corall[,] pearles[,] Crabbs eyes prepared of each a dragme & a halfe; Ambergrees halfe a dragme; Suger of flowers of oringes .13. ounces; Let these be made into Powder to be taken a spoonefull at a time, half an hower after every meale./

Let him be Purged as above againe & at night take ye same Cordiall as before./

[13r]

Leeches./
Two dayes after the last Purge apply .6: Leeches to the Hemerehoidale Veines, & draw .9: or :10. ounces of Bloud./

Epsom water./

Pharmacopoeia, 32). Coral was renowned more generally for its medical properties (especially in Antwerp) around this time (see Marlise Rijks (2019), 'Unusual Excrescences of Nature': Collected Coral and the Study of Petrified Luxury in Early Modern Antwerp. *Dutch Crossing: Journal of Low Country Studies* 43.2: 127–156 (134)).

[129] Pills of mastic were used as gentle purgatives and as strengthening medicines (Culpeper, *Pharmacopoeia*, 141).

After one dayes rest, Let him use the Epsom water beginning by a quart at first, & ascending every day of one pint more, takeing instantly upon them a spoon=full of this digestive Powder

Glyster./
Take Mallowes, violets, pellitory of the walle of each a handfull & a half; Rue, chamomile of each a handfull; Mellilot, St John wort, of each 3: pugills; Sage, rosemary, of each halfe a handfull; Bayes & Junipery berries of each .3. dragmes, make decoction in water to a pint in wch dissolve Catholicon one ounce[,] syrup of violets two ounces; castor dissolv'd in Canary wine one dragme./

Purging Pills./ Take Pilulæ cochiæ minores .18. grains[,] purpled Manna .12. graines, make .3. pills gilded, and take them betime[130] in the morneing .3. howers before his broth./

[13v]

Specific pills for Steele./
Take castor halfe an ounce, man's skull calcin'd, & wth oyle of vitrioll made into pastills[131]; rootes of wild valeriana; male Peony rootes & seed of each a dragme; Nutmegs, white amber prepar'd & Campher of each half a dragme; pound all these togeather into subtle powder, wch shall be made into a masse of pills, wth ye extract of Junipery berries made in waters of tylett[,] Lilly of ye valleys[,][132] black cherrys & Lavander.

Take of these pills half a dragme; of anima Hepatis & Hiera pigra[133] of each .6. graines, make fower[134] pills to be taken in ye morneing .3. howers before any meat.

Continue the use of this for a month augmenting the anima hepatis according to the operation./

[14r]

Sr Theador Mayernes Letter to the Lord Marquesse of Newcastell./

[130] At an early hour (*OED*, adv. 2).

[131] Pastilles.

[132] Culpeper held that lily of the valley strengthens "the brain and all the sences" (Culpeper, *Pharmacopoeia*, 60).

[133] Hiera picra. This drug was thought to be "an excellent remedy for vicious juyces" and "idle fancies" (Culpeper, *Pharmacopoeia*, 136).

[134] Four.

My Lord./

I see that yo^{u} are the same in respect of yor Health yo^{u} were at first when I had the ho=nour to know you, namely, Melancholyk Hypo=condriack,[135] troubled wth vapours, wch yor Stomach, the obstructions of the Meseraick vessels, and espetially yor Spleen, send continually to the Brayne, more or lesse Accordinge to the fault you Commit in yor Diett, and according as the Pas=sions of the Soule (whereof the tymes furnish yo^{u} wth subject Enough) doe move yor speritts, and Humours, I have spoaken of your Spleene, wch I hold particulerly considerable in this disease in the Center whereof Melancholik persons doe doe [*sic*] ordinarely heape a Leven[136] more remarkeable in its [14v] Activity,[137] then in its quantity, whereof a very little is able to cause great troubles, & to stir up, & produce divers meteors[138] accordinge to y^{e} diversity of Salts wch this Malignant sperrit meets in the little World this Conjunction causeinge extraor=dinary Boylings whence proceeds those Puffs wch the People, Sick Persons & Physitians themselves call winds, although they be not; When yo^{u} Lived contentedly yo^{u} felt these Harms, but more light=ly then yo^{u} doe now, being brought into the condition of yor fortune, neither to heare, nor see, any thinge that pleases yo^{u} (God will give an End to these) yo^{u} Complayne of Clowds wch fill yor Brayne, to the prejudice of yor Eyes, & of your Ears wch are importund[139] wth ringings. When there is fire on the harth, the Chimney wil be fil'd wth soot, wee must apply ou^{r} selves to ye causes, wthout troubling or amuseing our selves, wth the accidents, unlesse that by their violence they require perticuler care; yo^{u} aske my Advice concerning yor Diett [15r] And yo^{u} doe well for it is from thence you ought to hope yor Principall good; by the

[135] See Chapter 2, 21–22 and Chapter 6, 83–85.

[136] Leaven. Any substance that produces fermentation; a ferment; a causative agent of a zymotic disease (*OED*, n. 1b).

[137] Activity.

[138] Within the Aristotelian tradition, meteors were deemed to be affections of the earth (such as rainbows or volcanoes). Such phenomena might be understood, post-Paracelsus, as having to do not with elements (as was the case for Aristotle) but rather with salts, which were of course one aspect of the *tria prima*. The idea, then, seems to be that leaven produces changes in the body's salts, causing medically relevant phenomena, like gas, bloating, indigestion, etc. On the backdrop, see Roos, *The Salt of the Earth*, especially 10–107.

[139] Importune. To trouble; to pester, harass (*OED*, v. 4).

deminuition[140] of the Antecedent, & notable correction of ye conjunct causes; yo^{u} have had so many Counsells of able Phisitians thereupon; that to give yo^{u} new precepts, would be I believe to use vayne repetitions, yo^{u} are knowing Enough to observe what is good & what is hurtfull to yo^{u}. Take this onely for a Maxime that whatsoever heats yo^{u} or drys yo^{u} is very Prejuditiall to yo^{u}, and y^{t} on the Contrary yo^{u} must refreshe & moisten yo^{r}selfe, having reguard alwayes to open the obstructions of the vessels wthout any sharpenesse. Likewise whatsoever sowers in the stomack dus yo^{u} hurte Because this Vitriollick Spirits whereof all meate hath more, or lesse, Encrease yor naturall Vitriollique spirits whereof the seperation is made in the spleene, and by Consequence this Surplussage beyond the weight of nature destroyes the Stomach, Perverts the natural [15v] Functions,[141] & hurts all the others, And is the greatest cause of the Disorders whereof yo^{u} Complayne; yor Drinke is of greater moment then all the rest; Wine doth inflame yo^{u} and waxeth sower, & so doth Beare[142] & Ale, wch doe easily turne into Vinager; smalle Ale doth please one beter then Beare; the sharpnesse whereof may be prevented, if there [is] put into ye Barrell a good quantity of Scales of Iron, very cleane, and if in puttinge it into the Seller[143] yo^{u} cast Oatmeale into the vessel, so yor Drinke will never Corrupt, shall be alwayes quick and will become Phisicall,[144] but if I had to doe with a pliable Stomack, & without prejudication agaynst the Cold wch is so much feared in England I would Conclude to drinke nothing but water, so it be good, as that of Cesterne,[145] whereof there is no want at Paris & St. Jermynes,[146] at the least I would counsell to drinke two good glasses a day, in the morneinge fasting, & in the after noone two houres before super. If there is another sort of drinke besides

[140] Diminution. The action of diminishing or making less (*OED*, n. 1a).

[141] The reference here is to the Galenic natural faculties or vital functions of the body: that is, those activities of a living creature necessary for its life, or what Aristotle would label a ***nature*** (*physis*).

[142] Beer.

[143] Cellar.

[144] I.e. a medicine; good for physic.

[145] Cistern.

[146] St. Germain. The reference is to the western suburb of Paris, Saint-Germain-en-Laye, which had a spa that exploited springs on St-Germain hill.

[16r] Ptisan[147] of Barley & Lychoras,[148] the sweetenesse whereof doth temper all the sharpnesse wthin and wthout the Body, wch will refresshe, & by a longe usage haveing opened the Obstructions, will mer=vilously[149] comfort the naturall parts, for all the intentions that yo^{u} desire to be Accomplish'd in yo^{u}; mentioned in your Letter./

Fill a Barrell of Hogshead very cleane wth water of Cesterne, & put in it a good quantity of the aforesayde Iron Scales, leave this mixture in the vessell (wch will never Corrupt) to make yor Drinke, wch will be a kinde of Metheglin; To a galland[150] wch is 4. Paris Pintes, put an ounce of nefratick wood, half an ounce of eyther Coriander or Cynamond[151] broke into Little bits, Let it steepe at Least .24. howers before yo^{u} use it[;] strayne it and drinke it wth out addinge any Sugar to it wch as well as Honey would doe yo hurt, and the fewer sweet meates yo^{u} eat the better. If you would have it sweet put some Spanish Lychoris into the [16v] Infusion and I doe assure my selfe yo^{u} will Comend the use of this Liquor; I know all the World will crie out agayst this advise Espetially of mere water, as if it would ruine yor Stomack, but it is not the opinion of the Multitude that rules me, but experience is my reason[.] those y^{t} cannot perswade themselves to Receive benifitt from this advise, I allow them to drinke a little glasse of Sack at every meale whereof the heat is suffitiently Corrected by ye water, espetially beinge rawe, & the property of it is never to grow sower, so that of puer[152] & good Sack yo^{u} can never make vinager, yo^{u} will Phylosifise[153] here upon, & have the Judgment of yor Physitians to whose advise I much relie although the opinions of all the world are not alike; Yo^{u} speake to me of Sallads, wch canot be without Lemon, or vinager, Eate not much, & no Lettuce, they being hurtfull to the Brayne. [17r] Of Frute, Apples are the best, Strawberys and Resberys, very ripe & newly gathered, & Cherrys so they be very ripe, although they be something tarte, are good; but these frutes ought to be eate alone wth bread not cram=inge them upon the other meate &

[147] Tisane.

[148] Liquorice.

[149] Marvellously.

[150] Gallon.

[151] Cinnamon.

[152] Pure.

[153] Philosophise.

Aliments of divers natures, as is ordinarily done; Excellent Mellons & very ripe are good in their season. As for milke it should be alone, otherwise it doth Corrupt, & ofends the Stomack, & the head[;] I doe not know what Effect Butter milke pro=duces in yo^{u} if yo^{u} take it alone, to keepe your Belly open now & then it may be alowed, but you have otherwayes to keepe yor selfe loose as Broths, & Cooleing Glysters. Let us add to these tempringe things, Skim'd milke, as milke with all the Circomstances required, and Bathes of fresh lukewarme water. These Artickels may suffise, together wth the things that yor [17v] Physitians ordaynes[154] yo^{u} upon occasion, before I passe to what I conceive nessessary for yo^{u} at Present I must tell you that I condemne Extreamely the use of Tobacco, wch doth yo^{u} no good, seeinge it heates & dryes yo^{u}, as by its stupi=feing quality it hinders yo^{u} for perceaving the harme it doth you, yo^{u} have lived too longe in the Worlds, not to be resolved that yo^{u} must not feede upon smoake[;] however if you please to Continue, Leave Tobacco & use Tussilage[,][155] Bet=tony, & the greene pills of Pistashes, but it would be better to finde out some other Recreation to devert yor troubles, then to fill yor head & Lunges with winde, As for your Issues keepe them. Let the world talke, & hold you to what yo^{u} finde profitable, as many heads, so many minds, I have my Phanceys as well as others wch I shall not easily quitt, being no great freind of new opinions, Yo^{u} doe well & must Continue [18r] As Springe & Fall to Purge & Bleede, not onely in the Arme, but also in the Hemrodveynes[156] with Hors[157] Leaches, since nature showes you that waye, If the yeare was not to farr spent or if yo^{u} had advis'd wth me sooner, after the Universall things I had Counsell'd yo^{u} the use of my Steele for many weekes together which yo^{u} have so well aproved of heretofore as well as yor Wife when she was under my hands, and I am Perswaded that both she and you must make that yor Chefest[158] remedy for some yeares togeather, in the Spring and fall, But the heat drawing on during which you can use onely Minerall vitriolique, & Iron waters & the before mentioned Bath Iron is a great seacret for yor Cure. The waters

[154] Ordain. To order (a thing) to be made, provided, or equipped (*OED*, v. 15).

[155] Smoking the leaves of tussilago was reputed to cure lung diseases and be "good for inflamations" (Culpeper, *Pharmacopoeia*, 25).

[156] Haemorrhoid veins.

[157] Horse.

[158] Chiefest.

of Gorges[159] are very good, so are those of Pougues,[160] taken upon the Place doe penetrate more, those of the Spawe yo^{u} may have very [18v] Good at Paris, so yo^{u} Chuse them well[.] by this washinge used longe enough by Counsell you will abate the vapours wch doe afflict yo^{u} & will satisfy whatsoever the present state of yor health requires. To give you ease I must not give you my opinion of the an [*sic*] hearbe wch comes from China, and Jappan,[161] in that Country it is a Sadle for all horses whereof they make use as in other Places they doe, of Tobacco to Produce divers Contra=ry Effects. I cannot relate the use of it; nevertheless some Merchants who have Lived a longe while in the Indies have assured me that the Decoction of it drunk warme doth Mericells[162] Espetially to abate fumes, to keepe a man awake when he hath businesse, & to whet his under=standing when he hath need. I have taken of it my self wch hath done me neyther good nor hurt wch is no smale thinge; I understand that this Dragge[163] is very well knowne in [19r] France[;] yo^{u} may Informe yo^{r}self of those y^{t} use it, & doe as others doe, for I thinke it is not a misse that yo^{u} try it. This is all I can say to yo^{u} for yo^{r}self./

For my Lady she doth farre Exceede yo^{u} for matter of the Hypochondry. I have had hir in Cure of that disease heretofore wth good successe and the same way I led hir then, I would have hir goe againe the next Autumne, and so here after yearely, soundly Courtinge the Steele wch is onely Capable to doe hir good in the degree shee stands in neede of. I could Pre=scribe hir a hundred remedyes, & so to you but tis in vayne to use many where few may serve, notwithstanding my advise is, she should take waters for divers weekes, & if I had hir here I would have hir take those of Epsam, neere none such,[164] to the end to Purge hir Cleane before she use others, touching the evacuation [19v] Of vomits, wch she Prefers before those of the Belley, I am not wholey agaynst hir opinion, and seeing shee avoydes much Chollor y^{t} way is not inconvenient for hir

[159] Likely Gorges du Verdon. A river canyon located in the Provence-Alpes-Côte d'Azur region of Southeastern France.

[160] Probably Pougues-les-Eaux. An old water town in the Nièvre department in central France.

[161] Likely a reference to "China-Root".

[162] Miracles.

[163] Drug.

[164] The implication here is that there is nearly no equal to Epsom water.

Cure, but let hir remember, that the Stomack is a very pre=tious part, to wch shee must have a spetiall regarde for if the Hitching[165] fayle, farewell to all the rest. It would be better after she hath vomited to Endeaver to quit herself of that nesessity so much contrary to nature, in hin=dringe the Generation of those sharpe & Rebell Humors by some ordinary Drinke, very watry Even that wch I have ordained yo^u^ before, and Clensinge y^e^ first region, in wch is the roote of all Diseases, as by some stomack pills wch may be Prescrib'd hir, if she resolve to take them, Touching Conception, I know not, if in the estate she's in yo^u^ ought Earnestly to desire it, It is hard to get Children wth good Corage,[166] when [20r] One is Melancholy, and after they are got and come into the World, they bring a great deale of Payne wth them, And after that very often one looses them, as I have try'd to my great greefe and am sory to have had them[;] Be in good health & then you may till yor grownd, otherwise it will be but tyme lost if you enter that race frowningly; There are many perticulers where of I should be Enform'd before I can give a sollid Councell for generation, if my Lady goe to the Bougue waters,[167] after hir Body shall be well Cleansed, and she hath drunke the waters. The Bathes of Burboone[168] will not be unprofitable to hir, so that she use them to purpose, wth good Councell, to wch I referr my selfe, it being very dificult to conclud sol=lidly at this great distance, I humbly kisse hir hand, and wish hir the accomplishment of hir good desire./

[20v] Wee are here suffitiently troubled to strive a=gaynst Convulsive tremblings, wch trouble my Lord Henry yor Second Sonn in an Extraordinary manner[;] huge malignant vapours, rising from the low region of the Stomack, from the Mesen=terys, and from the Spleene, strike into the Heart and prick the Nerves, wch start & move unwilling=ly, sometimes one Part, sometimes another; make=ing a rownd Passinge from one Arme & one Legg to the other, makeing the knees knock one agaynst another, and the head & neck to shake, wth stronge shakes, & all w^th^out

165 Hitch. To fasten, especially in a temporary way (and against force acting in one direction) (*OED*, v. II.5a). The point is that if the hitching fails, the horse will run amock, and, by extension, if her stomach fails, her health will be in turmoil.

166 Courage.

167 Waters of Bourgogne. The water was from Canal de Bourgogne, which is in Burgundy in east-central France.

168 The Roman Baths of Bourbon-Lancy, which is a commune in the Saône-et-Loire department in the region of Bourgogne in eastern France.

touching the Judgment or the Sence & being past, Appeare not at all, Till they returne, wee must take the Brayne be not Affected,[169] for feare of an Epilep[^s]y, I have seene others, & by the grace of God Cured them; and I hope here to doe the same by his Mercy; Assure yor self that I will neyther spare my eare nor know=ledge to serve you in this Perticular, which [21r] Deservedly yo^u^ take so much to Heart, I have seene the advise yo^u^ have sent, and I hope wee have no need of so great a reamadyes[170] [*sic*], if wee must goe to them, we have those that are not inferior to the best they can propose to us, for the pre=cipitate made wth Mercury & Gold, I esteeme it lesse, then Mercury drawn from Gold, wth out addition of any other Mercury, wch I my self have made, haveing Prec[^i]pitated it along wth the fire, wth out any Corrisive or addition what=soever. I have used of it Prosperously, But af=ter all I doe farre esteeme Imperfect Mettalls[171] then perfect, for Physick; & have no great Con=fidence in Gold for human Bodyes. But when it is in the Pursse, doe not trouble yor self[;] very often great reamidyes fayle where some light ones doe the feat; Keepe me in the honour of yor favour & belive me

My Lord

Yo^r^ most Humble Servant
The: Mayerne./

Chelsey the .24.th of
May. 1648./

[21v]

A Letter from Doctor Davisson[172] to the Right Honble the Lord Marquesse of Newcastle./

My Lord./

[169] I.e. take it that the brain is not affected.

[170] Remedies.

[171] Metals that were considered less valuable than gold and silver (e.g. lead, tin, iron, or copper).

[172] William Davisson (1593–1669).

I have seene & Considered yor two Letters one from yor Lordships Sonne & the other from Mr Benoist,[173] And though I have no Perticular instruction of the Accidents that have befallen him, yet so farr as I can Conjecture by the relation, the universall Shakeing that waken=eth him from his sleepe may be some Epilep=ticke[174] Convulsions, or something that belongeth to that Sicknesse, well often wormes[175] generate in young Peoples Bodys, and nourished to an Ex=traordinary bignesse may Cause those simptomes and greater; Therefore my Councell is, that the Physitians that have him under Cure apply [to] him such thinges that may not onely kill the Wormes, but also divert the dangerous Consequence [22r] That may Come after, He is in so good handes that I will not hazard to vent[176] my Merchantdise further, then to Confirme the Excellent directions that Doctor Mayerne hath given for his recovering and doe Esteeme Exceedingly his steele to Purge and Extenuate[177] thicke & Limous[178] obstructions in the Spleene, but after the use of this spleneti^{c}ke I doubt not but the next course will be taken in giveing such things, that may

[173] Mark Anthony Benoist (active mid-seventeenth century), a longstanding Cavendish family retainer.

[174] While epilepsy was known to the ancients (see e.g. *De morbo sacro* in the Hippocratic corpus), the *OED* lists William Shakespeare's (1608), *King Lear* (London), vii. 79, and Lewis Machin and Gervase Markham's (1608), *Dumbe Knight* (London), iii. sig. E4v, as the earliest English uses of the adjectival form.

[175] It is not always clear what is meant by "worms", especially those in the body related to diseases. At the time, a worm could (like today) refer to a slender, creeping, limbless animal of the genus *Lumbricus*. But it might also signify any animal that creeps or crawls, including crocodiles (*OED*, n. I, 1 and 2a). As far as we know, there is no full work on early seventeenth-century ideas of worms, though do see Alanna Skuse (2014), Wombs, Worms and Wolves: Constructing Cancer in Early Modern England. *Social History of Medicine* 27.4: 632–648 and Ian MacInnes (2012), The Political Worm: Invertebrate Life in the Early Modern English Body. *The Indistinct Human in Renaissance Literature*, eds. Jean E. Feerick and Vin Nardizzi (Basingstoke), 253–273. For a slightly later period, see Julie Grissom (2014), *Parasitic Worms in Early Modern Science and Medicine, 1650–1810* (Dissertation, University of Oklahoma).

[176] Vend (to sell).

[177] Extenuate. To render thinner (the humours or concretions of the body) (*OED*, v. 3b).

[178] Muddy or slimy (*OED*, adj.).

kill the wormes and Purge that humour feculent[179] & melancholicke wch can be done by nothing more Conforme to reason & Experience, then by the use of the Powder of Mercury & Gold Precip[^i]tate together onely with the degrees of fire, ten graines of wch I thinke should be given Sixe weekes together thre dayes full, and foure dayes voide, but if it Cannot be had promptly, dulcified Mercury may be used for that Effect. And for the symptome that overtaketh him in the night, my advise is [22v] That use should be made of the narcoticke Sulpher of Hartmannus[180] wth the additions such as yor/𝓛/ shall finde set doune[181] in a Paper here inclosed, and he must keepe this remedy Constantly by him, takeing of it ten graines in the morneing three howers before any meat and at Evening as much when he goeth to bed And if it happeneth that he findeth any Parox=ysme in the night, those that are about him must put a dose of this remedy into his mouth and try to make him swallow it. But wth all this I should most willingly Counsell him to abstaine from Melancholicke thoughts, for assuredly his evill is nothing but a Child of Melancholy. His meat should be seasoned wth the Powder of froggs dried in an oven, & putten upon his meat as if it were nutmegge. And if the Paines could be taken to gather a great quan=tity of Livers, Hearts & Lights of froggs, & fry them in an oven with Peony flowers, the weight of a Golden Crowne of them taken wth y^{e} Conserve [23r] Of flowers of sauge[182] & Rosemary, betwixt every meale would be a soveraine reamedy and Antidote to that disease. I send yor/𝓛/ as much of the sulphur of Hartmannus wth my ad=ditions as I have presently by me, and should be Extreamely glad to

[179] Feculent. Containing or of the nature of fæces or dregs; abounding with sediment or impurities; thick, turbid. Also, laden or polluted with filth; foul, fetid (*OED*, adj. A, 1).

[180] A reference to sulphurated potash, which is a narcotic when taken internally. This particular recipe is named after Johannes Hartmann (1568–1631), who was the first chair in *Chymiatrica* at the University of Marburg (see Appendix 8D.III). His *Praxis Chymiatrica*, which was first published in 1633 but revised and expanded throughout the century, contains a recipe for "narcotic sulphur" in a section titled "Laudanum sine opio" (laudanum without opium) (*Praxis Chymiatrica*, Chap. II, Sec. 14, 44–45 and 316–317). This is the "narcoticke Sulpher of Hartmannus" in Pw V90. The recipe can also be found in the English translation by an anonymous "lover of Chymistry" ((1670), *Bazilica Chymica, & Praxis Chymiatricae, or Royal and Practical Chymistry* (London), Bk. III, 16–17) and Hartmann (1635), *Tractatus physico-medicus de opio* (Wittenberg), Theorem VII.3, 70–75.

[181] Down.

[182] Sage.

have any thinge that might be steadable[183] to my Lord yor Sonne being wth all my heart & affections

My Lord/

Yo^{r} most humble
& most obedient Servant
Davisson./

[23v]

An Excellent Receipt of Doctor Davissons

Take of Castorium an ounce & a halfe, Misletoe growinge upon Oakes, the Skull of a man not bury'd, Red & yellow Sanders of Each an ounce, Livers of greene froggs dried & pulver=ised, Seedes of male Peony, of each .6. dragmes[,] flowers of Tilliett & male Peony,[184] of each thre handfuls; Mace, Gallingall,[185] rownd Pepper, Long Pepper, Mummia[186] of each thre Dragmes, of little & great Cardamams,[187] of Each two dragms, of white Amber, an ounce & a halfe, Saffran,[188] Dung of white Peacocks of Each an ounce, of all these drawe the Tinc=ture wth Spirit of wine well rectified[;] Straine the Tincture through a Paper and Evaporate [^it] in Marys Bathe till it be wasted to the half, Then yo^{u} shall add to it Salt of [24r] Pearles, and Corrals, of each an ounce, Sulpher Narcotick of Hartmannus three ounces, then Continue the evaporating of it till it come to the Consistance of Hony/

When yo^{u} doe give this, you must add to each quantity you give at once, two drops of oyle of Camfor./

[24v is blank]

183 Serviceable, helpful (*OED*, adj.).

184 According to Culpeper, peony "helps women not sufficiently purged after travail, it provokes the Terms, and helps pains in the Belly, as also in the Reins and Bladder, Falling-sickness, and Convulsions in children" (Culpeper, *Pharmacopoeia*, 8).

185 Culpeper maintained that galangal "strengthens the stomach exceedingly, and takes away the pains thereof coming of cold or wind, the smel of it strengthens the brain, it releeves faint hearts, takes away windiness of the womb, heats the Reins, and provokes lust" (Culpeper, *Pharmacopoeia*, 7).

186 Also known as "mummy". Ingredients gathered from the dead were a constant feature of European medicine (see Richard Sugg (2011), *Mummies, Cannibals, and Vampires: The History of Corpse Medicine from the Renaissance to the Victorians* (London)).

187 Cardamom.

188 Saffron.

[25r]

A Letter from Sr Theador Mayerne to the Lord Marques of Newcastle Concerning the health of his Lady…

My Lord.

There is no neede y^{t} you should thanke me of ye care I have hade of my Lord your Sonne, who is now thankes be to God in good health; which I hope to see last long to your Comon Contentment. The taske of y^{t} Cure was very heavy But I believe that to Cure my Lady Marquesse your wife will be yet harder. Not so much for y^{e} nature of y^{e} Disease, which is Rebellious, as for y^{e} disposition of the Patient, who will not willingly submit to y^{e} Councell of her Physichans, be they never so good & so skilfull.

There is no neede of any Physician to describe a Sick=nesse if yo^{u} undertake to doe it as it appeares by your expresing so well the indisposition of my Lady, which is enough knowne to me, haveing had formerly thee [25v] Honour to Administer Phisick unto her La:^{p}[189] And I dare say not wthout a Considerable ease of all her Indispositions, which as I Perceave last yet, And are to be lessened or taken away by y^{e} same Remedies which she did use then, whereof Steele was y^{e} Prin=cipall. In ordinary diseases of this nature wee give of it for forty dayes, But in rooted & inveterate ones for severall monthes, And I have given it often for the space of a whole yeare with most happy successe, But for all that this is not now y^{e} season to begin it in since y^{e} heat of Summer is neare at hand, dureing which other remedyes are to be used to moisten that evell[190] black humour which hardens ye Belly, and dyes the excrement, Her La:^{p} hath beene Purged & Let bloud very much, and without doubt to much, since it hath beene done by her owne directions, as often at least as by ye advice of her Phisicians, a thing to be noted but not immitated. It is not my minde to forbid her all manner of Purgations, Shee hath neede of sume having respect to y^{e} whole body, which are to [26r] be Active, but used seldome, for by too often scoure a ketle it is at last worne out to holes, But y^{e} most necessary Purgations are those caled usuells, wch are for ordinary use, And doe not suffer any gather=ings to be made in y^{e} first

[189] While the exact date is unclear, this must have been before Cavendish left England in July 1644.

[190] Evil.

Pasiages[191] but carryes away y^{e} humours as soone as they are bred. The onely remedyes to be practised in this season till Autumne after some Purgation of Sena, Rubarbe, Agarick, Syrope of Roses, Compounded, Fumetera[192] and y^{e} like, wch wee call gentle Physicks, whose suffitient workeing is to be Procured by y^{e} quantity are y^{e} waters of Spaw, which where you are yo^{u} can easily get fresh & well Conditioned; It is ne=cessary when yo^{u} use them to be taken as to provide as well agaynst y^{e} Accident y^{t} may happen, though I believe none will agaynst such as my Lady might imagin, of which I expect some, so well I know y^{e} disposition of her minde in Point of [26v] Sicknesse, somtimes these waters trouble y^{e} head with vapours Causing in some a Light Payne, in others a kinde of giddy-nesse, But all passes a way after y^{e} water is evacuated, which works for all this such effects in y^{e} Entrailes as wee in=tende; The remedy to it is by frequent Clystors taken in y^{e} after noone if my Lady finds her selfe ill; But if her La:^{p} be troubled with heate as her so black excrements doe infal-labl shew she is very much; The refreshing Bath onely Luke warme; with Mallowes, violets, sallow vine,[193] nenuphar,[194] Plantine,[195] both terres-triall & aquaticke, & quantitie of rose-cakes will doe very well and may be taken in y^{e} afternoones of the very dayes she takes y^{e} waters in[.] This reme=dy is of so great Consequence, that I wish my Lady would make it familiar to her in y^{e} great heates of Summer, even as y^{e} Romanes[196] were used to doe who did Bath everyday. The waters are to be [27r] Drunke .30. or .40. dayes, even in a Longer time according as it shall doe good, If they should not worke, they may be helpt mixing in them some Manna or Lax=ative syrope of Roses, or syrope of violets, or Laxa-tive syrop of Apples, or eating .2. or .3. ounces of Perfumed Concerve of roses one houre after she drinkes the water which is not to be dune every day, But every third or fourth day. Concerning y^{e} quantity of water which my Lady is to drinke, she must Consult with ye strength of her stomach,

[191] Passages.

[192] Fumitory.

[193] The drooping branches of the sallow (or, more particularly, the willow tree), which are vine-like.

[194] Culpeper insisted that nenuphars "stop lust" (Culpeper, *Pharmacopoeia*, 8).

[195] Plantain.

[196] Romans.

And if she finde not ease emediatly,[197] she must not leave all as I feare she will, but she must be Constant, or else she shall not be cured; which I tell her La:^{p} without flattery, after she hath drunke ye waters till Autumne comes she may have made some distilled water of Milke wch shall be also Cordiall, to make Juleps with syropes of Apples, strawberryes or Rasberryes, or Cherrys, whereof she shall drinke a good draught once, or twice a day according as [27v] She shall finde her selfe hott, or thurstey,[198] And according as the black humour shall abound, I Pre=scribe no leting of bloud because I believe she hath her selfe done enough already, neverthelesse if during this time her Hemrodes or Piles[199] should swell I am not agaynst the Applying of Leeches unto them Provided y^{t} it be done with discretion in y^{e} afternoone and y^{t} by y^{e} meanes of y^{e} vapours of warme water, so much bloud be drawne as shall be judged necessary by the Physicians through whose handes it is abso=lutly needfull to Passe, every one being to be believed in his own Trade, In y^{e} Autumne we are to speake of somwhat els, for then we must come to stronger Purgations even to y^{e} black Ellebore, without banishing total y^{e} white one; if neede be, especially in Clysters, as at y^{t} tyme they may both be Prescrib'd. Afterwardes we shall come to Steele taking all-wayes care not to set y^{e} house on fyre for melan=choly People ought to be repressed[200] & moistoned[201] [28r] If one intend to Cure them; I know not whether my Lady have any Issues, It would doe her La:^{p} good If she had some in her thyes, foure fingers above her kneese, But fayre Ladyes who have dayntie[202] husbandes doe not willingly admit of y^{t} subjection; Yo^{u} would obliedge me to [give] to yo^{u} Perticuler Receipts wch at this distance I cannot well doe, As I will never doe any thing but well when it Concernes yor service; In this Councell I have sayde enough, But I am afrayde[203] that my Lady will doe nothing at all, or shall doe it by peace-meales according to her

[197] Immediately.

[198] Thirsty.

[199] Haemorrhoids.

[200] Repress. To control or counteract with a particular treatment or remedy (*OED*, v. 1 3b).

[201] Moistened.

[202] Dainty. There are various meanings to this word, but the most likely ones here are valuable, fine, handsome; choice, excellent. It might also imply being held in estimation, honour, favour; esteem, regard; affection, love (*OED*, adj. 1 and n. 1).

[203] Afraid.

Custome; Let not her La:^{p} be offended with my freedome whereof I am in Possession towardes all those to whome I have vowed some service. I need not Protestations to make yo^{u} believe how much I honour yo^{u} & how much I desire to please yo^{u}; Suffer me in this Place with all humility to kisse my Ladyes handes whoe will doe me ye honour to let me know how she likes this advice of myne, as yo^{u} will also. [28v] Be Pleased not to deny me ye favour to be believed by yo^{u}

My Lord

Yo^{r} most humble servant
De Mayerne

Chelsay the .22.th of
May 1649

[29r]

A Letter to the Lord Marquesse of Newcastle Concerning his owne health, from Sr Theador Mayerne./

My Lord

As Carelesnesse in Matters of health is blameable and dangerous, soe too much curiositie, wch breeds a thousand doubts in y^{e} minde, is superfluous and full of unquietnesse, It is an old Proverbe, y^{t} he that lives Physically, Lives Miserably,[204] being Con=tinually in doubt, And stopt at every step by difficulty as Conceaved without any necessitie[.] Praysed[205] by God[206] that by letting of bloud, and y^{e} use of other good remedyes, yo^{u} are Cured of yor feaver which I be=lieve was not very great upon those difficulties yo^{u} described to me in asking my advice about about [*sic*] many things, wch I doe not hold of very great Consequence. [29v] I will

[204] To live physically here means to live via physic: that is, according to the rules of medicine. While it was evidently an established aphorism by the time that Mayerne was writing, the earliest known print example of this phrase is in George Cheyne (1725), *Essay on Health and Long Life* (London), which was published over seventy years after Mayerne penned his letter. The 1818 edition of Samuel Johnson's dictionary attributed the phrase to Cheyne (*A Dictionary of the English Language*, Vol. 4 (London), "Physically", def. 2).

[205] Praised.

[206] By God. Expressing dependence upon or grateful recognition of God's providence; after post-classical Latin *gratia Dei* (or "by the grace of god") (*OED*, "god", n. P1, a (a)).

tell yo^{u} what I think yor bloud is hot And it is most Certaine y^{t} Sorrell,[207] Borage, Buglasse and other such refreshing hearbes are very good for yo^{u} & very needfull, And I advise you to use them specially in yor brothes, adding to them Endive, Cicory, Percelaine,[208] Dendelyon, ye Leaves of violets, & of Strawberrys, As for salads and other rawe hearbes, yo^{u} will doe well to forbeare them, for though I will not say y^{t} yo^{u} are old yet we know very well y^{t} yo^{u} are none of y^{e} youngest[.] Oranges & Limondes are good to sharpen the stomack & serve for sauce, not for nourish=ment, use not yo^{r}selfe so much to suger,[209] that yo^{u} may not want it sometimes in yor meates as y^{e} fashion of yor Country is. I have told yo^{u} before now, and it is very true, that there is [in] suger an Acid spirit so Corrosive (as is also in honey) that it is able to disolve Gold, And when [30r] It is separated from those other substances which Compose & make up this sweet salt (wch is done by the natural heate[210] in our bodyes) it Produces very ill effects, And it is no mockery to say that it Corodes & kottes[211] y^{e} teeth, which is most true, besydes it is easily converted into Choller the Concequen=ces whereof are in a longe time very hurtfull[;] whay is better then buttermilke, and in those North=erne Countryes where there is no

[207] Sorrel roots were considered "profitable against the Jaundice" (Culpeper, *Pharmacopoeia*, 7).

[208] In terms of its medical virtues, Culpeper said that purslane "cools the blood, liver, and is good for hot diseases, or inflamations in any of these places, stops fluxes, and the terms, and helps all inward inflamations whatsoever" (Culpeper, *Pharmacopoeia*, 22).

[209] I.e. do not use so much sugar.

[210] The natural—also animal, or innate, or vital—heat is a common element of Renaissance and early modern physiology, which was often thought to correspond, in some way, to the heat of the stars (Aristotle, *De generatione animalium* II.4, 736b29–737a6). As such, it was invoked to help explain the ability of living bodies to perform their necessary and teleologically organised functions like generation and nutrition and was integral to any understanding of the vegetative soul. Throughout the seventeenth century, many of these vital processes, which often involved the transformation of one substance (like food) into another (like blood), began to be understood in chymical terms. Significant in the English context are thinkers influenced by Harvey (including Boyle and Charleton), who reinterpreted some of his ideas about the blood in chymical terms. The fullest account of this remains Robert Frank (1980), *Harvey and the Oxford Physiologists: A Study of Scientific Ideas* (Berkeley).

[211] While the spelling is clear, the meaning is less so. The most likely reading is "coats", though the *OED* gives the first usage of "coat" to mean a layer of any substance covering a surface as 1663 (*OED*, n. II.9). Another possibility is "cuts".

good oyle fresh butter is ye most Plesant. Casia[212] refresheth ye Liver & ye kidneyes, wch I feare yo^{u} heate to much and may be taken without any subjection imediatly before dinner drinking som broth upon it to desolve it in ye stomack. Possets are a good invention of ye English & serve for broth & apozeme adding to them such hearbes as are put in Claryfyed[213] whay, & sweetening them with sirop of Pippins, or of Pearemanes, Apples are good for all sauces [30v] Either raw, when they are very Ripe, or rosted And for Capers & Currands[214] they are a most excellent remedy to those y^{t} are troubled with ye Spleene[;] instead of Currands, Raysens of the sunne[215] may be used, And y^{t} was ye great secret of Inigo-Johnes[216] against ye Melancholly, fermentations of windes of his Spleene, Straw-berryes & Cherryes very ripe & not sowre,[217] refreshe y^{e} kidneys, And are somewhat Cordiall; but remember y^{t} yo^{u} must not make a hotch-Potch[218] of all these things at once in yor stomack; It is much beter to use sometime one of them, sometime another severally without any thing els[e] in some houres before & after; Chief=ly y^{t} which comes from Milke admits no fellow=ship without danger of Corruption, If yor loose=nesse, wch I believe Cured by this time, should Con=tinue, Purge yo^{r}selfe with a little Cena, Rubarbe [31r] And Tammarins,[219] But not w^{th}out caling a Physician whoe will advise yo^{u} according to y^{e} Present necessity, I can tell yo^{u} no more but y^{t} yo^{u} are worth a Physician & a halfe, And as for the Admirable Receipts of y^{t} Noble Man made Duke of Northum=berland[220] by y^{e} Pope,[221] if yo^{u} were here I should be glad to see them, But in y^{e}

212 Cassia.

213 Clarified.

214 Currants.

215 Raisins of the sun are simply sun-dried grapes (see *OED*, "raisin", n. P2).

216 Inigo Jones (1573–1652), the great early modern English architect.

217 Sour.

218 Hodge-podge. It was sometimes used at this time (as now) to mean "A clumsy mixture of ingredients" (*OED*, n. 2).

219 Tamarind.

220 Robert Dudley (1574–1649).

221 It is unclear what exactly Mayerne is referring to here, since Dudley was never confirmed as the Duke of Northumberland, let alone by a pope, though he was favoured by the Grand Dukes of Tuscany.

most Part of such things Parturiunt monthes[222] &c. I wish you a perfect quietnesse of body & minde and all the Contintment[223] y^{t} yo^{u} can well desire, being fin alla morte & più se piu puote[224]/

My Lord

Yo^{r} most humble and
Most obedient servant
De Mayerne

Chelsay y^{e} .28. June
1649

[31v]

Sr Theadore Mayernes Receipt for the Hemærhoides[225] or Pills.[226]/

Takes rootes of y^{e} great scrophularia[227] .6. ounces; water Bettoney, orpin,[228] rootes & Leafes, of each 3. hand fulls, young houndes tongue[229] rootes & Leafes .4. handfulls, water=Lillies, Mellilott, St Johns-wort, water-Bettony, Scrophularia, y^{e} flowers of all these .3. Pugills, fonngreek,[230] pad,[231] Cimford,[232] of each an ounce & halfe, crocus

222 This is (oddly spelt) Latin for "mountain labour", which is a reference to Aesop's fable "The Mountain of Labour", wherein a mountain, after much ado, gives birth to a mouse. See Chapter 6, 95.

223 Contentment.

224 "Fin alla morte, e più, se più si puote" is an Italian phrase, which translates as "until death, and longer, if longer is possible". This phrase can be found in many early modern texts, perhaps most famously in Ludovico Ariosto (1543), *Orlando Furioso* (Venice), 242.

225 Haemorrhoids.

226 Piles.

227 According to Culpeper, scrophularia "helps the Piles or Hemorrhoids, and (they say) being hung about the neck preserves the body in health" (Culpeper, *Pharmacopoeia*, 24).

228 Culpeper held that orpine "cleers the Skin; inwardly taken, it helps gnawings of the stomach and bowels, ulcers in the lungues, bloody flux, and Quinsie in the throat" (Culpeper, *Pharmacopoeia*, 17).

229 Hound's tongue was thought to fend off "Hemorrhoids" and to be "good against burnings" (Culpeper, *Pharmacopoeia*, 5).

230 Fenugreek.

231 Pad (shorthand for paddock). A toad; later also a frog (*OED*, n. 1, 1).

232 Camphor.

martis, Aparins[233] three ounces, Juice of Eldere[234] a Pound, Swines grease. 3. Pounds[,] new ale, a Potle[,][235] black flyes caled beetles a hundred, boyle these to=gether till all y^{e} Liquor be spent, then strayne the grease, and take of y^{e} purest of it .12. ounces, oynt=ment of Popplar .6. ounces, Album graecum gatherd in y^{e} dog dayes,[236] 2. ounces, Suger of saturne an ounce, Let these be mix'd & agitated in a Leaden Morter 5. or 6. houres & be made an oyntment to anoynt the fundament[237] if it come downe, & so put it up a [32r] gaine. And allso to anoynt y^{e} hamorhoids, or Pilles Applying upon them a bolster of Cotton spread with the same oyntment./

Sr Theadore Mayernes Advise to the Lord Marquesse of Newcastle.

For ye Curiosityes[238] of meates yo^{u} should eate, It is no great matter as long as yo^{u} are well, for yo^{u} may Eat of what meat so ever as long as yo^{u} are well, soe yo^{u} be temperate in y^{e} quantity which onely can hurt yo^{u}; yor naturall strength being such as will easily surmount little errors in yor diet, soe y^{t} it be seldome,[239] & y^{t} yo^{u} Content yor selfe with a mediocratie.[240] If yo^{u} finde yor selfe over=charged, fast And when yo^{u} finde yo^{u} have an Appetite, eate & drinke Excercisce moderatly and for y^{e} rest let nature doe it, the minde hath a great Com=mandment & powre of y^{e} body, & therefore keepe yor minde as Cleare & well pleased as yo^{u} can.

[32v-36v are blank]

[37r]

233 According to Culpeper, "Aparine" or "Goose-grass, or Clevers", is useful for "the bitings of venemous beasts, keeps mens bodies from growing too fat, helps the yellow jaundice, staies bleeding, fluxes, and helps green wounds" (Culpeper, *Pharmacopoeia*, 14).

234 Juice from elderberries.

235 Pottle. In contrast to its standard usage as a synonym for a "pot" (as in Transcription, 110), here the reference is to a measurement for liquids.

236 Dog days. The hottest part of the summer, associated in ancient times with the heliacal rising of the Dog Star in the Mediterranean area; formerly considered to be the unhealthiest period of the year and a time of ill omen (*OED*, n. 1a).

237 The anus or rectum (*OED*, n. 2).

238 Curiosities (i.e. delicacies).

239 I.e. so long as it be seldom.

240 Mediocrity. A middle course of action; moderation, temperance (*OED*, n. 2).

The Duke of Northumber=landes Receipts for my Lord Marquesse of Newcastell./

The Duke of Northumberlands great Cordiall Invented by him selfe, And gi=ven with admirable successe who beleeves it the greatest Cordiall in the world and the surest remedy against all Poysons and the Plague. In wch cases it hath ben tried often with Admirable Successe; By coman=dement of the great Duke of Tuscany.[241]

℞.[242] Roman Treacle, Mithridat,[243] Confection of Alchemes, & of Dianthos,[244] of each, give six ounces. [37v] The great Dukes oleum[245] Contraverva; The Duke of Northumberlandes' oyle of Sulpher camphorated[246]; oyle of yellow Amber,[247] of Mastick[,] of Rosemary, of Juniper gumme, of Cinnamone, of Mace, of Cloves, of Nutmegs, of commonsalt, of Turpentine, & Spirit of sulpher, of each ʒiiii.foure Dragmes.—Magistery of Perle and of Corall of each. ℥i. one ounce, Tincture of Gold in Powder, Liquor of Silver, Crocus of Mars of each .ʒii. two dramms, Rinoceros=horne, Unicornes-horne, Bone in the Staggs=heart,[248] Contrayerva-roote in Powder of each ʒiii. thre dragmes; Tincture of Saffron, of Corall,[249] of Peoina,[250] of wormewood[,] of

[241] Ferdinando II de' Medici, Grand Duke of Tuscany (1610–1670).

[242] ℞. The initial letter of classical Latin *recipe* (*recipere*; to receive). In medical prescriptions, as the opening instruction, it means: "take" (*OED*, "R", n. III, 15b).

[243] Mithridate. This recipe (like much from Dudley) is both impressive, because of the expense and rarity of its ingredients, and confusing, because its all-inclusive nature means that it contains, for instance, treacle and mithridate, which are expensive panaceas that serve the same purpose.

[244] Species Dianthus. According to Culpeper, this causes a "cheerful mind" and "strengthens such as have been weakned by long sickness" (Culpeper, *Pharmacopoeia*, 123).

[245] Latin for "oil".

[246] I.e. having been treated with camphor.

[247] Culpeper relayed that yellow amber "prevails against moist diseases of the head, it helps violent Coughs, helps Consumptions of the lungues, spitting of blood, the whites in women, it helps such women that are out of measure unwealdy in their going with child, it stops bleeding at the nose, helps difficulty of urine" (Culpeper, *Pharmacopoeia*, 32).

[248] Medically, stag's heart was reputed to be "as great a strengthner to the heart as any is" and to resist "pestilences and poyson" (Culpeper, *Pharmacopoeia*, 31).

[249] These are two different tinctures.

[250] Peony.

Contra=yerva roote of each .ʒiii. three dragmes, of orientall-bezarstone, of the stone of the [38r] Porcospina,[251] of orientall bolus, of the powder of Vipers, of Diphoretike Antimony of each. ℥ß halfe an ounce.—Of Calcined=cristall, of Saphires[,] of Emeralds[,] of Rubies[,] of Amethists, of Jacinthes,[252] of Topasses[253] of each .ʒiiii. a dramme & halfe; of christalline Tarter,[254] of sal-prunella, of Lignum-aloes[,] of salt of worme-wood, of each .ʒiiii. foure dragmes. Incorporate all together./ and it is made/ The Dose from a ℈ß. to .ʒß. from halfe a scrupell to halfe a Dragme./

This Cordiall hath wrough Admirable Effects in all Pestilentiall diseases and in Cases of Poyson, taken Inwardly.[255]/

[38v]

The Preparation of Creame of Tartar./ of ye Duke of Northumberland./

2.[256] The maner of my Prepareing Tartar is this[:] first I pound it small. Then, in regard of the great quantitie of Earth wch is in all sorts of Tarter, I wash it in a Bason[257] of water, & rubbe it very well betweene my handes for halfe a quarter of an houre. Then poure of the water, wch is both black & muddy: & put to it an other Cleane water & doe as before, And Continue thus washing it for three or foure times, or rather till you see ye water Come cleane from off ye Tartar[;] Then Put it into an Earthen Pan, with store of faire water, And Boyle it for a quarter of an houre, Then taking ye Pan from the fire [39r] Let it Coole, and there will Presently

251 "Porcospino" is Italian for "porcupine", and porcupines are one of numerous herbaceous animals that produce stones, or bezoars, that have long been popular in both Western and Eastern medicine.

252 Jacinth was thought to "stop loosness" (Culpeper, *Pharmacopoeia*, 7).

253 Topazes.

254 I.e. white (or cream of) tartar.

255 An "inward" or "internal" medicine is taken internally (especially by ingestion), whereas an "outward" or "external" medicine is applied to the skin or wound as, say, a cream or ointment.

256 The number "1" has been omitted. Since there is no evidence of missing pages, it may well have been skipped in the transcription process. Another possibility—which is suggested by the fact that several of the numbers are scratched out or purposely obscured—is that the scribe was innumerate (on innumeracy at this time, see Keith Thomas (1987), Numeracy in Early Modern England. *Transactions of the Royal Historical Society* 37: 103–132).

257 Basin.

rise as it were an ice upon y^e water the wch I take off with a scummer[258] & it is y^e flower or best of y^e Tartar; Then I set y^e Panne to boyle againe and doe all as before, And Continue y^e same Course till there arise no more ice or Congeale=ment upon y^e water. Then I dry it, and pound it into more fine Powder, Passing it through a fine searse./

The Manner of Prepareing the Duke of Northumberlands great oyle of Sulpher, a great and safe Remedy./

3. Take first y^e Balsome of Sulpher as it is usually made by the oyle or spirit of Tur=pentine,[259] where of three partes (or 4.) put [39v] upon one Part of Sulpher in Powder, fetch=eth out by digestion y^e red Tincture of it Which is an Excellent Remedy for the asthma & Diseases of y^e Lungs, But to exalt it much higher Put it into a Retort & by degrees of fire distill it over; And yo^u shall have a Perfect oyle of Sulpher somewhat reddish, and much more Penetrating & effectu=all then y^e former Tincture. This serveth not onely for y^e Diseases of y^e Lunge, But most powrefully allso for Pleurices, for the Running of y^e Raines in either Sex[,] For y^e whites of women; and for weepeing in men through relaxation of y^e seminall partes by often Gonorehaas[260]; The ordinary dose to Continue sometime wch is .7. or .8. droppes, But in a Pleurisy where a present effecasious cure is requir'd, yo^u may increase it to .20. [40r] Droppes, It is also Excellent in Maligne feavers and if ye feaver be very great yo^u may desolve in it by gentle heat, a fourth Part of Capher[261] for y^e disease in y^e seminall vessels, It is to be taken some dayes together, Morneing and Night first & Last, .7. or 8. Droppes at a time, The frequent use of it Preserveth health and giveth strength to y^e inwarde partes notably. It is to be taken wth any Appro=priated Liquor, serving for a vehiculum./

The Duke of Northumberlands way to make Milke of Sulpher

4. Take two Pound of y^e best & purest yellow Sulfer and three Pound of deflegmed[262] Roman vitrioll, that wth deflegiming is[263] become browne.

[258] A shallow ladle or sieve for removing scum or floating matter from the surface of a liquid (*OED*, n. 1).

[259] Turpentine was used as a purgative (Culpeper, *Pharmacopoeia*, 29).

[260] Gonnorhea.

[261] A caper.

[262] Dephlegmated.

[263] I.e. has.

Pound them to Powder & Passe them through a [40v] Thinne searce and Put them into a Body of glasse; which set in Sand upon y^{e} furnace and sublime it up with moderate fire. The first y^{t} sublimeth is naught & stinking therefore Cast it away; But that wch sub=limeth the second time into y^{e} head, is the good therefore save that Carefully; And for this end yo^{u} must have two blinde heads, That when yo^{u} take off one to scrape the sublimed sulfer out of it, yo^{u} may at y^{e} instant put on another or if you leave y^{e} Body uncovered, so as Ayre may Come in The Sulfer in ye bottome will take fire: And do thus till no more Sulfer Sublime; Then take all that is Sublimed up, and grinde it small, and put it into a new body of glasse well luted[264] and poure upon it duble its weight of Excellent oyle of tartar of white wine, and mingle them well wth a [41r] Spatule[265] of wood. Then make this Composition boyle very gentley, till yo^{u} see ye oyle of Tartar that was white become of a yellow orange Colour. Then take ye body from ye fire and when the matter is Cold decant off ye Cleare and upon ye sediment, put new oyle of tartar and boyle againe, & decant, and do this till you see y^{t} ye Sulfer giveth no more tincture to ye oyle of Tartar; Then Joyne together all yor finest oyles, and in a new body of glasse evaporate wth moderate fire halfe ye Liquor and then take it from ye fire and when it is gone pretty Cold poure to it a Bottle of pure white wine by Little, & Little and yo^{u} shall see a pure white milke fall downe to ye bottome of ye body. When all is setled poure off ye liquor, and keepe ye milke by it selfe; and poure more white wine to the Liquor, to y^{e} [41v] End y^{t} if there be any more milke in it, it may Precipitate downe, and so yo^{u} loose none of it, And do thus till yo^{u} have recovered all y^{e} milke of it. Then Joyne all y^{e} milkes together, and wash them first wth white wine, letting it then setle and poure y^{e} thine Liquor from it; Af=terwardes wash it in y^{e} same manner w^{th} .3. or .4. warme waters till yo^{u} perceive y^{e} water Come off Cleare & wth out any tast. Then put it into purslane[266] Plates, that it lay but thinne and in a Place free from Dust, Covered slight=ly wth Papers, let it dry in the Sunne; And then keepe it for yor use./

[264] Lute. To coat with lute (clay or cement), especially to cover (a crucible, etc.) with lute as a protection against fire; to close or stop with lute (an orifice or joint); to stop with lute the cracks or joints of (a vessel). The coating or covering made of this clay (*OED*, v. 2, 1).

[265] Spatula.

[266] Porcelain.

This is an admirable medium for Tisikes[267] & all diseases of ye Lunges./

The dose is .6. or .7. graines in water of Scorzonera or any other good Cordiall and Pectorall water or Syrup./

[42r]

The Duke of Northumberlands Aurum potabile,[268] or Tincture of Gold which is an admirable Medicine./

5. Take perfect Spirit of wine .4. times rectified every time from his owne salt drawne out of the feces left in distilling ye Spirit, Put of this into ye bottome of a duble filosophers egge[269] (made as my Margent)[270] And in y^e^ upper egge (Fig. 7.1) hange by thrides[271] of Gold wyre thinne Plates of Pure Gold; and shut up y^e^ glasse Hermis seale. Then set it in moderate heat of Maria Prophetissa bath[272]; And y^e^ Spirit of y^e^ wine rising, and Circu=lating into y^e^ upper egge will worke upon the Sol,[273] and dissolve its Tincture, & drop downe with it like bloud into y^e^ rest of the Spirit of wine. Continue this till it be all Tincted a deepe red as any Blood; Then let yor Bath Coole [42v] And take out yor glasse and breakeing it open take out y^e^ tincted Liquor; which yo^u^ may eit=her give so to .20. droppes or distilling away y^e^ Spirit of wine till yo^u^ have yor Tincture of Gold either like an oyle, or in Powder; yo^u^ may give thereof .3. or .4. graines, ever to be in some litle quantity of some noble Cordiall, and it worketh miracles

[267] Phthisic. According to a section "Of the Tisike" in Stephen Batman's translation of Anglicus Bartholomaeus's *De proprietatibus rerum* (a thirteenth-century encyclopedia), "Tisike is consumption and wasting of kinde humour of the bodye, and commeth of whelkes and of botches of the lunges: and sometime of a reume dropping from the head into the lunges, and smiteth the lunges ofte, and thirleth them and maketh holes therein, and whelkes and botches, as dropping of raine perceth a stone" ((1582), *Batman uppon Bartholome: His Book De proprietatibus rerum* (London), unpaginated).

[268] Literally, drinkable gold.

[269] For a fuller discussion, see Chapter 6, 97–98.

[270] Also margin. The space on a page between its extreme edge and the main body of matter (*OED*, n. 1a). For the image in the margin, see Fig. 7.1.

[271] Threads.

[272] Mary's bath.

[273] Soul.

Fig. 7.1 Robert Dudley's Philosophers' Egg

in all ocasions. Even des=perate Feavors, when y^e Partie hath bene given over by y^e Physitians, I have Cured Soddainely.[274]

The Duke of Northumberlands best and true Preperation of Scammony.[275]/

6. Beat into Powder of y^e best scammony one Pound, Put it in a Cucurbite; & infuse upon it two poundes of y^e Infusion of [43r] Roses made according to art. Mingle all exceeding well and shutting y^e vessell very Close, set it for two dayes in balneo Mariæ[276] moderately heated, In wch time, the infusion of Roses will have Imbibed y^e Essence[277] of the Scammony. Poure off the tincted infu=sion and poure new upon y^e remaineing sedi=ment, but in lesse quantity then before, digest againe as before, and

[274] Suddenly.

[275] Culpeper warns that scammony gnaws patients' "bodyes as fast as Doctors gnaw their purses" (Culpeper, *Pharmacopoeia*, 29).

[276] Mary's bath.

[277] An extract obtained by distillation or otherwise from a plant, or from a medicinal, odoriferous, or alimentary substance, and containing its characteristic properties in a concentrated form. In pharmacy, it is chiefly applied to alcoholic solutions containing the volatile elements or "essential oil" to which the perfume, flavour, or therapeutic virtues of the substance are due (*OED*, n. 9a).

when it is tincted, poure it off to ye first, as formely yo^{u} did and Continue doing thus till yo^{u} see that the Scammony affordeth no more tincture to the Infusion of Roses, Then put all yor tincted Infusions joyned into a glasse limbeke and Distill away till ye residence[278] become of ye thicknesse of honey; wch then dry in ye shade in severall glasse vassells, putting but very litle in each; The dose from .3. graines to .10. [43v] And even to more in strong Complexions, But I alwayes give with it about a Dragme of Creame of Tartar./

The Duke of Northumberlands great Astringent

7. An excellent easy, safe and dayly approved remedy, of the Duke of Northumberlands to stoppe securely all fluxes of Blood; either by a vaine broken inwardly, or by the Hemoroides, or bleeding at the Nose or any other part of the body either inward or outward as woun=des. And never fayled in hundreds of Persons that it hath ben proved[279] upon; It also stoppeth the too abun=dant Courses of blood of women when fit to be stopped,

[44r]

A great and admirable remedy.

Take two ounces of exellent opium, and beate it grossely[280] in Powder wch put into an Earthen glased[281] dish that can indure the fire, or Purslane,[282] and set it over a chafing dish of Coales, But let the fire be temporate, and no more then can well melt the opium, and whiles it is melting; stiring Continually with ~~some~~ Some Spatule, or other fitt Instrument of Iron, to hinder[283] y^{t} the melted opium stick not to y^{e} dish; And Con=tinue doing thus, Continually stiring it till it melt no more; wch is a signe y^{t} it is sufficiently Calcined, and all y^{e} untuous[284] volatile poisonous part of it evaporated away. And this will require about two houres time,

[278] Matter that settles as a deposit in a liquid; sediment (*OED*, n. 2, 1a).

[279] I.e. tested with obvious success.

[280] Grossly. Of grinding, pounding, etc.: coarsely (*OED*, adv. 4).

[281] Glazed.

[282] Porcelain.

[283] I.e. to keep/stop the melted opium from sticking.

[284] Unctuous. Of the nature or quality of an unguent or ointment; oily, greasy (*OED*, adj. 1a).

and then it will be reduced to ye weight of one ounce & .2. Deniers[285] from two ounces y^{t} it was at first, The opium [44v] Being thus Calcined, grind it upon a Painters Marble stone, whereon he grindeth his Colours (that is to saye, such a one) wth a litle distilled vinager, And it will be imbibed Presently in ye pouder[286]; and never loose its virtue.—

The Dose of this Powder is from .7. to .10. graines either in a reare[287] egge, or in a Pill of venice Treacle, It may be taken at any time, when the necessity requireth; But if one Can Stay till then, it worketh best when one goeth to Sleepe upon it, usually it stoppeth bleeding in .4. or .5. houres, But if it should happen to fayle ye first time, This remedy is so safe as at y^{e} .5. houres ende yo^{u} may give another dose of .8. graines (wch is y^{e} ordinary) And at the ende of .5. houres more, a third dose if it be requisite; And in extreame necesity one may safely proceede to a fourth dose But hitherto it never missed Cureing at [45r] The third takeing; And seldome at the first[.] It is allso a secure remedy in y^{e} extreame tor=ments of y^{e} stone or gravell; In wch by eas=ing y^{e} Paine of ten times nature overcometh y^{e} Disease. It avayleth[288] in like manner in y^{e} Cholike[289] takeing a glister[290] before it./

The Vitriol and Spirits of Silver of the Duke of Northumberlands./

8. Calcine one Part of ☽[291] with a Parte of Sulfer in gentle fire .6. or .7. times, letting them burne of themselves wth out blowing or forcing y^{e} heate. Then fetch out y^{e} vitriol by often ebullitions[292] in water, and new Calcinations of y^{e} remaineing bo=dy, and filtrations of y^{e} waters, and evaporating of y^{e} pellicula[293] and then letting it precipitate in Christalls:

285 A French coin, the twelfth of the sou; originally, like the Roman denarius and English penny, of silver; but, from the sixteenth century, a small copper coin (*OED*, n. 3, 1).

286 Powder.

287 Rare: that is, a lightly cooked egg, left soft.

288 Availeth.

289 Colic.

290 Glyster.

291 The alchemical symbol for silver.

292 The process of boiling, or keeping a liquid at the boiling point by the application of heat; the state of bubbling agitation into which a liquid is thrown by being heated to the boiling point (*OED*, n. 1a).

293 Pellicle.

and do thus till yo^{u} have al ye Silver Converted into Christalline vitriol [45v] In ye same manner as is used in extracting the vitriol of ♀[294] but this requireth much more Calcination. Then out of this ⊕[295] Distill by force of fire the Spirit & oyle; and out of ye dead head, draw ye fixed Salt, with warme water and mix it wth ye Spirit, when it is perfectly depured[296] by frequent dissolution & Congelations[.][297] This is an admirable medicine for all diseases of ye head, as vertigos, Frensies[,] Lunacies, & all paines or distempers there. As also against hot feavours. The Dose from .4 droppes to .8. in y^{e} water of Scorzonera, or other fitt Cordiall water, wth a litle suger or agrodicedro[298] to dul-lify[299] y^{e} sharpenesse of y^{e} Spirit of vitriol of Silver. And if y^{e} season nor disease be very hott yo^{u} may take it wth a litle dulci=fyed greeke wine./

[46r]

1. The Duke of Northumberlands Receipts sent to Sr Kellam Digby[300] & in a letter for the Lord Marquesse of Newcastle, as followeth./

As touching the Infermity my Lord Marquesse of Newcastle writeth touching his owne Indispo=sition of vapors that ascende to y^{e} head, I my-selfe for many yeares Past have bene troubled with y^{e} same Infe[^r]mity; And for y^{e} same I doe Purge wth Certaine gentle Pills of myne twise every month & onely a quarter of an howre be=fore I goe to Diner, For soe they worke most gentley wthout any trouble; I take often Tobacko an[d] smoke in y^{e} Pipe at least six times a day, But not y^{t} day y^{t} I Purge, And with y^{e} Tobacko I put in wth it certayne Droppes of y^{e} oyle of Carebe[301]

294 The alchemical symbol for copper.

295 The alchemical symbol for vitriol.

296 Depure. To free from impurity, cleanse, purify (*OED*, v.).

297 The action of congealing or freezing; the process or state of being congealed (*OED*, n. 1a).

298 The Italian "agro di cedro" (or "sour cedar").

299 To render dull (*OED*, v.).

300 Kenelm Digby (1603–1665).

301 Carabe. Another term for "yellow amber" (Cotgrave, *A Dictionarie of the French and English Tongues*, unpaginated).

distilled, which is Amber, whereof six [46v] Droppes serve for six Pipefulls of Tobacke, And this Extinguesheth much the vapors that fly to the Head from ye Hypocondria; I take allso ye Balsum of Sulfer made wth ye Spirit-tuall oyle of Turpentine some six droppes therefore every time at least three or foure times every weeke, & this Purgeth well those vapors by urin, wch vapors Causeth noyses in ye head and by this meanes under God have Preserved myselfe. This verbatim the Dukes writing./

2. The Duke of Northumberland his Receipts./

The Pills wch I take when I goe to Dinner [^y^{t}] worke so gently, is breefely thus for one piesse.[302] I take half a deiner[303] wch is twelves graynes of pilule Cocci[304] for y^{e} head, and mix with it .6. graynes of my scamony pre=pared with y^{e} fume of sulfor burning in a Chafing=disshe as yor Lords.^{ip} knoweth by Doctor Coracckeno[305] [47r] his Booke written thereof, And .6. graynes of my Antimony Prepared with sall niter as is written in the same Booke in all maketh .24. graynes, which mix=ed together in [the] forme of Past[306] as Comonly used, wch yo^{u} part into .5. or .6. Litle Pills to take in broth or in iustia[307] as used; But to take them a litle before his Lord.^{ip} goe to Dinner, not to Passe a quarter of an houre, & to eat at dinner; & to drinke Compirably[308] well & rather lesse then he useth ordinarely, and if they worke to gently, & not suficiently, his Lord.^{ip} may add an other time two Pills more or Lesse acording to their working for his Complexion; And ye Balsum or oyle of sulfer may be taken in a spoonefull of any Liquor; either broth, or wine; or water./

[47v]

302 Piece.

303 Deaner. A half-shilling (*OED*, n.).

304 Pilulae cochiae.

305 A professor at the University of Pisa, Marco Cornacchinio, who published a more simplistic version of Dudley's powder, known as "Pulvis Warwicensis", in a 1619 work titled *Methodus qua omnes humani corporis affectiones ab humoribus copia* (Florence), which was dedicated to Dudley.

306 A paste.

307 Dubious reading. It could be the Latin for "juice".

308 Comparably.

3. Notes Out of Letters from Sr Kellam Digbey from Roome.[309]/

The Powder of vipers ten-times more Effectuall then the Troches of vipers; And by y^{e} great Ex=perience of y^{e} wonderfull good it doth in Restore=ing Natures, And even in makeing y^{e} outwarde sem=blance of People returne to a kinde of youth=fulnesse much more Powerfull then y^{e} troches. It hathe worne them[310] in a manner out of use, & out of Countnance,[311] You will finde y^{e} Roman Troches Different from y^{e} Paduan or Venetian Troches in Coller,[312] shape & bignesse, But they are esteemed farr beyond them; The vipers them-selves being accoun=ted better, And then being secure[313] to be right; whereas venice of late is accounted to sell falce[314] ware,[315] for if they weare[316] true & y^{e} quantity not inlarged by to great a Proportion of Bead,[317] or admixture of gummes, they Could not afford them at y^{e} rate [48r] They sell them at, But indeede y^{e} vipers have bene Exterordinary dear here this yeare, by reason of some strange effects they wrought last yeare up=on old decayed Persons .two. Men, & one woman which hath sett every body on fire to use this Me=dicine, and so y^{e} many byers made a dirth of y^{e} como=dity, I will send yor Lord.^{ip} a Printed discourse of their Vertue & use.[318]/

The Powder of viper & Troches is much better at Rome then at venice, or Padua because y^{e} mountaine vipers is beter then in Lowe & marish[319] places as also y^{e} heates being greater there it makes them the purer./

309 Rome. Between 1645 and 1648, Digby spent considerable time in Rome, in a failed attempt to convince Pope Innocent X to support the Royalist cause.

310 I.e. the troches of vipers.

311 Countenance. Estimation; credit or repute (*OED*, n. 1, III.9).

312 Colour.

313 I.e. certain; fully assured (*OED*, adj. II).

314 False.

315 A collective term for articles of merchandise or manufacture; goods, commodities (*OED*, n. 3, 1).

316 Were.

317 Bead amber.

318 Probably a reference to Marcus Aurelius Severinus's 1650 treatise, *Vipera Pythia, id est, de viperae natura, veneno, medicina, demonstrations* (Padua). While this was ostensibly published a year after Digby penned his letter, the ambiguity of the Gregorian/Julian calendrical systems makes it likely that Severinus's book had just been published (as Digby indicates) in 1649.

319 I.e. marshy (*OED*, n. B, 1a).

The Powder of vipers is made up with their hearts and Livers

4. The Virtues of the Powder of vipers and Troches./

Their Imediate efficasie, is agaynst all sortes [48v] of Poysons. But that wch they are most estee=med for here Is that by a longe use of such flesh of vipers; old Men grow young agen And strong Persons are kept in a perpetuall vigor; And Men growne Eunuches[320] by age become Priapus[321] agen; Indeede they restore nature strangely if they be Constantly used every Day for a longe time; as for a yeare or more

6.[322] Of the Virtues of the Powder and Troches of vipers./

I send yor Lord.^{ip} also a Paper lately Printed here of y^{e} virtues of vipers; In short their flesh is a greate Renovative & restorer of natures, Strengtheneth y^{e} stomack, & all y^{e} inward Partes; sharpeneth y^{e} sight, overcometh the Spleene; Reviveth y^{e} Spirits & strengthen=eth wonderfully y^{e} parts & actions serving [49r] To generation, The Powder & Troches are of like virtue; But y^{e} Powder much more Effica=tious: They will keepe well .3. or .4. yeares or more, if Preserved dry: The Treacle, and Metridate .30. yeares & upwards./

Of y^{e} Powder yo^{u} are to take a scrupell, or halfe a dramm at a time; Either in y^{e} Morneing fasting or goeing to bed after a Light supper in wine or broth or some Cordiall Conserve; Drinkeing a little wine upon it, as Rosemary-flowers or gilly-flowers, or of Roses, and to have a sensible great effect of it; It must be Continued so for .60: dayes, And if yo^{u} finde it not to strong, yo^{u} may take it morneing & night a scruple at a time./

That which yor Lord:^{ip} Likeneth to haires in yor Pouder of vipers is Part of ye back-bone which faleth into such thrides,[323] & wil not be beaton smaler. Some are of opinion it is y^{e} most effecative [49v] Part of y^{e} Viper, And therefore by no meanes put it away: They who seeke solid benifite and Restauration by this Medicament, take it a quarter or halfe a yeare wth out Intermi=tion every Day; But then exceede not a scru=pell, & but once a Day./

[320] Eunuch. A castrated person of the male sex (*OED*, n. a).

[321] A Graeco-Roman god of procreation and fertility, usually represented as a small, deformed figure with an enormous phallus, and later taken as a tutelary god of gardens, vineyards, etc. (*OED*, n. 1).

[322] There is no number "5".

[323] Threads.

7.

a. The Excellent Treacle made up wth true opobalsamum,[324] & all y^{e} Ingredients, y^{e} true one is very dear; The same rare Apothecary that maketh y^{e} Treacle, maketh also y^{e} Powder of vipers [^w^{ch}] he hath dun such miracles upon others, & him-selfe every yeare, he hath not a grayne left wth in. 3. monthes after y^{e} make=ing of it Excepting such store as his Concubin (for he hath no wife) maketh him keepe [50r] for himselfe for by meanes of it, & of his Treacle She findeth him a Cavalier of 25. yeares old, though indeede he bee. 40. If yor Lord.^{ip} will have any of these yo^{u} must bespeake in time for y^{e} viper season, & y^{e} Powder he maketh will be soone gon, And he hath now all his ingre=dients Compleat for makeing his Treacle and Metridate & intendeth y^{t} this shall be y^{e} Last time of ever his makeing[325] them wth such, For it is a wearisum & Chargable[326] diligence to get all the Drugges right, for which end he sendeth on pur=pose into Egipt, Arabia, & India[327] And y^{e} whole Coledge of Phisicians there have subscribed[328] y^{e} truth of all his Ingredients; This Excellent Treacle & metridate[329] is made but by one Apo=thycary, and he exceedeth Infinitely all others; he is old & cannot live Longe, and had bene dead Longe since but for y^{e} Continuall use of his owne [50v] Treacle, he maketh it of all y^{e} true Ingredients without suplying[330] y^{e} defects of any of them by such other Druggs as Physicians have a=pointed in stead of them when they are not to be gotten, for he sendeth on Purpose into Egipt for true Balsum, and into Arabia for y^{e} right Spices, & into India for other Druggs, And ha=th true all y^{t} other Pothecarys want; Though at an excessive rate; And he is a[s] curious in y^{e} makeing these two Compositions, as in Chusing their Ingredients./

[324] Opobalsamum was thought to be "excellent to take away a skar, and many other purposes: but it's very dear, and hard to be gotten" (Anonymous, (1657), *A Physical Dictionary* (London), unpaginated).

[325] I.e. the last time that he will ever make them.

[326] Chargeable. Burdensome, troublesome (*OED*, adj. I.1).

[327] For the importation of recipes from Egypt, the Arabian Peninsula, and India during this period, see Jenner and Wallis, *Medicine and the Market*; also see Chapter 5, 75–77.

[328] To indicate one's assent to; to support, favour, sanction, concur in (*OED*, v. I, 2c).

[329] Mithridate.

[330] Supplying. To make up a deficiency in (something) by providing or replacing what is missing; to supplement (*OED*, v. 1, I, 2b).

8. His name is Siginor Antonio Manfradi[331]; I have seene admirable effects of his Treacles be=yond all others, And surely it is ye greatest Medicin in nature to keepe a body in perfect strength & vigor, and to keepe back old age and decay; old men have told me Miracles of this who have used of it Constantly above [51r] 30. yeares; And they have believed themselves growne young by it[;] a Composition of it well made will Last good .60 yeares

~~8.~~
9. Treacle & Metridate are likewise Imediatly against Poyson[332]; But they are more valued here for their resisting of Putrefaction, for their digesting of Crudities & na[^u]tious humors, for their strengthening all ye Internall Partes espetialy ye stomack & ye head, for their in=gendring good spirites, for makeing good digestion and for giveing a notable vigor to all nature in a Person; they use them here, once, twice or thrice a weeke as ye Complexion admitteth[.] The makeing them as also ye Troches is just ye same as Gallen Proscribed, wch they wch observe Religiously & are accurate in ye Choyse of their Ingredients doe make perfect Compositions [51v] I may say Miraculus ones as ye Apothecary I have mentioned to yor Lord.ip But to Imitate him for it, Costeth to much, & taketh up to much time & Paynes; for sale ware[333] formerly these Compositions [which] weare onely made by ye Emperors And this man maketh them now onely for a de=light & ambition he hath in makeing such, for otherwise he rather loseth then geteth by these his Confections./

The dose of ye Treacle, & Metridate is from halfe a drame; to a Dramme, & but once a weeke in ye winter & spring./
In ye great heates it is to be wholy forborne

[52r]

The manner of how to make Troches of Vipers./

331 "Antonio Manfredi" also features prominently in Baldi's 1646 *Del vero opobalsamo orientale*, which, as the name suggests, was an apology for "opobalsamum". This text was published in Rome around the time that Digby was there on behalf of King Charles I (see Chapter 6, 86–87). It must have either been Digby's source, which seems likely since Baldi also dwells on the power of viper flesh, or he must have been in direct contact with Manfredi (*Del vero opobalsamo*, 99–100).

332 I.e. they are antidotes to poison.

333 Were.

Take Vipers & Cut of[334] their heads & tayles about the length of three fingers, & take out ye guts & fatt soe yt there remaine nothing but ye flesh; Boyle these vipers (so Prepared) in water, wth an hearb cal=ed Aretnuß,[335] & a Litle Salt, Boyle them untill they be tender, Then take of this flesh of vipers eight ounces, Then take of the finest wheaten Bred, or elce[336] of ye whitest Bisqueet-cakes[337] beaten into Powder and sivted[338] through a fine sive[339] two ounces; Let these two be mixed together, & then beaton in a Marble Morter very smale, untill you can make Litle Cakes of it ye bredth of a farthing[340][;] if ye Past be to drie you may take some of ye same Liquor, & moysten it in ye Morter, And after these Cakes be made, you may annoynt them wth a litle opabalsamum, or elce [52v] With oyle of Nutmegs, And so you may set them to dry betwixt two Papers, so yt the Cakes doe not tutch[341] one another Continually turneing them untill they be dry, And then you may keepe them for 3. yeares good being kept in a glasse very well stoped;

This is ye manner of makeing & keepeing of ye Troches of vipers That doe enter in ye venice Treacle/

The Troches of vipers are usually given to Cause Sweat; And ye dose is commonly .20 graines./

The Receipt to make Salt of vipers./

The Salt of vipers is made by Calcination as other salts are; in this manner Burne yor vipers to white Ashes; wch Ashes you shall put into a glasse botle and put to it of fayre water as much, that ye water may be .4. fingers above ye Ashes, Then Let it stand in some hot place .2. or .3. dayes or Longer or else untill ye water be Collered,[342] and then [53r] Separate ye water & keepe it in a glasse alone and lay ye water upon ye Ashes, This doe so long untill ye water have no tast, then take all these waters together & Boyle

[334] Off.

[335] This likely refers to the leaves of the aretnuß, which were often used medicinally.

[336] Else.

[337] Biscuit cakes.

[338] Sifted.

[339] Sieve.

[340] A quarter of a penny; the coin representing this value (*OED*, n. 1a).

[341] Touch.

[342] Coloured.

them until .3. Parts be Con=summed, And then let it stand; or else untill all be Consumed, And then yo^{u} have y^{e} Salt./

This is an Excellent Cordiall to Cause Sweat
The Dose is :15: graines, when neede Requires./

The Receipt of the Cooleing Julep or Diet drinke that Doctor Davison doth give in Fevors./

Take Barley-water two Pintes; Sirop of violets two ounces, Sirop of Limonds, one ounce[;] mix them together & use this, in Place of Beare to quench the thirst./

A Receipt of the Julep which Doctor Davison doth use in fevors[343] to loosen the Belly./

[53v] Take one ounce of Ivory, & one ounce of Harts=Horne, Raspe[344] them & put them in .3. Paris Pintes of water, & let them Boyle together untill the halfe be Consumed, Then straine it through a Cloth and Put it to y^{r} Liquor foure ounces of y^{e} best Joice of Barberys,[345] & one ounce & a halfe of y^{e} sirop of Pomgrannets, wch is in Latin S^{ir}. granator[346]: this is to be used instead of Beare

The Receipt how to make Cristall Minerals./

Take Salt-Peter of ye whitest as much as yo^{u} Please and let it melt in faire-water & Boyle it in a fit vessell & then straine it through a woollen Cloth & then set it in a Cold place .24. houres, then separate ye liquor from ye Cristall, & dry ye Cristell[;] The liquor yo^{u} shall boyle againe untill ye halfe be Consumed, & then doe as before, Thus yo^{u} may doe [54r] five or six times untill it be as white as may be, Then y^{e} Cristall being well dryed yo^{u} are to take a Crusible or an Earthen Pott, y^{t} will resist y^{e} fire, In wch Pott yo^{u} are to Cast in yor Salt-Peter & Let it melt[;] be=ing almost melted, add to it a litle flower of Brimstone and so let it burne together, & then Put it in some earthen vessell to Coole, & then it is made./

[343] Fevers.

[344] Rasp. To grate, file, or scrape with a rasp or other rough instrument (*OED*, v. 2a).

[345] Barberry was deemed to "purgeth Choller" and to be "a singular remedy for the yellow Jaundice" (Culpeper, *Pharmacopoeia*, 11).

[346] Latin for "syrup of pomegranate" (*sirupus granati*).

But if yo^{u} will have it Cristall, yo^{u} musts melt the same in water againe, And boyle it untill y^{e} halfe be Consumed, & then set it in a Cold place, & in the morneing separate yor Cristall & it is made./

It is good for all sortes of hot burning feavors to Coole y^{e} Body & to quench y^{e} thirst Being de=solved in barly-water with sirop of violets./

It openeth obstructions of y^{e} Liver, Milt and Kidneys, & easeth to make water, And is good for those y^{t} are troubled with gravell or y^{e} Stone, It is also an espetiall Medicin for y^{e} Paine in the throat, being desolved[347] in Plantin water, with [54v] Honey of red-Roses. For outward Applycations being desolved in Plaintin water, & Layde on with Linin Cloth upon a hot swelling./

Also ye Itch or Scabb it dryeth up & healeth the same in short time./

Doctor Floids[348] ways to Cure a loosenesse./

Take Cathalicum,[349] double, with Ruberbe[350]; 3. Drammes Pulpe of Tammerins, drawne through a Sive .1. drame & halfe[;] This to be taken in [^a] Bolus./

An other way./

Take halfe a dragme of Rhuberbe, and make into Six Pills with Sirop of Suckory, or, violets./

An other way./

An ounce, or an ounce & a halfe of Manna in a Pint of wheye./

[55r]

A Receipt how to Cure sore Eyes./

Take a quart of newe Milke, & Put it into a disshe and Cover it wth another dishe; & set it upon a gentle fire, y^{t} it may not Boyle, And y^{t} water which hangs upon ye upper dishe take & let it drope into some

[347] Dissolved.

[348] Robert Fludd (bap. 1574–1637).

[349] Catholicon.

[350] Rhubarb.

violl[351] or glasse, & then dissolve a litle white Suger-candy, And then with y^t^ water washe yor Eyes; And tis for Certaine a Perfect Cure./

[55v]

Receipts of Mr Basincotes[352] an Apothecary at Paris./

The true manner how to make Crocus-Metalorum and the use thereof./

Take of y^e^ best Salt-Peeter one ounce; And one ounce of the best Antimoney unprepared, being beaten into fine Powder, severaly, & well dryed Then mix them both together in an Iron-mortor or Laedle,[353] and holding them over the fire[;] yo^u^ shall stir them wth a red=hot Iron & they will burne & become hepar-Antimones[354] y^e^ which shall have y^e^ Co[^u]llor of Saffron. Then take this hepar & beat it into fine powder, and wash it three times in fayre-water, and let it stand [56r] one night to setle, and then y^r^ water being cast away, the Powder y^t^ remaine in ye bottome, you must drie; And it will serve for yor use as followeth[:]

This Powder is good for all sorts of feavors being given before ye fit, half an houre; Also it is good against the Plague, being given at the first begining;/

And it doth both Purge & Vomit; And often times by sweat./

It's also good for those that have their Stomackes full of tuffe[355] flegme./

This is to be given to such as are apt to vomit[;] You must take heede in giving [^of] this Powder to such Persons as are in Consumtions, or to those y^t^ have refluxions,[356] and also to those as doe now & then spit bloud; For there may follow many Inconveniences And also Death:

Use of this Powder in Substance, from .2. to .12. graines./

in substance, according to ye age & strength of ye Person./

[56v] But Commonly it is Infused in white wine, one day, two, or three; being often stired, and then Strayned

351 Vial.

352 Unfortunately, we have not been able to find any information concerning this apothecary.

353 Ladle. A pan with a handle, to hold molten metal for pouring (*OED*, n. 2b).

354 Hepar Antimony.

355 Tough.

356 Flowing back; an instance of this (*OED*, n.). The phenomenon in question is what we call "acid reflux".

Take of this Powder ye weight of two french-crownes and infuse it in an English Pinte of white-wine, for the space of two, or three dayes, in an Earthen or glasse-vessell, & in a warme place, & often stiring it, and then strayne it through a Browne Paper And yo^{u} may give of this wine, from halfe an ounce to two ounces, & a halfe; For such Diseases, as afore sayde, or as neede requireth or According to the Judgment of those y^{t} shall use this Medicin./

The manner of Prepareing Cristall Minerall and the use of it./

Take of the best refyned, or whitest, Salt-peter one-Pound; then wth .4. Pintes of water yo^{u} shall de=solve yor Salt peeter, & then boyle it untill halfe a Pinte be Consumed; Then strayne it through a Browne paper, and the strayned Liquor you must [57r] Boyle againe, till halfe y^{e} Liquor be Consumed, and then yo^{u} will have almost two pintes of boyled Liquor remayneing, Then set yor Earthen vesell in the Seller, or in any Cold place, for ye space of .24. houres and then yo^{u} shall finde in ye bottome fine white Cristall, like unto Suger-candy; The Liquor wch remaineth upon ye Cristall yo^{u} shall seperate and boyle it againe, and set in ye Seller as before, and the Cristall wch yo^{u} finde in the bottome of those vessells yo^{u} may dry before ye fire, or in ye Sunne./

Then in an earthen-vessell or Crusible, yo^{u} shall take .4. ounces of ye same Cristall & melt upon the fier, the wch being almost melted, yo^{u} shall kast[357] into the Crusuble a litle flower of brimston, into the salt-peter .2. or .3. times; And then yo^{u} shall have Lap: Prunella[358]; And to have it better this foresayde worke must be done .2. or .3. times./

[57v]

The vertues of Cristall-Mineralis

It is good for all sortes of hot burneing feavors for it is a great Cooler of blood; & quencheth y^{e} thirst; being desolved in barley-water, or Succo=ry water; with Sirope of violets; It doth open all sortes of obstructions, of Liver, & Milt, & doth Purge by urin,[359] And somtimes easeth sweat./

[357] Cast.

[358] Lapis prunella was also thought to help bruises, "the Rickets and other stoppings of the Liver; outwardly it is of wonderful force in curing wounds and ulcers, though festered, as also gangreens and Fistulaes, it helps broken bones" (Culpeper, *Pharmacopoeia*, 15).

[359] Urine.

It is good for all sortes of Inflamations, of the Throat, being desolved in Plaintin-water, and Sirope of violets, or Sirope of Mulberyes, & so gargill[360] y^e^ Throat therewith, y^e^ wch taketh away the Paine; It is also good for all sortes of hot swellings, being desolved in any Liquor, & so ap=plyed wth Clothes,[361] the wch dryeth up the same humor./

[58r]

The Preparation of Vitrioll or els[e] Calcination./

Take of the best greene vitrioll, as much as you Please, & Put in a Crusuble, or Earthen pott and set it on a Stronge fire, & stir it with a peece of iron untill it melteth, And then let it Calcine untill it bered[362]; Then take ye Pott from ye fire, & breake ye Pott, & save ye Calcined vitrioll; The which is to be used for Bleeding, by a vaine or at ye Nose./

The use of this Vitrioll./

Take of this Powder, & put a litle of it upon a litle Cotton with as much Bole-armoneck in Powder, and so apply to ye affected part; And it will stay[363] the Blood; The oyle of vitrioll is good for a kanker in y^e^ mouth, being mixed with a Litle Honey of Roses [58v] And so wash it .2. or .3. times a Day, wth a Cloth raped[364] upon a stick; And it will cure Infaliably./

The spirit of vitrioll is good for all kinde of bur=ning=feavors; Being taken in Barly-water or broth, or else Succory-water with Sirop of violets or Limonds[365]; so y^t^ y^e^ Julep be soureish; which to the quantity of .18. or .10. dropps; But may not be used soe freely in a Flux, except yo^u^ drawe y^e^ Tincture of Roses wth y^e^ spirit of vitrioll; wch is this man=er following./

Take an ounce of red rose leaves dryed, 20. drops of Spirits of vitrioll; Put y^e^ rose with y^e^ Spirit in an earthen Pott; haveing one Pinte of faire rayne or fountaine-water; And being boyling-hott, ease it into y^e^ pott where y^e^ roses are; And let them Infuse the space of .2. or .3. houres;

[360] Gargle.

[361] Cloths.

[362] Bere. To cry out, howl, roar (*OED*, v. 1). The implication here is that the green vitriol should calcine until the fire is roaring.

[363] I.e. block its flow.

[364] Wrapped.

[365] Lemons.

And then straine it through a Cloth; which being done;—If it be hot feavor that the Patient hath; yo^{u} shall put in .2. ounces [59r] Of Sirop of violets; If it be a Flux .2. ounces of Suger of Roses./

How to make Spirit of Salt, with the Virtues of it./

Take of the best french-salt .2. Pound; Calcine that in an earthin Pott; Then desolve y^{t} in ye seller with a Litle rayne-water; and then take ye Powder of red-bricks .6. Pound, & mix yor foresayde & salt with it, so y^{t} it may be of a sufficient hardnesse to make into Litle balls, the wch yo^{u} shall drye in ye Sunne the space of a fortnight, then put ye foresayde balls in an earthen Retort of a great bignesse, The wch must be put into a Furnesse made a Purpose[366]; And to ye retort yo^{u} must Iryne[367] a Recippent so y^{t} no ayre can come forth. But yo^{u} must put into yor Recipent a Litle Distilled-water; ye quantity of an English wine Pint. Yo^{r} Recipent must be twice so [59v] Big as ye Retort; And yo^{u} must Distill first with a gentle fire & then A[^u]gment a Litle, & a Litle untill ye fourth degree or else untill all be distiled which will be dun in ye space of .30. houres time./

This Spirit so distilled must bee Rectified twice over before it can be good./

The Virtues of Spirit of Salt

Take Spirit of salt .6. dropps, & mix it with Posset drinke, or broth, or white-wine, (If ye Patient have no Feavor) Is good for ye Stone; gravell or any other obstruction of ye Spleene, And also for ye yellow-Janders or beginning of Dropsies being taken every-day in Broth./

The Patient being first well Purged./

Allso for y^{e} Clening of y^{e} tooth; Take one drame of this spirit, & as much Joyce[368] of Limonds mixed together, and so rubing yor tooth with this water once a day, & they will be as white as Sno=we./

[60r]

[366] I.e. made to purpose.

[367] Iron.

[368] Juice.

The manner of makeing that water which Do: Dauvson[369] did Prescribe for Sr Charles[370] footmen, wherewith he was Cured of a great, & strange Coughe./

Take .2. or .3. Rootes of comfre, caled in Latin Cansolida: majorie[371]; the Leaves of Buge,[372] of Sanicle,[373] of scabiose,[374] Peruincke,[375] of Coults-foote,[376] Pulmoaria,[377] of Agromoine,[378] of Platin[379] Leaves of each one handfull. Boyle all these together in .4. Pintes of water, untill one Pint be Con=summed & then straine it, & add to it .8. ounces of Clarified Honey, Then boyle that in the same decoction, untill an other Pint be Consumed then strayne it againe, and see y^t^ it be well scummed,[380] And then it is fit for use./

[60v]

The Preparation of Divers sorts of Pultesses[381] for all sorts of hot Swellings./

369 Davisson.

370 Since William Cavendish's eldest son, Charles (1626?–1659), does not appear to have been knighted, the reference must be to the footmen of William's brother, the mathematician Sir Charles Cavendish (1595?–1654).

371 Culpeper thought that consolida maggiore counteracts "spitting of blood, Ruptures or Burstness, pains in the back, it strengthens the Reins, it stops the Terms, and helpeth Hemorrholds" (Culpeper, *Pharmacopoeia*, 5).

372 Bugle.

373 Sanicle was used to "clenseth wounds and ulcers" (Culpeper, *Pharmacopoeia*, 23).

374 Scabious was upheld as a protection against skin diseases including "Scabs and Itch" and "the French-pocks" (Culpeper, *Pharmacopoeia*, 9).

375 According to Gerard, the leaves of periwinkle "boiled in wine and drunken, stop the laske [laxness of the bowels] and bloudie flix [flux]" (*The Herball*, 895).

376 Coltsfoot.

377 According to Culpeper, pulmonaria (or lungwort) "helps infirmities of the Lungues" (Culpeper, *Pharmacopoeia*, 23).

378 Agrimony.

379 Plantain.

380 Scum. To clear (the surface of a liquid) of impurities or floating matter (*OED*, v. 1a).

381 Poultice.

Take March-Mallowes Leaves, Violet Leaves, Lilly, ~~Rootes;~~ Flauers,[382] Cammomile; Let all these be boyled together till they are soft, then strayne them from y^e^ Liquor, And then add to y^e^ Liquor foure ounces of Barley-meale, of Beane me=ale .3. ounces, then boyle these together, to a Pul=tis,[383] then add to y^t^ y^e^ herbs it selfe, being first beaten in a Mortor, and then add oyle of Lilless[384] and oyle of Linseede, And then Let them be well mix=ed together, & it is made./

It must be Applyed warme upon a linnen Cloth and this being dun twise a day will take away all sortes of swellings, & Ripen Impostumes

[61r]

A Pultis to ripen sore fingers or any Impostumes in short time./

Take .2. white Lille-Rootes, & .2. onions wrap these together in a Browne Paper wch must be wetted, & then cover in ye hot embers, and when they are Rosted, beat them in a Morter with an ounce of Leven,[385] & an ounce of Bassilicum.

Another Anodine[386] Pultis to take away all sortes of Panges of the Gout or any other Panges.

Take Crumes[387] of white-bred .4. ounces, one Pinte of Milke boyle these together continualy stir=ing them untill they come to y^e^ Consistance of a Pultise, then add to it .20. graines of Saffran, two yolks of eggs, & one ounce of oyle of Roses./

[61v]

The Physicall Proceeding in the Course of the Salt of Steele./

First of all yo^u^ must Purge./

The second day yo^u^ must take one dose of the Salt of Steele according to ye direction wch is written. The first is .4. graines; And there is .4. doses which is .16. graines in y^e^ 4. Papers. Every [^day] one Layer in the

382 Flowers.

383 Poultice.

384 Lilies.

385 Leaven.

386 Anodyne. Having the power to assuage or lessen pain (*OED*, adj. A, 1).

387 Crumbs.

Morning wth Broth or Posset-drinke, or water of distilled-Apples, and fast .4. houres upon it, with gen=tle Excersise. When those .4. Layers are taken yo^{u} shall take y^{e} second wch is .6. graines everyday./

The third is to be taken in y^{e} same manner; And are also .4. doces and in every doce .8. graines./

For y^{e} Last Papers there are .18. doces; every dose Containing .10. graines, wch is to be taken every day 10 graines so long as it lasteth./

[62r] Note that every third day yo^{u} must take of Aqua=pendentis-Pills[388]; 20. graines Dureing y^{e} hole Course And after y^{e} hole Course; yo^{u} must be Purged againe.

Note also y^{t} dureing y^{e} hole Course, yo^{u} must not eate any fishe, or Milke, or Any Frutes./

The manner of using Aquapen=dentis Pills./

They are to be taken y^{e} weight of halfe a deiner in .3. Litle Pills before yo^{u} goe to Supper or else y^{e} quantity before mentioned which is .20. graines./

And y^{e} next day will give .2. or .3. Stooles./

The Preparation of Creame of Tartar./

Take one Pound of white Tartar and beat it to fine Powder; then washe it, & let it setle a Litle [62v] And then Cast away ye water, and Boyle it in foure Paris Pints of water in an earthen Pott until halfe be Consumed, Then straine it through a Hipocras-Bagg, y^{t} the Liquor may be very cleare, Then Put it into ye Seller in an earthen Bole or dishe, & Let it stand one night, & yo^{u} shall finde yor Creame of Tartar in ye Bottome of y^{r} vessell[;] Cast away ye water & drye ye Creame of Tartar and beat it into fine Powder./

And yo^{u} shall take of this Creame of Tartar the weight of a French Crowne in Broth; in ye morning after yo^{u} have bene Purged./

The manner howe to Infuse one of the .12. Papers of Crocus Metalorum./

Take one of ye Papers, and Put to infuse in a wine glasse full of ye best white-wine wch is .4. ounces and so let it stand for .12. houres, and then strayne it through a Browne-paper, and it will be as [63r] Clean^{e} as before ye Powder was Steeped in it, to strayne yor wine through Paper, yo^{u} must

[388] Aquapendente's Pills, named after Fabricius ab Acquapendente (1533–1619).

have a funnell or glasse, and then take a Sheete of Browne-Paper, the wch yo^{u} must fitt into the funnell, and to Passe yor wine through./

My Lady Marquesse of Newcastells Receipt to Cure a Fluxe or Dis=sentorye.[389]/

Take a Pritty quantity of Milke, & Boyle it well .3. times over; and scume it every time and then Drinke it as hott yo^{u} Can indure it

Probatum Est[390]

Maryes[391] Medicin for the Same./

Take a good quantity of Milke a bout a Pinte & halfe, or more; And put a great white sheete of Paper in it, and Boyle it till halfe y^{e} Milke be [63v] Consumed; And so eate it wth a Scone for it will be very thick; And if yo^{u} eate y^{e} Paper to it will be much y^{e} better for y^{e} Paper will in a manner be Boyled to a Pare[392][;] This to be taken Morneing, & evening for .5. or .6. times; This Me=decin never fayles when all others doe./

Receipts given to my Lord Marquesse of Newcastle; By Doctor Nodin[393] at Paris./

The true Preparation of Crocus=Metalorum

Take of y^{e} Best Antimonie three-partes; Salt=peeter purified, two Partes, or, even quantity if it be not Purified; Beat them together to fine-Powder in an Iron-Morter, And in y^{e} same give them fire with a hot burneing-Coall, in a Chimny, or some other open-Place, for feare of y^{e} Smoake./

[64r] Let all be Coole & Put to Powder; And then put it in a bigg Earthen-Pott, wth a good quantity of water; Boyle it well, & also stir it with an Iron stick untill ye water becometh very Salt,[394] Then take it from the fire, and Put it in to a bigg earthen-dish, there Let it rest a

389 Dysentery.

390 "It has been approved/demonstrated". This common phrase signalled that a medicine has been tested and is effective (see Claire Jones (1998), Formula and Formulation: 'Efficacy Phrases' in Medieval English Medical Manuscripts. *Neuphilologische Mitteilungen* 99.2: 199–209.).

391 Unknown. There were several Marys in Margaret Cavendish's family (see 8D.IV).

392 I.e. to a shaving or peeling.

393 Théodore Naudin (1616–1675).

394 I.e. salty.

while, & then Powre out gently y^{e} water; The Powder which remaineth in the Bottome of y^{e} dishe is yor Crocus=Metallorum./

The true Preparation of the Spirit of Salt./

Take Sea-Salt whitened, and Purified as much as yo^{u} will, And Put it in a Stronge earthen Pott and set it in an open Furnace, with a good fire a=bout it and Cover it with a Tyle[395] or some Brickes and Let it lye there whilst it makes no noyse, Then this Salt being Cold, Put into fine Powder [64v] And add to y^{e} same three-times as much of Bolus-armonus[396] in fine Powder; Then put it into a glasse-Retorta[397] well-Luted, or in a stronge earthen one. Put this Retorta in ye furnace of opened reverbere[398][;] give fire at the beginning some-what Stronge to drawe out ye flegme, which is of no worth: And then Adapt to the Retort a large & bigg Recipient; Seale up thire[399] Joynts very well with ye commune Lut.[400] Continue the fire, & Increase from three-houres, to three-houres as long as any white Spirits will apeare./

At last urge ye fire to ye utmost for ye space of halfe an houre, And comeing no more Spirites end ye worke: Take good heede to not Let ye fire decrease when it is once well kindeled; If this Spirit be found to waterish [by] begin[^ing] againe the same worke it may be rectified; And y^{e} flegme of waterish Substance be seperated from y^{e} true Spirits; The wch is very sharpe./

[65r]

The true Preparation of the Salt of Steele./

Take a Peace of excellent Steele, & make it red-hott, and then tutch it with a stick of Brim=stone upon a Payle of water; By this meanes the Steele faleth by droppes, into y^{e} Payle of water; And being all Consumed gather it up & drie it, & Put to Powder, and Powre upon it very good &

395 Tile.

396 Bole Armoniac.

397 Retort.

398 Most likely a reverb (or reverberatory) furnace, which has a flame that was meant to isolate the material being processed from the fuel, but not from combustion gases (see Christophe Glaser (1667), *Traité de la chymie* (Paris), 44).

399 Their.

400 Common lute.

rectified Spirit of Sulfor, as much as neede will be to disolve it all; Let this be digested a litle while & at Last cast it into Stilled water So ye Salt of Steele gather it selfe in ye bottome of ye water; wch yo^u^ may take for yor use./

A Strength[^en]ing Electuary./

℞.[401] Pulpox interior. nucis indiex ʒ)iiii.[402] Pineol mund. in uino maluatico infusor pistach. recent. a. ʒ)ii.[403] rad. sayircon Condit. ʒ)iii.[404] Zinzibor viid. Condit.[405] [65v] nucis moschat. Condit.[406] a. ʒ)i Lumbor.[407] ~~Sim~~ cor=eineum[408] Cacertulor. a. ʒ)ß[409] uebetar cinamon inte=rior semin

[401] Abbreviation for "recipe", which is the second singular present imperative of *recipere* (Latin for "receive" or "take"). While the reading is relatively clear, the abbreviations, aberrant spelling, and unknown ingredients make the following recipe difficult to decipher. After consulting numerous paleographers and Latinists, we were unable to determine all of the meanings, but we have endeavored here to recreate as closely as possible the exact spellings, and in the footnotes suggest various interpretations. Note that a full stop (".") usually indicates an abbreviation, while "x" is occasionally used to abbreviate endings for noun declensions.

[402] "Four ounces interior pulp of Indian nuts", which is likely the coconut, but could be the areca nut.

[403] "Two ounces of pine-nuts cleaned in Madeira sack [and] poured into an equal amount of fresh pistachios". Here "pistachios" probably refers to turpentine, from *Pitacia Terebinthus*; also called Chian or Cyprian turpentine. Alternatively, it might refer to the pistachio nut from the tree *Pistacia vera*. "a." is an abbreviation for "ana", meaning "of each" or "of an equal amount".

[404] "Three ounces preserved satyrion roots".

[405] "Seven pennyworth preserved ginger".

[406] "Preserved nutmeg, each one ounce".

[407] "Lumbor" might refer to "lumborum", the genitive plural of "lumbus", meaning "the loins"; or, less likely, it might refer to "lumbrici", or earthworms, which were often used in medical recipes.

[408] The "sim" is blotted by ink, and could be crossed out. If the "sim" is crossed out, the "cor", combined with the letters on the next line, could render a strange spelling of "corium", which referred to leather, skin, or perhaps a peel: "corimeum". How to make sense of what ingredient is being referred to here, however, is not easy.

[409] "[...] half an ounce".

erueá. a. ʒiii.[410] cotie.[411] exter. citis siuiss ʒß.[412] terantur omnia diligentiss in mortar. in ar=mor.[413] & excip .S .4. ol. sanctrar & Syr. de cinamon[414] moscte. Ft. Electuarum instar.[415] Opeatx ad usum[416]

The use of this remedie is, First y^e Patient must keepe a very good diet, wch Consisteth first in the Choice of a drye & wholesome Ayre; or in y^e Corecting of y^e same by Parfumes, if it be other, Secondly in the Choice of Meat, & drinke, and therefore he may use of Mutton, veale, Capons,[417] Henns, Faisants,[418] Pa[^r]tridges, Pigeons, gicces,[419] cailles,[420] Larkes for y^e most Part Rosted, somtimes Larded of Sage, Rosemary, giriofle,[421] or Cinamont,[422] the young Connies,[423] & young hares are good too, And also Powder=ed, & Rosted-beefe; & new layde eggs; A bit of a good veneson Pye, or an neatstounge[424] is not a misse./

410 "[…] the plucked inner seeds of cinnamon […] each three drams". This is a dubious interpretation; the first word ("uebtar") is clear, but the meaning is not; the last word, "erueá", might be a form of "eruo-eruere", meaning "to dig up, to pluck". The cinnamon tree has a berry with tiny, black dicot seeds and has many of the same properties as the bark, which is more commonly used in food and medicine.

411 It might read "cotic".

412 "Take two and a half drams […] exterior and stir […]".

413 "Ground everything diligently by stirring in a mortar".

414 "Four drops oleum sanctum [dubious interpretation] and syrup of cinnamon", assuming that "S" means "stillae" for "liquid drops". "S" could also stand for "semicochleare", which means a half-spoonful.

415 "[and] let this be made into the form of an electuary", assuming that "mosct. Ft." is "miscreant fiat".

416 "Serve to/according to use", assuming that "Opeatx" is an abbreviation of "Operatum".

417 A castrated cock (*OED*, n. 1a).

418 Pheasants.

419 The reading is reasonably clear, but the reference is not. Given the context, however, it is likely "geese".

420 French for "quails".

421 Girofle (French for "clove").

422 Cinnamon.

423 Coney.

424 Neat's tongue.

[66r] Fishe is not good, but such as is Shelly,[425] or else they must be well fryed, & well seasoned; Let y^{r} drinke be of Beare moderate Stronge; wine is better; somewhat tempered a litle of Canarie or Muscadin with a test[426] will be very Conve=inent sometimes; halfe an houre before Dinner: Some thinke to Provoke [their] Stomacke by y^{e} use of tabacco; And Perhaps are not frustrated of their expectation; if they take it very soberly But it is Certaine y^{t} in this Case y^{e} Continuall use of the same is much Contrary. Thirdly in a moderat excercise and for wch belongeth the venetion[427] I esteeme [^to be] ye best to not Provoke it much; & to follow onely ye desires of nature, Forthly in the time of the Sleepe & wakeing, wch both should be moderate; yet if an excesse may be here Permitted that of Sleeping is farr lesse hurtfull then wakeing, wch I forbeare[428] expressly. Upon all things Mirth is necessary And let him Banish all other Passions but this./

Beside this generall Precept of Diet well observed if Nature do not satisfie to ye desire the Patient is two or three dayes before he will come to his end to take every morneing & eveneing goeing to bed, ye bignesse of a smale nut of this Electuary [66v] And to drinke upon it a litle glasse of Muskedine or Sack, The last day at .4. of y^{e} Clock in y^{e} afternoone, take a pritty quantity of Juce of Mutton, dresed wth new layd-eggs, wth a litle Nutmegg, & Amber. The Supper must be small./

At .9. of the Clock he shall take a double dose of the Electu=ary, with a litle of y^{e} said wine upon it, afterwards Let him expect wth Petience w^{t} it desireth; If not wth standing all this Nature is lasey,[429] & Slow wth other helps y^{t} verie Prentises[430] in this trade knoweth as well as y^{e} Master, he shall Receave Releefe & Comfort by anoynting the great toe of ye left foot with an oyle in wch Cantharides hath beene desolved./

[67r]

[425] I.e. shellfish.

[426] Toast.

[427] Venation. The action or occupation of hunting wild animals (*OED*, n. 1). The suggestion is that hunting is the best form of exercise.

[428] Forbear. To abstain or refrain from (some action or procedure); to cease, desist from (*OED*, v. 5). The implication is that the patient should abstain from staying awake for too long.

[429] Lazy.

[430] Prentice: that is, an apprentice.

Mr Phillips waters, with their severall virtues.

Beozer-Water[431]—Is an approved Antidote agaynst all Contagious diseases, as ye Plague, Purples, smale-Pox and the like, for preventing whereof two spoonefulls is to be taken with an indifferent draught of beare, to be dranke in y^{e} morning fasting; But to y^{e} infected there with three good spoonefulls, mixt with six of Cardus, or Angelica[432] water y^{t} is distild cold onely of y^{e} hearbe, And for want of those waters, in cleare posset drinke, wch by provo=keing a moderate sweate will Expell y^{e} mallignitie of the malladie the Patient being kept temporatelie[433] warme

It is likewise an experienced remedie in a violent surfet and when the stomacke is oppres'd with winde, cold flegme or any superfluity[;] it is Effectuallie digestive and in these cases to be drunke of it self without any other mixture

Spirit of Clary—Is an Excellent restorative in any kinde of weakenesse; But Cheiflie for that of the back, which it much [67v] Enableth, it recovereth those y^{t} are in (and Preserveth such as are incident to[434]) Consumtions, comforteth ye Heart, healeth the ulcerated Lung[^u]es[,][435] increaseth ye radicall humor, and is a great strengthner of women after their deliverances, to be taken morneing & evening; two good spoonefulls./

Spirit of Mints, Is an exceeding comfortable Cordiall, which rectifieth y^{e} infermities of y^{e} stomack, strengtheneth the retentive facultie[436] thereof, stayeth vomiting, Comforteth the spirits, preserveth y^{e} stomack from putrifaction, expel=leth winde and helpeth disgestive and is to be taken morneing & evening two spoonefulls./

431 Bezoar-water.

432 Culpeper held that "Angellica water the greater" "comforts the heart, cherisheth the vital Spirits, resisteth the pestilence, and al corrupt airs" (Culpeper, *Pharmacopoeia*, 63).

433 Temperately.

434 Incident. Apt to fall into; liable *to*; subject *to* (adj.1, I, 4).

435 Lungs.

436 The retentive faculty, which was one of Galen's natural faculties, was deemed central to nutrition insofar as it was thought to keep consumed food in its proper place during the process of assimilation (see e.g. Galen *De facultatibus naturalia*, III.1).

Treacle water, Is an Effectual remedie for y^e^ diseases of y^e^ spleene, and doth no less prevent then Expell y^e^ mallignitie of any manner of Contagion and when it is given as an Antedote, a spoonefull in y^e^ morneing suffi=ceth, but to y^e^ infected .3. good spoonefulls And y^e^ par=tie to sweat after it, And for y^e^ infermities of the Spleene .2. spoonefulls in y^e^ morneing and att evening when there is an ocasion for it./

[68r] Spirit of Saffron. Is a great Comforter of y^e^ Heart and helpeth y^e^ Passion[,][437] treambling & Pensivenesse thereof, it preventeth anie[438] Mallignitie that may oppresse it and suppresseth ye ill fumes arising from the spleene into ye Head and openeth the obstructions thereof and is an infalliable remedie against melancholy to be taken a spoonefull in y^e^ morneing and as much goeing to bed when there is ocasion, It is also good for women in Travell[439] in which case it hath bene frequently and Effectualy approved and in y^e^ Extremitie of the Paine to take .3. spoonefulls./

Spirit of Rosemary flowers. Is a Prevolent[440] preven=tion against infermitie of y^e^ stomack suppressing the mallignant vapours assending from thence into the head and is good agaynst y^e^ Appoplex, the virtigoes[441] & all diseases of that kinde, It doth also disgest y^e^ superfluous crudities of the stomack & expelleth wind^e^ being taken morneing & evening a good spoonefull./

[68v] Spirit of Roses./ For opening y^e^ Longues[442] and healeing y^e^ ulcerating of them hath beene approved[;] verie succesfull [at] Preventing the infermities thereof, and preserving them from Putrifactions, the breath from being corrup=ted[;] to be taken in y^e^ morneing fasting, at foure of the clock in the after noone and at night goeing to bed, a good spoonefull at a time/

[437] A fit or seizure; a faint (*OED*, n. 4b).

[438] Any.

[439] Travail. The effort and pain of childbirth; labour (*OED*, n. 1, 2).

[440] Prevalent. Having great power or force; effective, powerful; influential, potent (*OED*, adj. A.1).

[441] Vertigo.

[442] Lungs.

Adrian Gilberts[443] Cordiall water. Expelleth Poyson from y^{e} Heart and all infectious diseases, It helpeth the winde Collicke[444] & vomitting of meates, it's admirable for Consumtions[,] delivereth women & maketh them goe out their time[,][445] helpeth disgesteth[446] and restoreth na=ture decaied[,] purifieth y^{e} bloud and giveth a Lively ac=tion to y^{e} spirites, also it is Exellent in y^{e} time of the Measels & smale pox and for a surfet and for swoun=ings[,] for Rhumaticke Persons and cold ~~Persons~~ bodyes and may be given in a burning feaver and to Child new borne which sheweth tis good for all diseases, & hurt=full to none, The Doz[447] is a good spoonefull but in=case of Extremitie two spoonefulls may be given./

[69r]

Doctor Bispham[448] Receipt to make an excellent Balsum for any gre=ene wound, inward Bruse or hurt

Take of yellow wax halfe a Pound, Venice Turpentine one Pound, good Sallet-oyle three Pounds, red-sanders in Powder one ounce, naturall Balsum as much, or two ounces or as much as you shall finde sufficient for your intention, Spirit of wine halfe a pound, Mix alltogether according to Art, which is done after this manner, desolve your wax, Turpentine & oyle together upon a gentle fire[;] being desolved, Let them coole, continually stiring them with a spatle,[449] and before it be fully cold add to it your red-sanders made into an impalpable Powder, and with some of your spirit of wine make it into a thinn Pappe[450] or Past, which mixe in a morter with a spoonefull or two of y^{e} forementioned mixture of waxe, and by stiring them very well together, & being well incor=porated, put .2. or .3. spoonefulls more of y^{e} first mixture and in corporate them together, and y^{e} third time put 6. or .8. spoonefulls of y^{e} first mixture, and being

[443] Adrian Gilbert (1541–1628).

[444] Colic.

[445] The meaning here is obscure; perhaps the medicine helps women with their menstrual cycle, or with delivering babies.

[446] Digest. To divide and dispose, to distribute (*OED*, v. 1a). The meaning seems to be that the cordial helps to dispose of the decayed nature.

[447] Dose.

[448] Samuel Bispham (1612/13–1664).

[449] Spattle. A spatula (*OED*, n. 2, 1).

[450] Pap.

united [69v] as formerly you may mixe it very well with ye rest[;] wee doe this to open ye body of ye sanders, and because the thin Paste will not conveinently mixe with all the body of ye mixture at once; but rather runne into Litle Clottes, and be unproportionably mixed, Then mixe your Balsum with a Litle of that united body & after with more ye second & third time; Then add it to all ye rest, and mixe them well together by Continuall stiring, Lastly to your Spirit paper Litle of y^{e} mixture and stir it well therewith, & then put more & more of it till all this Balsum have sufficiently imbibed all the Spirit, Then Put it up in Gally-pott very close, and keepe it coole./ y^{e} doze for an inward bruse of this excellent Bal=sum, is a dramm, or a dramm & a halfe or .2. drames to be taken inwardly

An Exellent water of Doctor Bisphams, for old ulcers and Fistulas./

Take white vitrioll one ounce, Bole-armeneck .6. drag=mes, Camphor two dragmes, Powder them and Let them be well mixed wth one Pound of Plantin-water, & when you have ocasion to use it shake it well together, & therewth wash y^{e} sore, or siringe it./ a Pound, is an English Pinte

[70r]

For to make the Powder of Simpathy.

You must take Romaine[451] vitrioll and Lay it in ye sunne all ye Dogg-dayes, and then beat it to Powder, and apply it as followeth; Take of ye bloud, or Putrifaction[452] of ye wound upon a Clean Lining cloth, and mix as much Powder with it as you shall judge Proportionable; and keepe in a moderate warmeth, and binde up ye wound wth a cleane lining Cloth and open it not to ye Ayre under 24. houres and it will be Cured if ye wound be fresh But if it be an old Sort it must be Applyed twise or thrise; yo^{u} must not use womens lining[453] unlesse it have beene washed since worne or wearing; If there be severall wounds you must dresse ye bloud of each wound./

[70v]

451 Roman.

452 I.e. the scab.

453 Women's lining seems to have been used to wrap wounds. In this context, "lining" probably means underclothing (*OED*, n. 1, 1b).

Sr Thomas Cadamans way to make Pomander./

Take sweet Marjorum[454] & damaske Roses dry, of each 2. drammes, of yellow Amber, zyloaloes,[455] nutmegs, cloves, mace of each .3. dragmes, Bennjomin an ounce, storax 2. drames, Ambergreece, Muske, of each .20. graines, Cuiet[456] one graine, make all into balls according to Art, with gum-Arabic dissolved in Aquado an=gelic,[457] which whilst they are soft may be Pierced through/

The spannish way to make a Pound of Pastiels to eat, yo^{u} must beat to Powder a Pound of very white suger & passe it through a sive; and then add thereon a dram of Amber, & a dram of Muske, and y^{e} white of an egg, And if yo^{u} will you may Put .2. drames of each to a Pound of sugor/

[71r] The Spannish way to make a Pound of Pastoeles[458] to burne, Beat to fine Powder a Pound of Benjamin,[459] and Passe it through a sive, and then add to it .2. drames of suiet, & 2. drames of Amber And if you will yo^{u} may put .3. drames of each, or more./

[71v blank]

[72r]

A Letter from Mr Farer[460] to the Lord Marquesse of Newcastle Conserning ye health of Sr Thomas Glemham[461]

My Lord:

Sr Kellam Digby (my Noble and best freind) hath written to me soe effectually, Concerning Sr Thomas Glemham as he could not possible[462] in y^{t}

[454] Sweet marjoram was considered "an excellent remedy for cold diseases in the brain" that "helps such as are given to much sighing, easeth pains in the belly, provokes urine" (Culpeper, *Pharmacopoeia*, 13).

[455] Xylaloes (also "lign-aloes").

[456] Suet.

[457] Angel water. According to Culpeper, this "comforts the heart, cherisheth the vital Spirits, resisteth the pestilence, and al corrupt airs" (Culpeper, *Pharmacopoeia*, 63).

[458] Pastilles.

[459] Benjamin.

[460] Richard Farrar (active mid-seventeenth century).

[461] Thomas Glemham (1595–1649).

[462] Possibly.

kinde expresse his affection more to yor Lord.^{ip} then he did, And I were most blame worthey if I did not imploy to doe all y^{t} is in my powre at such a distance, And truly M^{r}. Endymion Porter[463] which I cannot forget hath by word of mouth dealt as earnestly with me as may be, in ye behalf of Sr Thomas Glemham, I have Conceaved (therefore) what I can, doe for y^{e} Pressent & what for y^{e} future, (y^{t} he survive)[.] This therefore (My most honored Lord) you may Please to know y^{t} it was not Possible for me upon .2. dayes warneing [72v] To make a mixture of such things as I must finde for a Perfect Cure, for they require a Longer time, not in Prepa=ring (for they are ready alwayes) but to apt[464] them for such a distance, & such a time (wch may be dun were it into Spayne) requires a Longer space; But I shal be ready against yor Lord.^{ips} answore to this (if yo^{u} Please to send any)[.] I take it for granted S^{r}. Tho: is [in] an ill Condition, much wasted with scurbruteary,[465] or hath beene (& must be) much Hy=droptick enough and if not mightely sweled yet not the better for y^{t}, I know his yeares & his Complexion though he knowes not me, yet therefore order shall be ta=ken to deliver [to] me y^{t} wch my medicines stand me in which I must send for .30. dayes (which will be y^{e} time for his Cure if God blesse it) wch will be twenty Poundes[.] I demaund noe more then y^{e} meanest give me for it, for it is a Chargable Medicin[;] I shall send as much of the extraordinary Medicine in a small glasse & a litle boxe as shall Perfect ye Cure, And what I send not, wch is ye decoction, I shall send ye way of it (wch is facile y^{t} any woman may make), And I shall set downe ye Preparations, [73r] of those cheife Ingredients I send him[;] they must be ordored and dosed, & all things Requiset[466] to ye Cure as playne as if I were there my selfe, And then yor Lord.^{ip} knowes what I can doe this way, And if God blesse ye Cure I will stand to his Curtesse[467] for ye good reward when God shall enable him, and y^{t} he Possesse his owne in quietnesse I know for ye Present a litle is a great matter, And were I able I would give to such a Person what my owne Condition admits not, and myne (my Lord) as I cann doe no more for any great way of Cure for ye Present soe I am not soe unmindefull but y^{t} I will doe all, in the

463 Endymion Porter (1587–1649).

464 Apt. To make fit, adapt (to), prepare suitably (for) (*OED*, v. 1a).

465 Scorbute.

466 Requisite.

467 Curtsy. To make a curtsy; to do reverence to (*OED*, v. 1a).

intrem[468] y^{t} I can to advise ye best y^{t} I can to stoppe the Prog[^r]esse of ye disease; I have therefore at the end of this Letter written a drinke which he may drinke at his Pleasure, And it is a good tast, Let not this Cleare drinke be despised, for although it is not Possible for it to operate to a Cure, yet it is so Proper for him, for it works altogether by urin [73v] That untill I can send ye other things (wch is as I heare from yor Lord.^{ip} for they shall be ready) there is nothing so Proper for him, not onely to prepare, but abate ye groweth of ye disease till ye other come, And in this I have done as much as lyes in my Art & Powre./

As soone as this arrives to yor Lord.^{ips} handes, Let him forbeare all manner of Purges whatere un-lesse Clysters (if his body be bound)[;] those he must use once in .48. howers for his Liver is weake, And all Purges make it weaker, Nor is there any true Cure of a Dropsey but by urin, wch is y^{e} way my great Medicine doth [^also] worke and nothing else, he must eat but one meale a day, and forbeare all drinke but this I send him[;] y^{e} making I here inclo=sed: he must eat nothing but rosted meat, noe fruite, Milke, Cheese, nor such thinges, and no [74r] Suppers at all unlesse a dryed biskett, these he must begin as soone as he hath this advise And now my good Lord I humbly Crave yor Pardon for my Prolixity & scribling[;] more I cannot wright[469] & lesse I ought not, and soe desiring al=mighty God to give a blesing for y^{e} recovery of that noble Knight, I will onely take y^{e} boldnesse to kisse yor Lord.^{ips} hands and shall ever acknowledge my self to be

My Lord

Yo.r Honors most humble servant

July 24th 1649

Richard Farrar./

[74v]

M^{r}. Farrers Receipt for Sr Thomas Glemham./

Take horehound,[470] hartes-tongue, Liver-woort and wormwood, of each one handfull (a good hand=full)[;] shred them well with a knife and put th=em into a Part of Barley meale and Let it be well Closed as can be,

[468] Interim.

[469] Write.

[470] Horehound was also used as a purgative (Culpeper, *Pharmacopoeia*, 107).

And bake it as yo^u^ would doe another Py[471] very well and then have in a readynesse foure English gallons of good Ale, (If none to be had, of good beare not to stronge) and when it is in ye Earthen vesell or stand taken hott from keele, presently breake in .3. or .4. peaces of the Pastey wherein yor hearbes are and cover ye stand and when it hath dun working (for y^t^ will make it worke againe) broach[472] it & let him drinke of it as he is dry, but drinke nothing else, till I send my other Medicin, (y^t^ it be required) and y^t^ [75r] he goe not to stoole in .48. howres take som slight Clyster onely to keepe him open body'd: This all for y^e^ Present which Let be carefully observed[;] I Pray and so God blesse him & it, & it to him Amen./

Doctor Farrers Decoction for y^e^ Dropsie./

Take a neck of Mutton (cutting of all y^e^ fatt & pare=ing it away it away [*sic*] very Cleane)[,] chop it very well in peaces, put it into a cleane new Pipkin & Powre upon it a Gallon (eight English pintes) of springe water set on y^e^ fire and when it is well scumed put in a litle salt to season it, then put into it of Parsley-rootes & Fennillrootes picked & slised, of bot[h] a smale handfull, and throwe in of Cloves and candy Mace of each a quarter of an ounce, then put [75v] halfe a handfull of Parsley well washed & pickt[,] 6. or .8. sprigs of time,[473] one sprige of Rosemary, and a litle sweet Marjorum, and Penny-Royall, a small handfull[;] then take .8. ounces of y^e^ best Currans,[474] well washed & pickt & stampe them in a Cleane Mortor to past & put them allso into y^e^ Pott with y^e^ hearbes And then all these being put in close or Cover y^e^ pott and let it boyle till it be halfe Consumed y^t^ is from 8. pintes to .4. (all this over a Charcoale fire Conti=nually watched & tended). Yo^u^ are to observe y^t^ when yo^u^ thinke by y^e^ opening of y^e^ Lidd y^t^ two pints are Consumed or wasted away, I say then yo^u^ are to put in as many cloves of Garlick as growes on an ordi=nary head of Garlick, each clove y^e^ skin being taken off and to put them into y^e^ Pott & Cover it close keeping it gently boyling till y^e^ other two pintes be Consumed, so y^t^ there be left but y^e^ one halfe (wch is .4. Pints) then take it off y^e^ fire & let it runn th=rough a strainer, not

471 Pie.

472 To pierce (a cask, etc.) to draw out the liquor; to tap (*OED*, v. 1, 4a).

473 Thyme.

474 Currants.

presing of it, And so poure [76r] It into some Cleane glazed Earthen Pott or vessell and set it in a cold place and when it is cold take of all y^{e} fatt y^{t} is on ye top of it be sure, And it must be kept alwayes in a cold Place not to close covered least it sowor,[475] This proportion will Last three dayes or there a bouts both for ye goodnesse in not sowring and allso for ye Proportion to drinke for a litle above a pinte a day, or a pinte & a halfe will be enough, ye wch he must take in time & proportion as I shall set downe by & by.

In ye morning at .6. of ye Clock before yo^{u} rise out of yor bed cause halfe an English wine Pinte of the decoction to be made warme in a Porranger[476] (and all ye fatt taken of)[;] make it as warme as yo^{u} can well drinke it, and upon ye point of a kin[477] so take of ye red Extract (wch I send yo^{u} in y^{e} box) the quantity of a nutmeg sixty graines by weight & no more and desolve it in y^{e} warme decoction w^{ch} [76v] you are to drinke presently, stiring it well with a spoone & it will desolve in y^{e} decoction, & instantly as hot as yo^{u} well can drinke it all at a draught Leave=ing none, and ly downe in yor bed, and sleepe if yo^{u} can after it, Eat nothing till noone, and then what yo^{u} Eat must be fresh, & Rosted, no fish nor salt meates[,] no butter, Cheese, fruits nor any such thing, but Rosted Mutton, Lambe, Veale, Chicken, Pullet,[478] and it matter not how Litle yo^{u} Eat, The Decoction wth what yo^{u} put into it of y^{e} Extract (though it be lit=le in quantity) is of a heigh nourishment[479] & virtue[;] If yo^{u} will yo^{u} may take y^{e} quantity of a small nutmeg of y^{e} Extract on a knifes point, & swallow it befor yo^{u} drinke y^{e} decoction & instantly drinke the decoction hott upon it; At foure a Clock of the after noone drinke another halfe pinte of the Decoction warme, And first as before written on a knives point swallow downe as much of the [77r] Red Extract as a small nutmeg (And so yo^{u} put none into ye decoction, for if yo^{u} doe ye one, you must not do ye other) And drinke ye halfe pinte of the Decoction instantly. And that for ye first .8. dayes I would have yo^{u} but drinke a

[475] Sour.

[476] Porringer. A small bowl or basin, typically with a handle, used for soup, stews, or similar dishes (*OED*, n. 1).

[477] Unclear meaning, though the reading is clear. It perhaps refers to the top of a kiln; less likely, it might refer to a bodkin, which is a long, sharp implement.

[478] A young domestic hen (*OED*, n. 1).

[479] High nourishment (i.e. very nourishing).

Pinte in a day at the two severall times afore set downe, And for yor dinner, eat more or lesse as yo^{u} can (it matters not) of such Meates as are described, and for ye first eight dayes yo^{u} may drinke a halfe wine pinte of such beare as yo^{u} like, or [^if you] have no good beare, as much good Rhenish wine with a litle suger in it, but it were better if possible to forbeare drinking at all which will be hard at first but Per dura & aperer sic itur ad Astra[480] But I assure yo^{u} after eight dayes you must drinke nothing but ye decoction, and ye former quantity of ye extract every time, wth or before ye decoction which I have thought good now to set downe in waight[481] justly wherein yo^{u} cannot erre, rather [77v] Then by guesse at ye quantity of a small nutmeg (as I writ before) And therefore yo^{u} are to waigh out of the Extract y^{e} waight of one dragme (wch is .60. graines) & swallow it and then drinke yor proportion of decoction So for y^{e} first .8. dayes two ounces of y^{e} Extraction will be enough, which is two dragmes every day, but after eight dayes yo^{u} must drinke .3. times a day of the De=coction, and take at each time one dragme of the Extraction, so it will be .3. dragmes of the decoction per diem, And .3. halfe pintes of yor decoction, at .3. severall times. vis[482]: in y^{e} morning betimes, at noone for yor dinner, And at night when yo^{u} goe to bed, But this Course yo^{u} begin not till .8. dayes be Compleat and Past with twice takeing each day as is set downe above./ Now haveing given this playne directi=on for yor ordering yor selfe wth yor decoction and Extract, and dyet: I must say somthing how you must order your self, If yo^{u} faile to goe to stoole once in .24. howres (wch I am sure yo^{u} will) when [78r] You Come to fall[483] close (after y^{e} first .8. dayes) to eat, nor drinke[484] any thing but yor decoction and Extract .3. times a day, I say when yo^{u} finde yor selfe bound y^{t} yo^{u} goe not to stoole in 24. howres then yo^{u} must take this simple Clystor./

480 "Sic itur ad astra" is a famous phrase from Virgil's *Aeneid,* Book IX, line 641, which translates as "thus one journeys to the stars". Here it means something like "with difficulty and suffering, immortality (or perhaps health) can be achieved".

481 I.e. provide the relevant weights.

482 Viz. (short for "videlicet").

483 Perhaps "feel".

484 The phraseology and grammar are odd, but the meaning seems relatively clear: when you are close to going to stool, do not eat anything, but rather take the decoction.

Let a Pinte of Milke be set over ye fire, and made but bloud warme, & wth a spoone disolve two ounces of Course[485] Clystor sugar in it and then strayne it. And add to it two ounces of Syrope of violets, and so made warme; about six a Clock at night let this be administered to yo^{u}, it will Coole & Refresh yo^{u}, And this must be dun (if yo^{u} goe not to stoole in .24. howres) continu=ally, yet one thing more (for I would omit nothing) yo^{u} must keepe yor selfe very warme, if yo^{u} take Cold (wch this Decoction & Extract will make yo^{u} apt to be) it will doe yo^{u} no good, but harme, for yor partes will be opened, And then Cold is as bad as drinke [78v] Both which are Mortall enemyes to yo^{u}[;] I Pray yo^{u} therefore have a speciall care of this, as of what also I write./

An other Letter from Mr. Farror to the Lord Marquesse of Newcastle.

Right Honble./

I have receaved yor Lord.^{ips} minde in the Papers yo^{u} wear pleased to send to me by this post and have written & sent by this Post, full direc=tions for S^{r}. Tho: Glemham soe playne as there can be no error if not since fully omitted by him selfe in observing, or those y^{t} make it in y^{e} Prepareing which every Cooke mayde,[486] much more any more skil-full cannot fayle in following y^{e} directions for I shall take y^{e} boldnesse to refer yor Honour to [79r] My Letter to S^{r}. Tho: for what Concernes ye estate of y^{r} Ladys health, My Lord I have well Pondred what yor Lord.^{ips} doth wright Concerning her, that shee is highly affected wth ye Spleene is aparant[487] the which must needs produce Melancholy in aboundance for ye want of her Courses,[488] y^{t} adds much to her affliction, none any good Lord, soe much I am realy a servant to your Lord.^{ips} (wth out least end or flattery I write it) that I shall set you downe two short directions, ye one for her Spleene wch is an Ellectuary, the other for an orderly (& not over hasty) bringing of them downe wherein I assure yor Lord.^{ip} although yo^{u} finde them playne in ye reading, and ye latter not so pleasing to fayre Laydes as being of Excrements of a Ramm, wch Shee needs not know, and then shee will not be troubled in ye Least at it, yet they are both so approved, as y^{t} it is not fit to be Communicated, and I Pray God blesse

[485] Coarse.

[486] Cookmaid.

[487] Apparent.

[488] The menstrual discharge, catamenia, menses (*OED*, "course", n. IV, 28).

them to her Lady.^{ip} y^{t} shee may finde a good effect[.] [79v] for what Concernes yor Lord.^{ips} singing in yor head, or kissing: I must Proposse my Lord y^{t} if it came from the vapours of y^{e} Spleene (as all y^{e} Learned Doc=tors have told yo^{u}) I know no way but by Cureing of yor Spleene, a hard taske I Confesse, But I assure your Lord.^{ip} y^{e} symple Electuary wch yo^{u} shall finde here inclosed of oyle, Capers & Currans, wth yor constant use, and a good dyet for .6. weekes to=gether wrought strange effects: Now if my Lady or yor Lord.^{ip} have not oyle, then I send yo^{u} it in vaine, The Ellectuary is thus made./

Take .8. ounces of y^{e} best Capers, y^{e} stalks pict of put them into a glasse[489] and cover them with good white wine y^{t} y^{e} wine stand a fingers breadth above them, soe let it stand .24. howres, then poure it Cleane of and put as much fresh white wine and let so [an]other .24. howres, this doe .3. or .4. times till all ye saltnesse be gotten from ye Capers, then let all ye wine draine out cleane from ye Capers that the Capers be dry, Then in a Pipkin stue[490] or [80r] Plumpe[491] .4. ounces of ye best Currans y^{t} are .17^{d}.[492] which must be first well picked very Carefully and washed exceedingly of y^{t} great care must be had and then plump'd in white wine (but not sodden to breake) And of these Currans thus Carefully pick'd[,] wash'd & plump'd waigh out .4. ounces & .8. ounces of ye prepared Capers and in a Morter (ye Currans and Capers very well dry'd first from ye wine if y^{t} be new) beat them together wth as much of ye best sallad-oyle, or oyle of sweet Almonds (wch is best liked) as will beat them into ye Consistance of an Ellectuary not to soft; of this take ye quantity of a walnut fasting, in ye morning betimes, & fast 3. howres after it, as much as .4. after noone (upon an empty stomack) and eat nothing .3. howres after this[;] Continued orderly with a good dyet forbeareing wine & saltmeals & fishe and Cheese & butter, will in .40 dayes strangely work even to admiration [80v] in some Persons it hath dun.

finis./

489 Long space in original.

490 Stew.

491 Plump. To make plump; to cause to swell (especially in early use, with water) (*OED*, v. 3, 1a). The method involved covering fruit in boiling water.

492 I.e. 17 pennies' worth.

For to bring downe y^e^ devices.[493]/

Take y^e^ dunge of a Ramm y^e^ quntity[494] of a handfull[;] let y^e^ Ramm be well hunted upon[495] & downe & it will come from him, And it must be taken fresh, and put y^e^ hand full of it into a fine Cloth wth a peny=worth[496] of saffran not ty'd hard,[497] And put it in to a Pinte of white wine & Let stand in it .24. howres then strayne it out of y^e^ Cloth into y^e^ wine, and divide it into .3. partes, takeing one third part every Morneing, for .3. dayes one after y^e^ other, & shee must stirr[498] after it, or excersize, this to be taken. 4. dayes before she exspects them, this hath rarely fayled to doe y^e^ offise[499] at once, but if not it must be taken agayne y^e^ second Month, My Lady needs not know what it is, though it is not so naucious[500] as it may be Conceaved./

[81r] And thus my Noble Lord I have endeavoured my best to give your Lord.^ip^ satisfaction; And I Pray God ble=sse it to yo^u^ & yours, and send yo^u^ all happynesse; And to Pray as hee; who is

My Lord

Yo^r^ Honours most [^humble]

Ri: Farrar./

The .10.th of August

493 The meaning of "Devices" here is obscure, but note that Farrar refers to taking the medicine "before she exspects them" and that, if it does not work in the first month, it "must be taken agayne ye second Month". Moreover, in the prior letter, Farrar refers to the "want of her courses", stating that he would provide instructions for "an orderly (& not over hasty) bringing of them downe". Given Farrar's advice and the context of this recipe, "Devices" may refer to menstrual pains and discomfort.

494 Quantity.

495 In this context, "well hunted upon" means "tired out" (see Charles Etienne (1616), *Maison Rustique, or, The Countrey Farme*, trans. Gervase Markham (London), 459).

496 Pennyworth. As much as can be bought or sold for a penny (*OED*, n. 1a).

497 The idea here is that the dung and saffron are put in a loosely tied cloth and steeped in the white wine.

498 Stir. Movement, considered in contrast to or as an interruption of rest or stillness (*OED*, n. 1, 1).

499 Office. That which is done, or is intended or expected to be done, by a particular thing (*OED*, n. 4a).

500 Nauseous.

1649

[81v]

Doctor Farrers Receipt for the Sterillitie in Women./

Take pure springe water & put therein as much of the hearbe Balme as ye water will Cover, And let them disgest over a gentle fire till ye Balme be full so it is papp, then strayn it very hard: take of this decoction a quarter of a Pinte, half a Spoonefull of naturall Balsum, caled Balsum of Peru, or Hyspanicum,[501] and warme them together still beating ye Balsum as it will swime in great dropps at ye toppe, therewith let ye Party make Injection into ye Matrix wth a Serenge[502] warme[;] doe this Morne & Night severall dayes./

Let her anoynt allso ye Navill with ye Balsum alone (The Navill is Centrum hominis[503])[;] shee may allso take a litle of it in a spoone, & so inwardly it hath a wonderfull effect.

Doctor Farrers Fortification a great secret./

Burne in a cleane Chimney, Oaken wood & no other wood with it, the fire is not plesant, but ye Ashes are rare and when you have a peck[504] of ye Ashes, or halfe a peck [82r] But yo^{u} must be sure no other wood be burnt dureing the time, nor no other Ashes mixed; then take those Ashes, wch must be burned white, not dead black Coales amongst them, I say makes a Ly, or Lixium out of them with distilled Rayne water, once distilled as women doe when they make Ly to buck[505] their Lining, wch is by Leting ye warme Rayne water runn through ye Ashes divers times

[501] Various plants that are common in herbals contain the term "hispanicum" (i.e. Spanish). In his *Herball*, Gerard listed, for instance, "Pseudospartum Hispanicum" (Spartium junceum, Spanish broom) (1318); "Origanum Hispanicum" (Oregano vulgare, Spanish thyme) (667); and "Dorychinium Hispanicum" (Lotus dorycnium, prostrate Canary clover) (1308).

[502] Syringe.

[503] Latin for the "centre of man". From ancient Greece through to the Middle Ages and Renaissance, much was written about the "primary" areas of the body, and their importance in understanding physiology. The point here is that rubbing ointment at the centre of the body will make it work faster and more effectively.

[504] Literally, a quarter of a bushel, or, figuratively, a considerable quantity or number; a great deal, a heap, a lot (*OED*, n. 1, 2a and 4).

[505] To steep or boil in an alkaline lye as an initial process in buck-washing, or bleaching (*OED*, v. 1, 1).

or severall waters, poured on ye Ashes, for the same water must not be poured above 2.[506] on the Ashes, And when you have made a Ly, then let ye water setle after it hath runn through ye Ashes, and Filter it Cleane through a Filter, then evaporate it in a fitt vessell of glasse till it hath a scumm at ye top then take it off, set it in a cold Celler, and then [there] will shoot from it a white salt like Dyamonds, 4. sqare[507][;] take onely of yt salt wch will seperate from ye brownish salt and use it as followeth/

Powder & mixe with it a Litle powder of nutmeggs a sixth part, And take ye quantity of yt Powder, as much as will Ly on a six pence .6. morne-ings together in a Litle white wine and fast .2. houres after, and [82v] Expect ye effect, use it more or lesse as you se[e] cause[;] [there is] no Chymick or Apothecary, but can make this if you give him ye Ashes, he needs not know what the Ashes are or to what use, but he can make or seperate ye right salt by my former expressed Method/ finis

Doctor Farrers Receipt for the Spleene in Man or Woman./

Take .8. ounces of ye best Capers, pick of ye stalkes cleane[;] put them into a broad glasse, or cleane glas=ed bason & cover them over yt ye wine be above the Capers a fingers breadth, let it so .24. houres, then part ye wine Cleare of, then put in as much more fresh wine, Let it stand soe 24. howres more, then powre yt of; agayne put on fresh & Let it stand 24. howres, this in 4. times doeing will drawe out all ye saltnesse from ye Capers, then let all ye wine drayne cleane out yt the Capers be dry, then take 4. ounces of good Currants plumped in white wine warme, but ye Currants must be well washed and [83r] Picked first & beate ye .4 ounces of Currants so pre=pared with ye .8. ounces of Capers soe Prepared and in a cleane Morter, of Marble with ye best salad=oyle you can gett, or else with oyle of sweete Almondes newly drawne beat it up into ye Consistance of an El=ectuary; of this take fasting ye quantity of a walnut and fast .2. howres after it and in ye afternoone as much as allso at night when you goe to bed. This use for 40. dayes, I doubt whether there be a better remedye for ye spleene, God blesse it./ Amen/.

[506] I.e. more than twice.

[507] Four-square. Having four equal sides; square (*OED*, adj. a). In this case, the reference is to the cut of the diamond (for a similar usage, see Elias Ashmole (1672), *The Institution Laws and Ceremonies of the Most Noble Order of the Garter* (London), 204).

In the time of takeing, forbeare salt Meats and wine, frute, Chese,[508] & Milke./

Sr Thomas Glemhams way to Cure an Ague and Cough./

Take a peace of a fine white leather, cut round a little broder than a Mans hand, Let it [be] pricked full of holes wth y^e^ point of a knife, then take of the best & Clearest franckeunsence[509] finely Powdred and wth a very hott knife spread it all over the [83v] Leather very well then take of y^e^ best venice turpentine & spread y^t^ very well upon y^e^ francun=cence reasonable[510] thick; All this being done let the Party apply it [to] y^e^ Navell some 2 howres be=fore y^e^ fitt comes and neyther eat nor drinke .3. howres before & after, binding fast on wth a nap-kin, this plaster is not to be taken off untill it faleth of it selfe, any body but a women with Child may take this.

If y^e^ first does it not, yo^u^ may Apply a second./

For the Cough./

Take Syrope of fugitus[511] and Syrope of Diacodium of each a like quantity[;] mix thes well together and Let ye Party take of it with a Liquoris stick well brused[512] at ye end soe often as neede shall require

Mr Rusts[513] way to make Almond butter[514]/

[508] Cheese.

[509] Frankincense.

[510] I.e. reasonably.

[511] The reference is unclear. "Fugitus" is Latin for "to flee" or "fly away", and it likely refers to a volatile syrup such as the "syrupus volatilis" that George Bate mentions ((1694), *Pharmacopœia Bateana: or, Bate's Dispensatory*, Vol. 1 (London), 784). There is also a long tradition, starting with Paracelsus, of using volatile salts in medicine, and these were sometimes called "fugitivus" and discussed in English as "fugitive" salts (see, for instance, Robert Boyle (1661), *The Sceptical Chymist* (London), 72–73 and 419).

[512] Bruise. To grind or break down (especially foodstuff) completely or partially with a pestle, millstone, or another implement; to pound, to crush (*OED*, v. 3).

[513] Calixt Rust (d. 1658/1659).

[514] According to the English physician Thomas Cogan, almond milk was good for hot diseases, while almond butter, which "cunning cookes" and the "learned Physitian" alike should know how to make, was "good for a stuffed breast" ((1636), *The Haven of Health* (London), 112). Agreeing with this evaluation, Margaret opined in a section "Of Coughs" in *The World's Olio* that "I should think Almond-milk should be very good; for

Take blancht[515] Almonds wth Cold water halfe a Pound[;] new butter washed very well in Rose or orange & flower water .3. quarters of a Pound or a Pound, fine rose suger, or wch is better ye white suger of Roses .4. ounces[;] [84r] Muske, & Amber greece more or lesse as yo^{u} desire the tast./

First beat ye Almonds in a Marble Morter till it comes to a very fine Past, but in beating the Almonds put a spoone full or 2. of Rose water to them to keepe them from oyling/

Then beat well ye suger wth ye Almonds untill all be very fine then add to it ye butter & beat it all together for a short time, & Passe ye butter through a spring & it may Ly in wormes

Mr Rusts way to make Barley Creame

Take of ye best pearle Barley halfe a Pound[,] boyle it halfe an howre in faire water, then ease away ye water & save ye barley, wch set to boyle againe in .6. Pintes of water, & let boyle till .4. onely remayne or until ye barley be very tender then strayne it through a clean napkin & first in a Marble Morter beat very well halfe an ounce of white Poppy seede with a spoonefull of Rose water to keepe it from oyling, then beat with [84v] them of y^{e} foure Cold seeds,[516] of each a quarter of an ounce wth a spoonefull of Rose water, Al=mondes blancht .3 ounces, halfe y^{e} boyled Barley or more at Pleasure, beat all these in order to [make] a soft Past casting in sometimes a few dropps of Rosewater & Lastly add of yor decoction Lately strayn'd out .3. Pintes, stur it well together & strayne it for=th very hard, And add to it of Manus Cristie as much as will suffice to make plesant./

This if it be well beaten will be in y^{e} forme and Colour of Creame./

The Ladye Brackleyes[517] way to make Almond Butter./

the French barley, that is boyled in the Water, is both cooling and clensing, and quenches out the Fiery Heat; and Almonds are healing and smothering" (193).

[515] Blanch. To whiten almonds, or the like, by taking off the skin; as is usually done by throwing them into boiling water (*OED*, v. 1, 2a).

[516] According to Culpeper, "The four greater cold Seeds" are "Gourds, Cittuls, Cucumber, Melons" while "The four lesser cold Seeds" are "Endive, Succory, Lettice, Purslain" (Culpeper, *Pharmacopoeia*, 79). Which of the two is being referred to here is unclear.

[517] Elizabeth Egerton [*née* Cavendish] (1626–1663).

Take a Pinte of sweet Creame & ye yolkes of .6. eggs new Lay'd, mingle ye Creame & egg well together then take halfe a handfull of Jordan Almonds[518] which have beene blancht in cold water, beat them in [85r] Morter & now & then put in a litle Rose water to keepe them from oyling, when they are finely beaten mingle them wth halfe a pinte of new Milke, and so put them to your Creame & eggs and mixe it well together, so strayne it into a Fayre skellet,[519] & set it over ye fire stiring it Continually till it come to a Curd, then put it into a canves bagg and hange it up y^{t} all ye whey may draine out when it is cold & hard, breake it wth a spoone or Ladle and so season it with suger & Rose water as yo^{u} Please./

Mr Rust's way to make Almond Butter./

Take of ye best new butter halfe a Pound, wash it well in Damaske rose water, then take of Almondes blancht a quarter of a Pound, Beat ye Almonds in a Stone Morter, & in beating of them put in by degrees .3. or .4. spoonefulls of ye best orange flowre water or Rose-water or if these be not liked yo^{u} [85v] May use .4. or .5. ~~dropps~~ spoonefulls of fresh Creame, the waters, or Creame is added to keepe y^{e} Almonds from running to oyle wch otherwise they would doe in beating: yor Almonds well beaten add .2. or .3. ounces of fine lofe suger,[520] beat them well together and Lastly add y^{e} Butter, And when they be all well mix=ed in yor Morter. It is done./

Yo^{r} Butter must be beaten in 2. or 3. severall waters I mean Rose-water, & Let it ly in y^{e} last water .4. howres./

Mr Rust's way to make white suger Candy./

Take what quantity of white powder suger you please wch Clarefye y^{e} ordinary way $w^{t}h$ y^{e} white of an egge, & haveing Past it through a gelly bagg boyle it up a litle above y^{e} haigh[521] of a syrop, or un=till yo^{u} observe

[518] A fine variety of almond, now coming chiefly from Malaga; also, simply Jordan (*OED*, n.).

[519] Fair skillet.

[520] Loaf Sugar. Sugar refined and moulded into a loaf or conical mass (*OED*, n.).

[521] Height. There are several such phrases in this manuscript, which do not refer to a strict measurement, but rather to the quality of a liquid: that is, to its viscousness (*OED*, n. I, 8). Philemon Holland, for instance, wrote in his 1601 *The Historie of the World, Commonly called, The Naturall Historie of C. Plinium Secundus* that a decoction must be "Suffered to seeth gently and leisurely to the height or consistence of honey" (Vol. 2 (London), 276).

litle skinns upon ye toppe of ye sugor then Powre it into a Cleane earthen bason, and [86r] Put litle sticks into ye Bason soe y^{t} they may rest by both ends upon ye sides of ye Bason, & soe y^{t} they may be just Cover'd with ye sugre; this done, tie a double paper over ye bason whilst ye suger is hott and set it by; And in a fewe dayes yo^{u} will finde ye suger in Part Cristalized about ye sticks & endes of the bason, Poure off ye suger y^{t} is not Cristalized and dry gently in a stove ye rest, And ye other suger y^{t} was powred of boyle it agayne to ye same haight as formerly & set to Cristalize as yo^{u} did ye other and thus doe untill all ye suger be turned into Cristall or Candy./

After y^{e} same manner in all things is made the browne suger Candy, wth Muscuvadoe suger./

The Earle of Chesterfields[522] Red Diaphorettick Powder./

Take of fine Rose three ounces, terra-Lemna[523] two ounces both these ground upon a stone wth Rose water and wine=vinager, & after dryed and Powdred; Redd=[86v] Corall halfe an ounce, white Corall .6. dragmes[;] of Emroides,[524] Rubes,[525] Saphores[526] & Jacinthes of each halfe an ounce; Pearle prepared, five dragmes, Cina=man two dragmes; Cloves, a drame, sorrell=seeds two dragmes, Cittron-pills,[527] white sanders of each a dragme and a halfe; Redd-sanders two dragmes, burnt Ey=vory[528] five dragmes, Risped Eyvorey,[529] three dragmes[;] Saffran a dragme, Unicornes horne, or if y^{t} cannot be hadd, y^{e} bones of a Harts heart halfe an ounce[;] of y^{e} Leaves of gold twenty./

Make all these into a very fine powder, & beat them up into a Past, with y^{e} gelly of Harts horne and make them into balls which dry in ye shade and keepe them for use./

[522] Philip Stanhope (1583/4–1656).

[523] Terra Lemnia was supposed to stop the flow of blood (Culpeper, *Pharmacopoeia*, 149).

[524] Emeralds.

[525] Rubies.

[526] Sapphires.

[527] Citron peels.

[528] Ivory.

[529] Rasped ivory: that is, ivory that has been scoured with a sharp implement.

The dose is from .20. to .30. graynes in sack or Cor=dialle water wth a drop of Clove gilly flowres/

To Children it is given from .10. to .12. graynes[;] as aforded[530] it is an excellent Remedy agaynst the Plague or any Pestilentiall feavor, Allso agaynst defallance[531] of spirits & is an excellent Cordiall [87r] At all times. This powder I have seene used upon his Lord^ps^ daughter[532] wth good success in fainting fitts./

The Countes of Kents[533] Powder.

Take of Pearles[,] of Crabs eyes, of w^ht^[534] Amber[,] Hartes horne, w^ht^ Corall, Contrayerva of each halfe an ounce, of ye black tipps of Crabs Clawes three ounce, the best East Bezoar six dragmes[;] make all these into very fine Powder, And wth ye Jelly of Hartts-horne make them into Past and so into Balls. This is a most excellent Cordiall.

The manner how to make Gascons Powder which Mr Rust had from ye Countesse of Arendell[535]

Take of ye black-tipps of Crabbs-clawes gathered in June, whit Corall & Pearle of each a dragme & halfe[,] Crabbs Eyes & white Amber of each a scruple; make all these into very fine Powders and beat them up into a Past wth gelly made of vipers skinns, or snakes skinns./

[87v] Taken in March or if these cannot be hadd with y^e^ Gelly of Harts-horne, And soe made into Balls as formerly, wch drye & keepe for use./

This Cordiall Powder with y^e^ last proscri=bed are given from .10 graynes to .30. in Cordiall water or burnt Clarret[;] agaynst y^e^ Plague[,] small Pox[,] spotted feaver & surfetts./

Contra-yerva.

530 Afforded (i.e. as described in the preceding recipe).

531 Defailance. Failing; failure (*OED*, n.).

532 The reference is presumably to one of Stanhope's daughters, Sarah or Elizabeth.

533 Elizabeth Talbot Grey (1582–1651).

534 White.

535 Aletheia Talbot Howard (c. 1585–1654). On Gascoigne's powder, see John K. Crellin (2019), Gascoigne's Powder: A British Prescription and Home Medicine, 1600s to early 1900s. *Pharmaceutical Historian* 49.1: 1–15.

Take of y^{e} Rootes called Contra-yerva, or snake=Rootes make them into fine Powder, & wth the Gelly of snakes skin & harts-horne make them into balls wch drie & keepe for use./ The dose is from .10. graynes to .30. in Cordiall waters or burnt Clarrett./

[88r]

The true Receipts of M^{r}. Rusts to make Mead, Metheglin and Braccet.[536]

Mead is made with ye Hoeney & Combes together broken into water, & in such a quantity that an Egge may swimm upon it,[537] then to [be] boyled very well & ye skin taken off as it riseth, then take it from ye fire & let it stand 24. howres that what Impuritie remaynes may have time to fall to the bottome: It wil keepe well .2. or. 3. yeares very well/

And for ye Meade y^{t} yo^{u} keepe longe yo^{u} are to put about three spoone fulls of yest[538] to it about a weeke before yo^{u} intend to drinke it & it will work & make it drinke very briskly.[539]/

But if yo^{u} be to drinke it Presently[540] skim it onely & then being allmost Cold put to some yest & let it worke in ye vessell as beare & then stoppe it wth Clay & after .4. or 5. dayes it will be fitt to drinke./

Braccet is made as followeth./

With stronge Ale wort & Honey in such a quantity y^{t} an egge may swim upon it, & boyled & skimed as the [88v] Former; And hange a bagg of such spices in it whilst it boyles as yo^{u} like best; Then add to a bar=rell a Pint of yest or thereabouts. And let it worke in y^{e} vessell, & when it hath done workeing stoppe it up & keepe it for use./

Matheglen[541] is made as followeth./

With water & Honey in such quantity that an egge may swim upon it, and then set it to boyle with all sortes of Comfortable[542] hearbes both hot &

536 Bragget.

537 I.e. that an egg can float on the liquid.

538 Yeast.

539 I.e. one should drink it quickly.

540 I.e. if you want to drink it right away.

541 Metheglin.

542 Strengthening or refreshing to the bodily faculties or organs; sustaining (*OED*, adj. I, 3).

Cold, As Borage, Buglosse, strawberry, & violet Leaves[,] nettle=buds, Sage, Bayes, Rosemary, Time, And all sorts of Sallad & Pott=hearbes, for thus y^{e} welsh make it and y^{t} in a good quantity; And havein[543] boyled [it] very well take it from y^{e} fire & strayne it out very hard into a fatt where yo^{u} must let it worke with a Reasonable proportione of yest, and as it workes up, beat it downe with a staffe & wth all stir it well together, and thus doe until it hath done workeing, wch will be a=bout .14. dayes; then powre y^{e} Cleare into a vessell & keepe well stopped for use./ These I had from a welsh Lady y^{t} hath often made them herselfe/

C: Rust./

[89r]

The Preparation of Juice of Liquoris or the Extract of Liquoris./

Take what quantity of greene Liquoris you please[,] rubb it Cleane with lining Clothes & [do] not wash it[,] cutt it in smale Peaces & bruse it very well in a stone Morter, then boyle it in a sufficient quantity of water untill neare halfe be boyled away, then strayne it through a Peace of new Canves in a Presse, and then put it into an earthen pan which plase upon a very gentle fyre of Charcoales y^{t} it may neither boyle nor simper but onely breath off gently untill it comes to the haight of Honey, then take a droppe or two out & put it upon a Putor[544] plate, And if (when it be Cold), [it] feeles stif, it is well, then take it off and make it into round balls or what other fashon yo^{u} Please./

Yo^{u} must be very Carefull to keepe yor fire very gentle towards y^{e} latter end for feare of burning wch it is very subject too; some doe add when it is almost heigh enough a quantity of suger Candy in Powder; others Manus-Cristie & Pen=ny dice.[545]/ but the symple is best./

[89v]

M$^{ris.}$ Alsbergers[546] way to Make Metheglin./

543 Having.

544 Pewter.

545 Penide.

546 Unknown. Active mid-seventeenth century.

Take Harts tongue, Liverwort, scabions,[547] five Leave grasse,[548] Endive, Succory, Strawberry Leaves, Agri=money, Buglosse, Borage, wilde Carrot topps, all these of each a small handfull, two sprigs of Rose=mary, mix sprigs of Hysope,[549] y^{e} like quantity of time,[550] three quartes of Hony, Boyle all these in thirteene quartes of water, for y^{e} space of an howre or till y^{e} fourth part be Consumed then take out the hearbs & Let it stand till it be Cold, then put a Litle barme into y^{e} Liquor, & so put it into yor vessel; After a Month or there about drawe it out; Cleane yor vessell & put it in agayne,/ this is to be made about Michellmas.[551]/

[90r]

Doctor Kings[552] way to make Meade/

Take five quarts & a Pint of water to one quart of Hony, to every gallon of Liquor one Limond, and a quarter of an ounce of nutmegs, it must boyle till the fume rise black—If yo^{u} will have it quickly ready to drinke squeeze in a Leamon when yo^{u} tunne[553] it up; yo^{u} must Let it be Cold before you tunne it./

M^{r}. Rust's opinion of Honey

Honey clenseth ye stomach from y^{t} tough flegme wch causeth Coughs, It opens obstrucktions in ye Lunges and is [a] remedy agaynst dificult breathing, It Preserves the Lungs from Empostumes[554] & is of great use to remove them when they are there, which it doth both by Clening and

547 Scabious.

548 According to Culpeper, "Five-leaved, or Five-fingered grass" is "admirable against all Fluxes, and stops blood flowing from any part of the body, it helps infirmities of the Liver and Lungues, helps putrified ulcers of the mouth, the Root boyled in Vinegar is good against the Shingles, and appeaseth the rage of any fretting sores" (Culpeper, *Pharmacopoeia*, 8).

549 Hyssop.

550 Thyme.

551 Michaelmas. The feast of St Michael, one of the quarter days in England, Ireland, and Wales; the date of this, 29 September. Also, a period or season around this date (*OED*, n.).

552 John King (bap. 1614–1681).

553 Tune. To put into a proper condition for producing some effect; to adapt to a particular purpose (*OED*, v. 2, 2c).

554 Impostume.

healing; It cures a Catarr[555] wch is a salt tartarus humour distilling upon ye Lungs & doth in time Corrode and inflame them wch sends a gennerall distemper through the whole body wch is then called by Physicians a hectick feavor

[90v] And indeed my Lord it is ye Balsum of nature distributing nutriment abundantly through ye whole bode; the way to take it my Lord, is spred upon bread & eaten every morning fasting/

But y^{e} best way is to take every morning fasting the quantity of an ordinary spoone full of y^{e} essence of Honey in a Large wine glasse full of mint water, or any other Cordiall water y^{t} is not inordonatly hott, And if y^{e} Patient be troubled wth a hot stomach, Pro=ceeding from a distempred Liver; It may be taken in Borage, or Buglosse-water, which are Cordiall & y^{t} Refreshing./

Those my Lord are my thoughts upon this subject which I humbly offer, & doe as humbly begg thare[556] aceptance with yor Honor, wch will be no small enco=ragment to improve my vallant[557] to serve yor Honor which shall always be with, those dutifull Reciepts that may become[558]

My Lord

Co: Garden[559] Octo^{r} 5.th 49

The most oblidg'd
amongst yor Hon^{rs}. servants
Ca: Rust

[91r]

Mr Rust's way to make y^{e} Essense of Hony:

Take of Pure Life Hony what quantity you Please which melt gently & take of ye skin y^{t} riseth, then mix it whilst it is hott with as much sea Sand (washed & dryed) as will make it into a stiff Past, wch put into a Cleane Pipkin, and Powre upon it of ye best spirit of wine untill it swim[s]

555 Catarrh.

556 Their.

557 Valliant. Having or possessing courage; especially acting with or showing boldness or bravery in fight or on the field of battle; bold, brave, courageous, stout-hearted (*OED*, adj. 2a).

558 To accord with, agree with, be suitable to; to befit (*OED*, v. III, 7).

559 Covent Garden.

above ye Matter .3. fingers then Cover ye Pipkin very Close, & let it stand warme 48. howres, then powre the Cleare through a Paper filter, & put fresh spirit of ye Masse as formely & Let it stand warme 48. howres and Passe through ye filtor as ye former, Then breath both this & ye former Passed over a gentle fire untill it Comes to ye haight of a Syrop of melted Hony which keepe in a glasse for your use; well stoped/

[91v]

~~Henry~~ [^]Thomas Michalls[560] way to boyle a Carpe/

Take a Carpe, & take y^{e} scales Cleane of, then open the beley[561] & powre in some Claret wine, & take all y^{e} blood & keepe it in a dish for y^{e} Sause; Then take Clarret wine & fayre water of each a quantity, put=ing halfe a handfull of Balme, Mint, as much sweet Marjorum; 5. or .6. bay Leaves, a Litle time[,][562] as much Rose-mary[,] a Litle Pennyroyall; Cut y^{e} fishe into peaces and when y^{e} Liquor boyles put in as much Salt as you thinke fitt, & fyshe will indure Salt well soe Let it boyle sufficiently/

The Sauce for it./

Take .3. or .4. sponne fulls of vinager, halfe a Pound of sweet Butter, & wash all ye buttermilke out in .3. or .4. waters, & put it into ye Bloud, alwayes stiring it that it may Caudle[563]; Then add to it a good nutmeg 6. cloves[,] .8. flacks[564] of large Mace, beaten very small altogether & soe put into ye Butter Continually stiring of them till they be al incorporated [92r] Then Cover ye dish y^{t} ye spice loose not their spirits[;] yo^{u} must put in ye spawne or Milt[565] when you put in ye Carpe, And when you dishe it up strowe[566] Litle ginger about ye brimes of ye dishe

Directions to make Barley Creame./

Take halfe a pound of Barley & wash in 5. or 6. severall waters, rubing it betwixt y^{r} hands, then let it ly in fayre water a day & a night, Then

[560] Unknown. Active mid-seventeenth century.

[561] Belly.

[562] Thyme.

[563] The recipe calls for the mixture to have the consistency of a caudle.

[564] Flakes.

[565] Fish semen (*OED*, n. 2).

[566] Strew. To scatter (rushes, straw, flowers, etc.) over the surface of something; to sprinkle (something granulated or in powder) over a surface (*OED*, v. 1a).

boyle in fresh water .7. or .8. times a quarter of an howre at a time then powre out y^{t} water till ye Brownes of ye barley be taken away, then boyle ye next as tender as your can, Soe dun take .5. or .6. Almonds & blanch them and put them into fayre water, then take them & beat them puting .3. or .4. drops of Rosewater to them then take out yor barley leting ye water drayne Cleane out and save ye rest to strayne out ye Pulpe, ye Barley being very well beaten amongst ye Almonds, soe yo^{u} make yor Creame thick or thin as yo^{u} Please, When yo^{u} have strayn'd y^{e} Creame yo^{u} must put Rosewater & Suger & y^{e} Juice of a Lemon

According to yor tast

[92v]

For the Boyling of Tey[567]

To a pint of fayre water of Antworpe measure[568] put a litle spoonefull of Tey, Let them boyle to=gether half an houre, after that Take it off & drinke it all times as you please, being a Litle warmed./

[93r]

A rare Secret called villemaigne[569] to Cure a Horse when he is prick'd

Take 2 ounces of Pine tree-seeds, 2. ounces of Gumme Elemy,[570] one ounce & a halfe of Turpentine of venice[,] 4. ounces & a halfe of flax redd with vermillon, halfe an ounce of Dragons bloud, All is to be put into powder then melted, afterwards put ye Terpentine to it, Then stir it & leave it a while upon ye fire; Being taken off from ye fire & halfe cold you must put to it halfe an ounce of round Birthwort, and as much of ye longe one in fine Powder & mixing all will make your Salve; When it is used you must melt it; And it is very good for all manner of wounds./

567 Tea.

568 Antwerp measure. This is a measure of length (and occasionally the volume of liquid) that is expressed in "ells". The measurement varies in different countries, but the Antwerp ell is probably the same (or very similar) to the Flemish ell, which is 27 inches (*OED*, n. 1, 1a and c).

569 Villemagne. A commune in the Aude department in southern France, probably where the cure originated. The French chymist Nicolas Lémery mentioned a similar "Plaister of *Villemagne* made upon soft Leather" ((1685), *Recueil de curiosités rares et nouvelles* (London), 249).

570 Elemi was used in plasters, ointments, and the manufacture of varnish, and deemed to be "good for fractures of the skul, as also in wounds" (Culpeper, *Pharmacopoeia*, 29).

To make an oyntment of the thicknesse and Colour of Honey good to lay a charge[571] on a Horse beaten in his Limbes, or foundred[,][572] to disolve swellings, to breake them, or suddenly Cure them

Take 2. pound of Muttons Tallow, one pound of sweet suet, one pound of Turpentine of venice, one pound of burgundy=Pitch,[573] & one Pound of black-pitch, 2. ounces of Cumin, 2. ounces of oyle of Bayes, one duzen of bolearmenak=[93v]Stones,[574] halfe a pound of Honey, halfe a pint or there aboutes of oyle olive, with as much aquavitæ, and 2. or .3. pound of flowre, which is to be mix'd with as much wine & made like papp, stiring it Continu=ally till it begine to growe thick, Then you shall put to it y^e^ two sortes of Pitch, y^e^ Tallow & y^e^ Suet till they be melting with boyling After which yo^u^ must add all y^e^ rest, & Continue y^e^ boyling for a quarter of an howre longer/

An oyntment to make Horses hoofes growe
And to keepe them freshe/

Take a Pound of Mutton Tallow, 2. Pound of Hoggs=grease, one Pound of Terpentine, halfe a pound of bee=wax, halfe a pinte of oyle olive, boyle all together that it be of y^e^ thicknesse of oyntment when it is Cold./

A Balsum well try'd for all manner of Cold fluxes & sciaticas & all kinde of Bruses as well of y^e^ flesh, as of y^e^ nerves./

Take flowres of Sage, of Rosemary, of Marjorum[,] of Lavender, of Time, of S^t^. Johns-wort, of dwarfe-Elderne[575] And in Case you finde not flowres, you must take ye toppes of them, & dry either of them [94r] In the shade, Because ~~Either~~ the Sunne takes away their force, of each sort of these flowres, or toppes, yo^u^ must put a good hand-full in an Earthen

[571] Charge. To load; to cause to bear, hold, or receive (*OED*, v. I, 1a); in this case ointment is to be "loaded" on an injured horse.

[572] Founder. Chiefly of a horse or its rider. To stumble violently, fall helplessly to the ground, collapse; to fall lame (*OED*, v. 4).

[573] Burgundy pitch was regarded as excellent "against old aches coming of former bruises or dislocations" (Culpeper, *Pharmacopoeia*, 49).

[574] Bole Armoniac Stones.

[575] Culpeper held that Dwarf Elder "wasts hard swellings, being applied in form of a pultis; the hair of the head being anointed with the juyce of it turns black; the leaves being applied to the place, help inflamations, burnings, scaldings, the bitings of mad-dogs" (Culpeper, *Pharmacopoeia*, 18).

botle wth two quartes of aquavitæ well rectifi'd, as also halfe a glasse of Pismires water distill'd; The botle is to be well stopped And according as yo^{u} will make more, yo^{u} must put pro=portionably more flowres & more Aquavitæ, which the better it ~~doth~~ is rectified, y^{e} better it doth penetrate[;] you may keepe this Balsume as longe as yo^{u} will, If you keepe it well stopped; but a fore you can use it it must be 15. dayes old, And when you use it, yo^{u} must rubb with a lining Cloth before ye fire ye Paine full part that ye Pores may be open'd And soe warme the Balsum that it be not Cold, & no more for it would with great heat evaporate away, And after ye part is rub'd with it it is to be Cover'd with a Linin dip'd in ye same balsum

To Cure a broken-winded[576] Horse/

Take .3. pound of Brimstone in powder, 3. pound of Steele dust, put in an Earthen pott soe much of ye brimstone as will Cover ye bottome halfe a finger thick, then put upon it of ye Steele dust halfe the thicknesse of the Brimstone, And upon ye Steele ye Brimstone agen, And soe Continue to put led, upon led, till all is in & that the last led be of Brimstone. Then set it on fyre with an iron red-hot thrusting it to ye very bottom, And to make it burne ye better, put a peace of paper burning [94v] at the toppe when it is well burnt, you must pound it in an iron Morter, or brasse Morter, & sift it through a fine cames,[577] And put a spoon full of it amongs the oates, which yo^{u} must wet a litle, that ye powder may stick to it, And so give it to ye horse for some dayes/

[95r and 95v are blank]

[96r]

The Manner how the Spaw Waters are to be taken at Antwerp by the Right Hon.^{bl} the Lady Marquesse of Newcastle

My Lady must first of all be Purged, And afterwards take every morneing the waters as followeth; The first day her La^{ip} shall drinke one pint, ye second day a pinte & a halfe, The third day a quart, The fourth day three

[576] Broken wind. An incurable disease of the organs of respiration in horses, caused by the rupture of the air-cells, which disables them from bearing fatigue (*OED*, n.).

[577] The reading is reasonably clear, although the meaning is not. It could perhaps be "chemise": that is, a kind of undergarment worn by women that could have been used as a filter (camis is an old spelling, from the Latin "camisia") (*OED*, n. 1a).

Pintes, The fifth day two of those glasse=botles that the waters are sold in, And so every day to increase; And the more her La^{p} drinkes of it, If her stomack can indure it, & nature puts it well out,[578] The better it is./

Every time her La^{p} Drinkes these waters, Shee must now and then eat some Anyseeds[579] Confits, And if they should not passe well in ye begining, as Commonly they doe not 4. or .5. drops of Spirits of vitrioll are to be put into every Glasse full of Water./

During the takeing of these waters her La^{p} must use some moderate Exercise, And may also now & then rest her selfe But not Sleepe/

The Water haveing done working her La^{p} may dine and according as She hath a Stomack eate well of any Meat of easy disgesture,[580] As Partridges[,] Capons, Henns, Chickens, Mutton, All of them rather rosted [96v] then Boyl'd, And somtimes minc'd,[581] But at Supper her La^{p} must eate very litle, And that eyther Eggs, or Prunes. And must drinke at Meales either Renish=wine, or Spaw=water alone, or with wine/

The drinking of thes waters may be Continu'd thus .3. 4. .5. or .6. Weekes According as her La^{p} finds it be=nificiall, And all y^{e} while Shee must take Care that She be not Bound/

As her La^{p} must be purged in y^{e} beginning of this Course of waters, soe it will be well to take a purge in y^{e} midle of it, & another at the Latter end./

What other things should be done according to the Accidents that may happen is to be Left to the more perticular advice of ye doctor.

The Spaw-waters are sold in Glasse-botles, Containing each of them .3. pintes or there about and care must be taken in ye buying of them that they be very full & well stopped; A botle being once open'd, it cannot be kept for ye next day, And if ye water be not very Cleare, it is not to be drunke./

[97r]

The Lady Throgmortons[582] way to Stew a Carpe.

[578] Put out. To drive, force; to expel (*OED*, "put", v. 5a). I.e. nature drives the substance out of Margaret's stomach.

[579] Aniseeds.

[580] Digesture. The process or faculty of digesting (*OED*, n.).

[581] Minced.

[582] Elizabeth Throckmorton, Lady Raleigh (1565–1647).

Take the Carpe & wash and scrape it very Cleane, Then open it & take out the Gall[583]; Then cut the tayle and save the Bloud with a Spoone, and with wine wash the Belly that you may have the more bloud; Then take three parts wine, and one of water with a reasonable quan=tity of salt, And then put it to boyle in an new earthen Pott; when it boyles put in a Spoonefull of vinager that it may skumm[584] the cleaner, and when it is Cleane skumm'd then take one nutmegg, half a quarter of an ounce of Cloves, & a litle mace, all powdred small togeather with a little Sugar, one great onion whole, or two lesser[585] and soe set it to stew altogeather, and when it is al=most enough, put in the quantity of an Egg of Butter, And when it is ready to be taken up, then put a [^clove] or two of Garlick into it, or else rub the dish with Garlick.

[97v]

Robert Atkins[586] his medicin for a Cough.

Take a quart of milke & let it be made into a posset, and Let y^{e} Curd be very well taken from it, Then put into it the quantity of two Spoonefulls of Raddle and let the Patient drinke a Pinte of it when he goes to bed as warme as he can take it, and soe likewise in the morning, keeping him selfe warme in his bed, for a night or two; This has beene approved on by divers. And it Cures likewise all Swine that hath the Murraine or Cough./

A Medicin for y^{e} Runing of the Raynes

After the Patient hath beene very well purg'd take a good quantity of milke, 3. or .4. quartes, Put therein soe much Isinge=Glasse[587] cut in small peaces, that it may the sooner desolve, and let boyle till you finde all the Isinge-glasse desolved, and growne to such a thicknesse that when it is Cold tis like a Gelly, then let the Pati=ent drinke thereof a Pint in y^{e} morning, & another at night warm'd; This hath beene well aproved off./

[98r]

583 Gall. The gallbladder and its contents (*OED*, n. 1, I, 2a).

584 Scum. To clear (the surface of a liquid) of impurities or floating matter (*OED*, v. 1a).

585 I.e. smaller.

586 Robert Atkyns (bap. 1621–1710).

587 Isinglass.

A Medicin for the Stone./

Take a Thunder=boalt stone & beat it into very fine powder and take as much of it as will ly upon a grot,[588] and put it into halfe a pint of whit wine; And let the Patient drinke it every morneing & night; This to: Chiffins,[589] & his brother doth know divers that hath beene Cured of that disease./

[98v]

Sr Walter Relighs[590] Receipt for Stomack Pills.

Take two ounces of Aloes, of Mirrh & Mastick each halfe an ounce, beat them severally into fine Powder, then mingle ye Powders together, and put to them two dramms of Saffron; Put all these into a Glasse viall with a narrow neck of about a Potle bignesse, Then powre upon them as much Spirit of wine as will Cover them & be about an inch above them, Let y^{e} Glasse stand neare y^{e} fire two dayes or till y^{e} Spirit of wine be deepe Colloured, Then powre of the spirit of wine soe Colloured into a=nother Glasse & put fresh Spirit of wine upon y^{e} Powder as at first; And when it is deepe Colloured powre it off agayne; Continue to doe thus till y^{e} Spirit of wine will Receave[591] no more Tincture from y^{e} Powders; Afterwards Distill this Spirit of wine soe tinctured in a Glasse still in Balneo well Luted, And Yo^{u} shall have all the Spirit of wine come Cleare into y^{r} Receavor, which by ye way is an Excellent Plague or Stomack water, And that wch is to be ye Medicine will remayne in ye Still like an Honey; wch put into a Masse & keepe in a Pott till you have an occasion to make it up in Pills of which you may take one, or two at a time of ye bignesse of a Peaso.[592]/

[99r]

588 Grot. A fragment, particle, atom (*OED*, n. 1). The meaning seems to be that one should take the smallest fragment of thunderbolt-stone.

589 The reference is to two members of the Chiffinch family of Staplehurst in Kent: Thomas Chiffinch (1600–1666), a confidant of Charles I and Charles II and a painting connoisseur, and his younger brother, William Chiffinch (c. 1602–1691), a courtier and Royal official. The Chiffinches, who followed Charles II to the Continent, would likely have been acquainted with the Cavendishes.

590 Walter Raleigh (1554–1618).

591 Receive.

592 Peso.

A Letter from Mr Rust to the Marquesse of Newcastle./

My Lord,

I Receaved yor Hon.^rs^ Commands for some Crocus=metal=lorum, which I have sent; But not in those quantities yor Lord.^ip^ directed, Because they are too great; for whereas yor Lord.^ip^ has directed .30. graines in a paper, for a dose, wee usually give But .3. 4. Or 5. graines at the most according to the strength of the Patient; But the most usuall, & safe way of takeing of it, is as followeth.

Take of Crocus-Metallorum two ounces[,] beat it to fine Powder and put it into a quart Glasse; And fill ye Glasse with Sack and let it stand warme, a weeke, or .10. dayes, alwayes shakeing the Glasse, twise, or thrice a day; And after it hath stood warme .10. dayes, set it by, And when it is setled very Cleare, Give of the Liquor, onely an ounce, to ye weakest; An ounce and a half to the stronger; And two ounces to the strongest; And let them drinke posset-drinke after they have had two or three vomets; And keepe from sleeping untill the vomet hath done working; And if yor Lo^ip^ be pleased to use the Powder wth out infusion, you are onely to mix the quantity yor Lo^ip^ please to use in a small draught of Posset-drink, or white wine and so give it. The first & comune way is Called, Aqua ben=idictæ Roulandy[593]; And this worketh more gently then y^e^ Powder. This with my most humble servis is all from

My Lord
Yo^r^ Hon^rs^ most obedient
Sarvuant
Ca: Rust

Co: Garden London
Jan: 1654

[99v]

An approved medicine agaynst melancholy, given mee by an Nun, att Antwerpe

Take of Sena Cleane picked the weight of 4. Shillings, of Liquoris the weight of .10. pence, of Cinamon the weight of 4. pence; Bruse your Liquoris and Cinamon then put them with y^r^ Sena into 38. spoonefulls of

[593] Aqua Benedicta Rulandi is a purgative described by Hartmann. For the recipe itself, see *Praxis Chymiatrica*, Chap. II, Sec. 14, 6, and, for an English translation, *Royal and Practical Chymistry*, Bk III, 2.

water, there to infuse an howre, Then set it over a gentle fire of Charcole, that it may not boyle, but simper,[594] till all the grosse[595] substance be sunke to the bottome, Then take it off the fire and Cover it till it be almost Cold, Then strayne it & take thereof .3. mornings .10. Spoonefulls at a time blood warme, Put to every 10. spoonefulls, 3. spoonefulls of syrrop of vinager, and the weight of 6. pence of Creame of Tartar, Let the potion be so hott that the Tartar may be dis=olved, Put halfe a spoonefull of Balme water or Cinamon water into it before you drinke.

Lye in yor bed an houre after yo^{u} have taken it neither sleepe nor sweat, But ly warme, fast 4. howres after it, Then take broath made with veale & Chicken and thickened with y^{e} Crust of white bread, Put therein large mace, a peace of nutmeg, the roots of fennell & Parsley, the root of balme, one branch of Rosemary & another of time,[596] Boyle the broath so stronge of meat[597] that it may gellye when it is Cold.

The Perfect Cure is to take this medicine 14. times, one or two dayes distance betweene each .3. Potions.

[100r]

Doctor fraysers[598] receipt for The colick and the spleen

Take the leaves of agrimony a handfull[,] of the tops of wormwood, of leser centory[,][599] caladine[600] of each as much as your may take with three fingers[;] set them to steepe in a pinte of renish wyne 24 hours[;] ad thereto as much sugar as is suficient and drinke everie morning fast=ing five or six ounces for five days together/

Take conserve of the floures of borage[,] bue=glas[,] violets[,] prim-roses and roses of each an ounce, conserve of floures of sage and rose marie of ½ an ounce[,] of citroune pills presser[^ved][,] of preserved mirabolonns[601] of each six drams, of species diarodan abatis[602] five

594 To simmer (*OED*, v. 1, 1).

595 I.e. coarse.

596 Thyme.

597 I.e. with lots of meat.

598 Alexander Fraizer (1607?-1681).

599 Centaury.

600 Celandine.

601 Mirabelle. A small, sweet, yellow plum, eaten chiefly in France (*OED*, n. 1a).

602 Diarrhodon abbatis.

drames, with a sufice=ient quantyty of scabiosa[603] sirope make an opiate whereof everie morning fasting and at night going to bed, take a drame and a halfe and that after the use of the aforesaid medecinall wyne/

[100v]

Purge with this folowing potion

Take rubarbe one drame[,] seana leaves three drams[,] thamarinds[604] halfe an ounce[,] coriander seeds a drame[,] cinomont[605] a scruple[;] set them to steep one night in a suficient quantity of Juce of oranges, strayne them and in three ounces of the liquore desolve six drams of mana, of sirup of roses laxative an ounce, mingle all these and make a postion/ 100 days after take the folowing pils/

Take of aloes rosate[606] a scruple, of pils mastichina[607] halfe a drame, salt of wormwood 15 graynes[;] forme therof 7 pils, and take them in the evening a quarter of an houre before super which is to bee continued five days drinking everie morning a draught of broth wherein a drame of christiall of barberum[608] is desolved/

[101r] having ended the pils use the folowing wyne for five mornings together three hours before diner

---

A pouder for the Collick and stone
y'lord conaways[609] receits

603 Scabious.

604 Tamarinds.

605 Cinnamon.

606 As its name suggests, aloes rosata is aloes "washed with juyce of roses" (Culpeper, *Pharmacopoeia*, 29). According to Culpeper, these aloes are "a gallant gentle purger of choller, frees the stomach from superfluous humors, opens stoppings, and other infirmities of the body proceeding from choller or flegm, as yellow Jaundice &c." (Culpeper, *Pharmacopoeia*, 140).

607 Pills of Mastic.

608 While the reading of "crystal of barbarum" is clear enough, the exact reference is not; it may refer to rhubarb (*Rheum rhaponticum*), as this shrub does leave behind a calcium oxalate phytolith when burned.

609 Edward Conway (bap. 1594–1655).

Take wilde time[,] garden time[,] mead[,] parsley[,] garden parsley[,] ash keys, watercreses, red mints[,] stone cropp,[610] gromwell[611] of each a like quantity, lay them like hey[612] and beate them to pouder, let the patient [^drink][613] thereof in poset drink, or caudle and it will help him of the collick presently *

For the collicke and Stich[614]

Take a small wooden dish and put therein cold ashes and lay upon them hot embers without any greate coales and thereupon strow[615] a good quantity of comminseed[616] and sprinkle it over with malmsey soe doeing till the dish bee filld, then cover the dish with a faire linon cloth doubled and lay it hot to your side where the colick lyeth, use this twice or thrice and it will help you, if you cannot get commense[^e]d take red sage[,] cammomille or sage or peney Royall or time of each a like quantity and use it as the other/

[101v]

For the wynde Collick

Take a handfull of camomile and a hand full of betoney and halfe a handfull of saxafage[617] and peneyroyall, bind them together and make porage in a pipkin with a litle meat once a weeke and eat them in the morning fasting/

For the wynd Collick

The fume of tobacco blowne into ye fundament is a good remedy/

For the Collick

610 Stonecrop is interpolated in a different hand; probably William Cavendish's.

611 According to Culpeper, "Gromwell-seed, provokes urine, helps the chollich, breaks the stone, and expels wind" (Culpeper, *Pharmacopoeia*, 28).

612 I.e. in a heap.

613 "drink" is interpolated in a different hand; again, probably William Cavendish's.

614 Stitch. For "stitch" as a synonym for "colic" at this time, see Hannah Woolley (1677), *The Compleat Servant-Maid; or, The Young Maidens Tutor* (London), 55.

615 Strew.

616 Cumin seed.

617 Saxifrage.

Take the kirnells of ashen keys ripe, ye roote of philipendula[618] each a like quantity[;] beat them small, take as much as will ly upon a six pence and drinke it in broth or cardus or camomile poset drink as much as the payne requireth/

Another for the Collick/

Take akornes[619] when they are redy to fall and lay them by till you have ocasion to use them then grate or scrape halfe of one and take it in a spoonfull of beere ale or white wyne which is best, and drink a good draut after it, but wyne is to hot[620] for a child

[102r]

An exelent Medecine for the Collick it hath cured many in great extremity

Take an old root of a hartichoke[621] the older the beter[,] wash it and take out the pith, then slice it and put it into a quart of white wyne, let it boyle to a pinte then being in bed drink one halfe[;] the other halfe at going to bed the next night

To Make Collick water

Take six quarts of the strongest white wyne you cann get and distill it in a limbeck, take a handfull of of [*sic*] sage balme, philipendula, saxafr[^a]g[,] bean flours[,] wild time[,] pelitory of the wall[,] germander of each a good quantity with one gallon of samphere[622][;] crush them all in peeces, and infuse them one night[;] before you distill them put to them one peck of may floures if they bee new gathered but if they be dry a gallon will serve, put in a quarter of a pound of aniseeds brused, and as much liquoras[623] shred, then destill them alltogether and when you are troubled take a spoonfull at a time, this is well aproved

[618] Filipendula was deemed to "provoke Urine, case pains in the bladder, and are a good preservative against the Falling sickness" (Culpeper, *Pharmacopoeia*, 6–7).

[619] Acorns.

[620] In terms of temperament, wine was considered hot (see, for example, Batman, *Batman uppon Bartholome*, 328–329).

[621] Artichoke.

[622] Samphire.

[623] Liquorice.

[102v]

For the Collick and exelent for y^e stone

Take a galon of white wine and put it in a limbeck, put of sage, fenell, balme, phill=ipendula, saxafrage, beane floure or bean cods[624] of each a handfull, wild time[,] pelitory of the wall, germander of each a good quantity, and a galon of sampere,[625] soe cut them in peeces and put them in the pot where the white wyne is[;] put thereto anis seed and liquoras each a quarter of a pound and put thereto a peeck[626] of may flours[;] let all these steep together one night and soe distill it in a limbeck and the water will apease the paine of the Collick

For the Collick in the small guts

Time made into pouder 3 drams with oximell and a litle salte[;] it purgeth by stoole tough and clamy flegmes and sharp and cholerick humers, and is good against the windynes of the side bely and stone and geniture[627] alsoe/ ***

[103r]

For the Collick in the small Guts

Take mithridate with a litle white wyne or with a decoction of camomile 4 or 5 houres after diner if the patient bee natureally loose in his bely or made loos[628] by glisters ***

For the wynd Collick

Take mountayne parsley[,] dry it in the shade, then hang it up in a bagg in the chimney, then a[^f]terwards put it in a bakers oven after the bread is drawne and being dry beat it into fine pouder, and keep it to your use, give it a spounfull at a time in white wyne poset/

For the Collick and Stone

624 Pods (*OED*, "bean", n. C2).

625 Probably "samphire".

626 Peck.

627 Genitor.

628 Loose.

Take elder floures[,] dry them and make pouder thereof[;] take as much of it as will ly on a shiling soe drinke it in a glass of sack/

Sir Thomas Alisburies[629]

For the Collick in the stomack

Take a pinte of cleare poset drinke[,] boyle therin of camomile floures and peletory of the wall each a quarter of a handfull, anisseeds and coriander seeds prepared each a spounfull then strayne it and put to it two ounces of diacatholicon,[630] course sugar two ounces, the yolk of an egg, and strayne it againe then use it

[103v]

The pill Mastick is verie good for the stomack Collick

Sr John Mints[631]

Receipt Remedie for a Tertian and Quarteran[632] Ague:

Take too figs and slice them and sprinkle them well with the pouder of white helebore[633]: lay the figs to the wrest[634] to that holow part wch Joynes both the inside and out side of the hand, bind it on fast with a linon cloth, aply it one houre or too before the fitt, let it ly some days till it dry of, the Quartern is cured with the same thing used the same way ***

Mr Foscues[635] Receipt For an ague

Take a quarter of a pint of clarret wyne[,] three spoonfuls of the best aqua vita[,] the Juce of three good oranges or too if they bee verie bigg: sweeten it with a spounfull or too of sugar and let the patient drink it when the cold fit begins[;] use it twice or thrice [104r] and make a plaister

[629] Thomas Aylesbury (1579/80–1658). The name appears to have been added later.

[630] Culpeper distinguished "Diacalaminthe Simple" from "Diacalaminthe Compound", the latter of which he maintained "heats and comforts cold bodies, cuts thick and gross flegm, provokes urin & the terms in women" (Culpeper, *Pharmacopoeia*, 121).

[631] John Mennes (1599–1671).

[632] Quartan Ague.

[633] Hellebore.

[634] Wrist.

[635] Unknown. Active mid-seventeenth century.

with mithredate upon a paper, and aply it to the region of the stomak halfe an ounce ~~wh~~ iss [*sic*] suficent/ *

For the stoping of the ears

Take of rosmary, betony, time, rue, each a handfull, put all these into a pinte of sack ~~and~~ [^or] a pinte of strong beare, let it all boyle gentlly till halfe bee consumed, then strayne it out and put the liquore into a bason, and being milke warme, spoute some of it into the eare with a sirringe, doe this soe often till it bringeth forth the cloted eare wax which hindereth the hearinge, then stop the eare with a litle black wooll touched with cy[^u]et,[636] keep it stoped soe close three or foure dayes to prevent the taking of could[637]/

Sir Thomas Alisburies receipt

[104v]

Doctor Morleys[638] receite for the papall or vaticall pills/

Take Amomi veri, aut sacsedan: anisi,[639] mastich,[640] Gingeber,[641] acori veri,[642] cardamom, cinamon, Zedoar,[643] macis,[644] nucis mosch,[645] garyophilop, croci,[646] cubeb,[647] lignialoes,[648] turbith, manna, agaric, sennea, cassice lign,[649] quinque mirablanoū,[650] of each one a scruple, rubarb one ounce, aloes too ounces and six drams[,] card(~~ij~~)=dui [*sic*]

636 Likely "suet".

637 Cold.

638 George Morley (1598?–1684).

639 True amomi, or substitute aniseeds.

640 Mastic.

641 Ginger.

642 Sweet Flag.

643 Zedoary.

644 Mace.

645 Nutmeg.

646 Saffron.

647 Cubebs.

648 Lignum aloes.

649 Cassia lignum.

650 Five myrobalan.

sancti,[651] scordi[652] of each half a drame[,] cum siropo rosarū soll[653] fiat massa pil[^u]cap[654]

[105r]

Doctor Prigions[655] receit for The Collick in the stomacke

Take of pills of mercury, of stomacke pills with gums[,] with pills of hiera,[656] with agarick of imperiall pills[,][657] of aloes rosata, of extract of rudius, each halfe a dram, salt of wormewood distiled as much as is sufficient[;] make six pills out of each dram and take too or three twice a week in the mornings[;] one[658] such days as the pills are not taken use a dose of this folowing electuary

Take imperiall water,[659] aqua mirabilis[660] of each foure ounces, fine sugar six ounces, boyle them according to the consistance of a sirope[;] add thereto of London electuary[,] of sasafras too ounces, species of aro=maticum rosatum[,] of london diamber[661] each too drams, Juce of

[651] Carduus sanctus (another name for Carduus benedictus).

[652] Water germander, *Teucrium scordium*. According to Culpeper, "it provokes urine and the terms, opens stoppings of the liver, spleen, reins, bladder, and matrix, it is a great counterpoyson, and easeth the breast oppressed with flegm" (Culpeper, *Pharmacopoeia*, 9 and 23).

[653] With syrup of roses solutive. Culpeper held that this "loosneth the belly, and gently bringeth out choller and flegm, but leaves a binding quality behind it" (Culpeper, *Pharmacopoeia*, 110).

[654] Made into a mass for pills.

[655] Francis Prujean (bap. 1597–1666).

[656] Hiera picra.

[657] Culpeper maintained that the imperial pill "clenseth the body of mixt humors, and strengthens the stomach exceedingly, as also the bowels, liver, and natural spirit" (Culpeper, *Pharmacopoeia*, 141).

[658] On.

[659] Culpeper held that imperial water "comforts and strengthens the heart against faintings and swoonings, and is held to be a preservative against consumptions and apoplexies" (Culpeper, *Pharmacopoeia*, 66).

[660] It was thought that aqua mirabilis "heats cold stomachs", prevents "apoplexies, and restores lost speech" (Culpeper, *Pharmacopoeia*, 67).

[661] According to Culpeper, diamber "heats and strengthens the brain, causeth mirth, helps concoction, cherisheth the Animal, Vital, and Natural Spirit, it strengthens the heart and stomach, and resists all cold diseases" (Culpeper, *Pharmacopoeia*, 122).

kermis[662] an ounce, of confection alkermis halfe an ounce, leavs of gold twelve, of london laudanum too grains, make this electuary for the use aforesaid the dose being to drams

[105v]

Sir Thomas Alisburies Leter to the Most Honored Lord Marquis of Newcastle For the Collick in the stomack:

Most Honorable Lord/

wee are all very much affected with your indisposs=ition of health but have hope that you have ease before this come to you, my wife thinkes nothing better for you then glisters if the like hapen againe, the receit of one is this under written, alsoe to drinke in those fits warme posset drinke boyled with cammomille which though it bee bitter in the tast yet in the efect it will make you a requitall, alsoe you may aply to the place afected a toast of bread well moistned in aqua vita and verie warme and lay it to the pitt of your stomack, this under corection[663] & of beter Judgments is her opinion having had experience of it with good sucses, now my lord give mee leave to tell you that this ill afect=ion may bee caused thorough[664] the usuall drinking of the smalest beare, drinke in the morning a good draught of wormwood beare or wine and fast and walke after it; [have] another at your meales or els[e] some other good wine which I am confident will doe you much good and prevent these fitts for it is not now as when you came first to court[;] your stomake now doth require hoter drinks then before[;] if you like not [106r] wormwood wine use wormwood beare which there is very good in antwerpp[;] you have the harty prayers of all for your health never excepting

My Lord

you^u^r Lordships most
humble servant
Thomas Aylesbury

Breda Febb y^e^
8th 1654

[662] The juice of the so-called "kermes-berry" (or "alkermes").

[663] Under correction. Subject to correction; a formula expressing deference to superior information, or critical authority (*OED*, "correction", n. 1b).

[664] Through.

Sir Thomas Aylesburies receit for the Colick in the stomacke [^, A Glister][665]

Take a pint of cleare poset drink and boyle ~~it~~ in it cam=momile floures and peletory of the wall each a quarter of a handfull[;] anysseed, coryander ~~seed~~ prepared each a spoonfull then strayne it out and put to it of diacath=oholicon too ounces[,] course sugar too ounces and the yoke of an egg[;] straine it againe if you will and soe take it

-- *

if you take some times the pilulea masticum[666] finaly they are verie exelent for this purpose/

[106v]

A Letter frome Mr Rust an Apothecary att London aboute the Collick in my Stomack.

My Lorde.

Your greefe Inn my aprehention prosedes frome a bundance[667] of coller, remassed[668] Inn your Stomack & parts Ajatient[669] & nott withoute some obstruction In the Spleen, & these[670] are setled In to Fitts, Sommthinge Inn nature of ann Ague for ther Is no Agewe,[671] butt their bases or originall Is Coller[672] & Itt Is this sharpe Humour, that gives those gripinges[673] or fitts, which your Lo:ps calls a Collick & frome the fume off this coller, together with the obstructions off the Spleen, Is your heade Molested with these Vapors you Complayne off

That which I woulde advise if your Lo:ps have strength to beare Itt, Is a gentle vomitt, off ann ounce off the Infutionn off Croqus Metalorumm to bee takenn ann houre, before the fitt, cometh & to drinke verye

665 "A Glister" appears to have been added later.

666 Mastic pills.

667 An abundance.

668 Remassed. To come together again in a mass or masses; to reassemble (*OED*, v. 1).

669 Adjacent.

670 I.e. the choler.

671 Ague.

672 Choler.

673 Gripe. To affect with "gripes"; to produce griping pains in; more specifically, to produce pain in the bowels as if by constriction or contraction (*OED*, v. 1, 8a and b).

plentifullye posett drinke, after Everye vomitt, & two dayes after to purge Gentlye with Sena & Ruburb, as your Lo:ps hath donn & after this to take this Glister twice or thrice a weeke untill your paynes bee Gone.—

[107r] Ande to take Morninge & Eveninge the quantetye of a wallnutt of Conserve of Borage flowers, & wood forwith mixte together, this will Coole & temper the Coller;—Rosted Pippins also are very good, Iff your Lo:ps Suffer the Grouth[674] off this Coller Itt Will proseed, to a weaknes & Indispositionn Inn the Limms,[675] & discoulleringe the Skinn towards a yellowe—

your Lo:ps shall doe well to observe wether you have Inn your walkinge or other Exercises a dificultie off breathing, or anye dull payns In your side, if you have Itt maye bee Fearde your violente Exersise maye have donn hurte, butt iff your Lo:ps bee free frome these Simtomms[676] you maye conclude your presente greefe [comes] onlye frome Coller, & your trouble from your heade desendente[677] frome that—All hott thinges are bad, so are oyleye & fatt thinges for the one doth Inflame & the other Increases the Coller & so contrariwise all Coole thinges doe temper & allaye the heate off Itt butt regarde muste bee had to your Lo:ps Age leaste[678] Inn corectinge the Heate, Wee doe nott putt oute the fier,

[107v]

<u>The Glister.</u>

Take ann ounce off Cenna, baye beries, & Juniper Beries of each ann ounce, Chamomell flowers & Melilott flowers, of Each a handful, March Mallowes topps, a handful[;] boyle these In two pints of water untill halfe bee boyled a waye, then strayne Itt, & putt to Itt <u>2:</u> ounces off Sirop off violetts & give Itt,—This Glister Will give <u>5:</u> or <u>6:</u> stooles & purge Gentlye, as well as a medcin,[679] & maye bee donn Inn the house[;] Thus Inn obedience to your Lo:ps Commandes, as ~~your~~

Your Moste
Humble Servante

[674] Growth.

[675] Limbs.

[676] Symptoms.

[677] Descendant.

[678] Lest.

[679] I.e. the glyster will work as well as a medicine that is taken orally.

Ca: Rust.

Strann[680] Maye[681]
31: 1654

[108r]

a letter frome M^r^. Rust the Apothecarye thus.—

My Lord

Those coolinge herbs Mentionde are verye good, Inn broth, & whenn theye are used in whaye,[682] a litle fumitorye maye bee added,—Spinige[683] Is ann Exselente Sallett, Stued,[684] so are Lettis, Rosted Aples are verye good Eaten Colde att all times butt Espetialye goinge to bed,—Bledinge maye bee Saflye used butt nott to Exsed 3: ounces att a time & care muste bee had that colde water bee nott to frequentlye used for feare off coolinge the stomack to much to bedwarde[685] whenn ther will wante Exercise to digeste Itt butt rosted aples are then Safe & I amm Confidente woulde helpe much to temper & Alaye the heate & sharpness off the Coller,

Your Lo:ps Moste
Humble Servante,
Ca: Rust.

[108v]

A Letter frome thatt moste famous, Lerned, & moste Exselante Phititian Sr Theodore Mayherne aboute the Collick Inn my Stomach—

My Lord

you set mee too greate a taske in desiring my counsell for your infirmity in English of which language I dare not brag, and I doe not love to doe

680 Perhaps a reference to the Strand in London. This would make sense since (as shown in Appendix 8D, "Biographies") Rust appears to have been based around St Martin in the Fields—where he wanted to be buried—which is on the northern side of the western extremity of the Strand.

681 May.

682 Whey.

683 Spinach.

684 Stewed.

685 Bedward. Towards bedtime; just before going to bed (*OED*, adv. 2).

any thing that I cannot doe de bonne grace.[686] but since you will have it soe it must be soe:

According to your relation, y^{e} diseas seems to mee to bee a vitious cholicke: an accident very painefull. and mighty trouble some: In many verie long and tedious: in others dan=gerous, bringing either some contractions of y^{e} armes. or a palsie in the lower parts [109r] of y^{e} body: I have seen some in it taken with con=vultions and falling sickness with a succes[687]: though some have escaped out of that calamity and extremety: It cometh from a hot distemper of y^{e} intrailes: y^{e} most part of y^{e} liver: sometimes of y^{e} spleen: in y^{e} melancholy bodys by which greate store of cholar: for y^{e} most parte yelow: some times blacke being gathered: is rejected upon ye guts: which are imbibed thoroughly with these burning humers: as if a limon cloth were diped in watter high coloured wth saffron when nature is strong: it is discharged by ye confinnite[688] of ye mesentery upon ye root of it: and from thence upon ye sinews or marrow of ye back bone: and causeth ye above mentioned simtomes of contraction or resolution: for ye most parte: this is ye fruite of an evill and irigulare diet alone: in others ye passion of ye minde: acting fierc^{e}ly upon ye body and upon its humers doe contribute much to y^{e} generating: of y^{e} materiall causes: and doe extreamly increas the malignity of them[.] the cure of this diseas is not very easie particularly wheare y^{e} [^infirmity] cause is not [109v] ordinary, as in this [*sic*] countryes: in france and in germany: it is to common: y^{e} people particu=larly in france: having a great deale more choler then flegme: it hath been knowne by y^{e} ancients and well remarked by Paulus Agyneta[689] the caus of this torture is hott and y^{e} phisitians: which doe perswade y^{m} selves these windes doe come of a cold humor are much deceaved: and will hardly ease theire patient, the methode wch you have to follow is: to bee purged of this superfluous humor well thoroughlly: ye remedy which I finde very effectuall is this: take too drams of very good rubarbe: put it in small pouder and add to it as much sirup: of violets as will serve to make a bolus: wch must be takn in waters: as it is well knowne to you: and imediatly after drinke upon it ye disolution

[686] "De bonne grâce" (French for "with good grace").

[687] A "melancholicus succus", which was another name for a "scirrhus" (see Alexander Read (1636), *The Chirurgicall Lectures of Tumors and Ulcers* (London), 75–80).

[688] Confinity. The position of bordering on something else; neighbourhood, contiguity, adjacency (*OED*, n.).

[689] Paul of Aegina, the seventh-century Byzantine Greek physician.

of 2 ounces of exelent mana of Calabria melt=ed in a suficient quantity of cammomille water[;] I hope the effect and operation of this medicine shall satisfie you:/

[110r] Purge with it three or foure times if need bee leaving one or too days betwixt every purge to rest and give nature time to recollect[690] his strength: some times y^e^ deseas is soe stubborne that it requireth a phisick of more strength to bring forth a green humor which is like verdegrice[691] and is ye cause of the intolerable paines ordinary in this kinde of cholicke: this purgative is mercurius dulcis[692]: a very safe powder whear of[693] twenty grains are to be mixed: with as much or twenty five graines of ye pills wch wee call cochias minores: foure hours after a thin broth is to bee taken as after other purging medecines like wise[;] this purging phisick must bee given againe three or foure times leaving at ye least 3 days betwixt every exhibition[694]: and that day take heed of cold and goe not out of your chamber: in case of great dolours[695]: which are used to come by fits[696]: use glisters only to ease y^e^ paine and not of a violent operation: those y^t^ are made with sack and oyle of wallnuts doe well: and may bee will keep long with great ease/

[110v] usually in case if need be of evacuation by y^e^ said glisters they due[697] use pure=gative yea coloquinte[698]: but I take some milder confection as a catholicon, which I due spurre[699] with y^e^ infusion of crocus met=tallorum: and these glisters doe very well stupifying things & provoke sleepe: [they] are often times prescribed: by y^e^ physitians as

[690] To collect, gather (*OED*, v. 1, 1a).

[691] Verdigris.

[692] According to Culpeper, there are two ways to make "Mercurius dulcis": "Mercurius dulcis Sublimatus" (sublimated through fire) and "Mercurius dulcis precipitated" (precipitated in aquafortis and water) (Culpeper, *Pharmacopoeia*, 186). The exact method used here is unclear.

[693] Whereof.

[694] The administration of a remedy (*OED*, n. II, 4).

[695] Dolours. Physical suffering, pain; also (with plural), a painful affection, a disease (*OED*, n. 1).

[696] I.e. which often accompany fits.

[697] Do.

[698] Culpeper warns that coloquint causes one to purge "violently, and therefore are not rashly to be medled withal" (Culpeper, *Pharmacopoeia*, 46).

[699] Spur. To stimulate (*OED*, v. 1, I, 2a).

laudanum which is ye best: and other opiate medicines: but I give of them as litle as I can: because they are more for the credit and quiet of ye phisitian: then to y^e good of y^e patient: take away y^e cause and y^e effect will cease: yet nesesity hath not law[;] one time is not custome: some of our authers in the greate torments have ministred a great draught of watter cold from y^e conduit: which remedy being found in reason and authority: I can com=mend as most efectuall by my owne exprience and this drinke according to y^e exigence of the disease 3 or foure times in twenty four houres: have better opinion of y^e phisitians amongst whome you live: that greate and famous City whear you are, [111r] Should bee more infortunate[700]: if it hath not as good as any parte of y^e world: in ordering your rem=medyes above mentioned: use their counsell but above all helps[.][701] you have been sufi=ciently purged: y^e watters of spaw drunken phi=sically and in order are most exelent: with all y^e ca[^u]tions[702] nesesary in y^e taking of y^m: you can have there y^e whole yeare: wheare you are now and have used them many years: they will keep more then an yeare: which weare better transparted[703] then ours of townbridge[704] or knest=row[705] in yorkshire: I have seen y^e use of y^m in contracture and palsies: and in y^e lingring of this cholick: most effectuall and certaine in many patients: the halfe bath: which wee call semi cupium[706] doth extreamly ease the paine of y^e cholick: whatsoever it bee: but must bee used discreetly after due purging otherwise in melting ye humors often times it doth hasten ye palsy data tempore pusant[707][;] if after all ye remedies have been administred rationally and methodically some paines under ye navel doe continue or by fits cometh [111v] And if some weaknes afflict ye sinews

[700] Unfortunate.

[701] Long space in original.

[702] Cautions.

[703] Transported.

[704] Perhaps Trowbridge, the county town of Wiltshire, England.

[705] Unclear reference; perhaps the market and spa town of Knaresborough in Yorkshire, England.

[706] Semi cupium. The first word looks more like "sepui"; it is quite possible the scribe did not know this term nor the Latin from whence it originated.

[707] From Ovid's *Remedia amoris* I.131: "Temporibus medicina valet: data tempore prosunt, Et data non apto tempore vina nocent" ["Medicine is healthy at the right time: given at the proper time, wine is beneficial, otherwise, given at the wrong time, it does harm"].

in y[e] arme: and in y[e] thigh: the unction[708] for y[e] belly is made with galbanum[709] parcelli[710] but purge well: and qualifie[711] well y[e] humer[.] you shall have noe need of soe many remedys[;] this is all and is enough for a full answere to y[e] letter: and for y[e] absolute evre: with a good diet: in which all heating things must be avoyded: I pray excuse my rude Stile which had been beter in another Idiome: but you and I shal be satisfied if y[e] end of all bee your perfect cure and recovery: wished heartyly by mee

My Lord

Your most humble servant

Chelsey
March 29
1654

[112r]

S[r] Theodore Mayerns Advice

S[r] I was with S[r] Theodore and shewed him both your leters: of which heath[712] taken much care: for hee refused all busines till hee had answered them which I can asure you: being present with him he did with some laboure and industry: you may please to take it word for word as I receaved it from him: for hee was pleased to command mee to write as hee dictated. Yr[s]:/

S[r] y[e] continuall use of tobacco is most prejudiciall to y[e] patient who should not bee bewitched by such a stinking comodity: w[ch] doth spoile y[e] braine and stupifie it: with prejudice to y[e] faccultyes[713] and weakens

[708] Unction. The action of anointing or rubbing with an ointment or oil as a lubricating or preserving substance (*OED*, n. 4).

[709] Culpeper held that galbanum "hastens both birth and afterbirth, applied to the navel it staies the strangling of the womb, commonly called the fits of the mother, helps pains in the sides, and difficuty [*sic*] of breathing, being applied to it, and the smel of it helps the vertigo or dissiness" (Culpeper, *Pharmacopoeia*, 29).

[710] Perhaps "parcelly", which refers to something being divided "by parcels, parts, or portions" (*OED*, adv.). But it most likely indicates a portion (or parcel) of the galbanum. This use of "parcel" is quite common (see, e.g., Gerard's *Herball*, 366 and 545).

[711] Qualify. To bring into, or keep in, a normal or proper condition; to control, regulate, modulate (*OED*, v. II, 13).

[712] He hath.

[713] This is most likely meant in terms of the Galenic idea of the faculties: that is, the various powers and parts of the body that sustain life.

extrealy ye sinewes in him: it were good to keep it for a necesity in a sicke ayre and a moist and fogy weather: but a tobacconist is not easiely perswaded against his ill custome for which I am sory[714]: It is strange to mee if seana hath made ye patient sicke: if hee thot[715] a solid medecine will worke beter efects upon ye stomacke let these pills bee sent him

take gum of guaick[716] too drams: fine aloes 3 drames[,] pure ambergreece one dram: let these bee [112v] Poudred[717] very fine and with a suficient quantity of balsum of perue[;] make all up into a mass: first let too scruples bee taken a quarter of an houre before a light supers: then according to y^e efects of this medicine: increas or dimin=ish y^e quantity: let it be taken too houres after super going to bed: when hee takes not his pills: he may take foure or five drops of y^e balsume of perue: in a litle muscadine[;] this medicine we advised henry y^e son for y^e same distemper./

But especially to alay vapours I doe com=mend y^e electuary of sasafras: to be taken [the] Quantity of a nutmeg at all times when y^e winde trou-bleth: y^e vapours that trouble y^e head have their originall from y^e lower region which must bee kept cleane;

Botle beare is starke naught[718] and [sends] fumes mightily too y^e braine: sack is not good[;] claret wine at meales if it be pure is very good: for putting y^e head in cold water is dangerous this could[719] season:/

[113r] The auther himselfe is of this counsell: who was celsus[720][;] hee prescribes [^to] doe it in hott weather y^t may be dryed best in sumer[721]: experimentum per=iculosom saith Hipocrates[722]:

[714] Long space in original.

[715] Thought.

[716] Guaiacum.

[717] Powdered.

[718] I.e. bottled beer gives no benefit or reward.

[719] Cold.

[720] Aulus Cornelius Celsus, the Roman encyclopaedist.

[721] I.e. one should only put one's head in cold water in the summer because, when it is hot, the hair (and face) dries quicker.

[722] This is a Latin translation of a quotation from the Hippocratic corpus, the full version of which is: "Ars longa, vita brevis, occasio praeceps, experimentum periculosum, iudicium difficile" (Hippocrates, *Aphorisms*, I.1). While it roughly translates as "life is short, the art is long; opportunity fleeting, experiment dangerous, judgment difficult", the exact meaning of this passage has been much debated. For some of the medical aspects

For my lady tis another mater: ye second water was counsailed her[723] against melan=cholick sowernes: which comes into ye stom=ack from ye spleen to which all liquors that doe growe sower are contrary: sack if it be pure can never be turned into vinagare and in a hipacondriack it must be alayd wth much water: but to ye other party her husband:/

Juniper beryes is very good for this windy dis=position: they are very comfortable and may be taken prepared in maligo[724] sacke and amber greece: steeping them and drying them: and chewing and swalowing them from 10 to 20 berries every morning drinking three or foure spounfulls of a kinde of wine or Liquore made of ye same beries very ripe [113v] With watter only as they meale[725][;] with Liquore [it] is extreame comfortable[726] and this I have tryde offten with good succes to expell windy vapours:

For Isues[727] in ye thighs they may bee spared: since those on the shoulders doe soe good operations: a thousand thanks and an assurance of service for ye bookes: which shall be kept as a monument of ye patients love and favoure: this from ye Doctor

Sr

you will finde in my last that amongst other thing: Sr Theodor hath advised you the use of that balsum: but not in soe great a quantity: wch is very iregulare but I shall accquainte[728] Sr theodor with your disire in this last: for I wrote at large concerning ye former:

for your sirop you may please to take 2 ounces in a small draught of poset ale [which] is enough for your selfe: but for my lady as much more: or you may please to begin with an ounce and a halfe only and inlarge the dose in ye next as you [114r] finde cause: this sirope is somwhat in ye nature of a gelly but will easily disolve in any warme liquore: drinke posset drinke: as you doe with other phisicks: keeping your selfe in motion as much as your can whiles ye phisick workes

of the experimental part, see Evan R. Ragland (2017), "Making Trials" in Sixteenth- and Early Seventeenth-Century European Academic Medicine. *Isis* 108.3: 503–528.

723 I.e. Mayerne counselled her to take the second water.

724 Malaga.

725 I.e. eat.

726 I.e. juniper berries with liquor strengthen or refresh the body.

727 Issues.

728 Acquaint.

your Servant C: Rust

[114v]

Sr Theodor Mayerne Advice For My Lady and my Self

It is noe great wonder that glistering sand of a white collour: which my lady voideth[729] in her urine: since her constitution being sub=ject to to many obstructions of y^{e} naturall conduits it is nesesary that this tartarous matter ingendered in y^{e} stomack then finished and seperated in all y^{e} wayes to y^{e} kidneys and y^{e} blader: bee voided by nature: thorough[730] y^{e} conduit ordained to this purpose: sub paean of[731] congealing of stone: or by ading of mater by ye coagulation of it: ye crudity in ye first digestion or ye imperfection thereof is ye mother of most deseases: therefore I knowe not if my Lady doth well to drinke nothing but water and perhaps it would be more convenient too add unto it a fourth parte of good, milde and comfortable old claret wine: I say french not of those strong Liquors[732]: which upon hope [115r] of comfort doe spoyle ye liver: and doe noe good to ye kidneys, wch have noe need of[733] babions[734] wth diueretick[735] under ye pretence of cleansing: for feare of a diebetick passion[736]: since it seemeth that my lady drinketh much and maketh much water, if she will not admit wine for her ordinary drinke she may use ye infusion of china roots: lentisque[737] of spaine and true lignum

[729] Simple present indicative form of "void". It means, in this context, to rid, to make free or clear, of (or from) some quality or condition (*OED*, v. 1b).

[730] Through.

[731] Subpoena. On pain of (*OED*, adv.).

[732] I.e. claret (a French wine) is better than the stronger wines from other places.

[733] Long space in original.

[734] Long space in original. Likely "dation", which is a word for "dose".

[735] Diuretic.

[736] Diabetic. Characterised by an excessive outpouring or effusion (*OED*, adj. B, 1). A "Diabetic passion" led to one "almost insensibly evacuating all" (Richard Allestree (1673), *The Ladies Calling* (Oxford), I.i.11).

[737] Pistacia lentiscus (or "Pills of Mastic").

nefreticum[738]: wch drinke in time[739]: without any sensible alteration of nature: will produce great efects for ye general: In ye season spaw water will be exelent: both to cleans: to coole and comfort ye liver and kidneys: and wash away all ye rags[740] which may bee found in ye ways betwixt ye mouth and ye blader: ye last time that I spake to to [*sic*] my lady about her health[741]: shee did heare[742] patiently resolve to doe: but that resolution did quickly pass and she did nothing: and perhaps will doe soe now: her Ladiships occupation in writing of bookes: with a sedentay[743] life is absolutely bad for health, I have given her at other times steell liquore: wch hath done [115v] Exedingly well with good succes: it weare not amis that shee should goe to it againe in ye later spring and it can be given in ye waters of spaw exelently: for ye rest shee wanteth not phisitians whear she is: to give order to ye preparation and purgation of her body: health is subject to daly mutabilities and there is noe security to deale with a body at a great distance: I doe beleeve this will sufice: for ye present necesety: wch ye lady hath to be advised for her cheife: I would try for a year to take as litle phisick as he could[744] and to fly drugs miserere viuit qui medica viuit[745]: let him rule his hours[,] his meales and his exercises and all all [*sic*] ye action of his life: soe that nature being in a rationall: easy and constant way: by a disposition in a habit to be well: and let him not hearken to soe many flying phisitians: vis all ye world y^{t} speaketh to him: of which every hath a receit: without ground or reason for ye most part: wch are not worth ye hearing: he hath a good constitution [116r] of body[;] let him not spoyle it: in overdoing all things as it is a very familiare ye populor erroure in this nation: the dose of y^{e} ~~stomack~~ balsum is according to ye stomack: I have given from 8 to forty drops and to

[738] Lignum nephriticum (or "Nephritic Wood").

[739] This could mean either "in season" or "early enough; not too late" (*OED*, "time", n. P3, k, a.i and a.ii).

[740] A piece or mass of hard or rough stone (*OED*, n. 1, 1a).

[741] This encounter occurred during Margaret's return trip to England between late 1651 and early 1653.

[742] Here.

[743] Sedentary.

[744] A number of words are blacked out to the point of illegibility here.

[745] "Fly drugs" means, in this context, to "stay away from drugs". This is reinforced by the Latin, which is an inverted version of the English quotation used by Mayerne above: "one lives miserably, who lives by the physician" (Transcription, 135).

a dramm alsoe with profite: in necesary case[s] with one or two good spoonfulls of good canary wine: there is another balsum caled balsum de tolu[746]: which is very good: but I remember that hipocrates saith: sape optima medecina est non uti medecina[747]: this is ye counsell of a freind and a servant: ye liquore of Juniper berries can bee made any wheare: very easily ye wch being being [*sic*] only nessesary to make it: berries and good water: with some ~~water~~ yeast to procure ye fermentation: Thankes be given for ye booke: wch must be given to ye booke binder againe to perfect them: becaus in them something is wanting[;] he doth promis to doe it honestly: then they shall be Renewed by your Servant *[748] (Fig. 7.2).

[116v]

Certane notes out of severall letters by the advice of Sr Theodre Mayerne to Mr Rust an apothecary at London as followeth:

Sr Theodore mislikes ye bathing of your head in cold watter especially in winter and hee sayes y^{t} ye auther himselfe who first devised it[749]: absolutely forbids it in winter: but hee adviseth to purge as hee directeth: at least twice a month ye yeare throughout: and to have Isues: Sr Theodore likes well of your taking creame of tarrter.[750]

For ye wine of Juniper berries wch Sr Theodore adviseth: was further medicine for ye stone and he saith: as exelent a remedy to cleans ye uritories[751]: as to prevent windy vapours

[746] According to some at the time, this balsam "cures most diseases in human Bodies, particularly helps, all Pains coming from cold, chiefly pains in the Stomach, want of Appetite, corroborating and strengthning the whole Body" (R.B. (1680), *The Life of Oliver Cromwel* (London), unpaginated).

[747] "Often the best medicine is not to use [any] medicine". The provenance of this aphorism is not entirely clear, though some late nineteenth-century authors attributed it not to Hippocrates but Celsus (see William Hamilton (1866), *Discussions on Philosophy and Literature, Education and University Reform* (Edinburgh), 254 and J. Warburton Begbie (1875), Address in Medicine. *The British Medical Journal* 2.762: 164–171). We have not been able to confirm this attribution independently.

[748] There is an insignia at the end of this letter, for which see Fig. 7.2.

[749] A reference to Celsus.

[750] Tartar.

[751] Ureter.

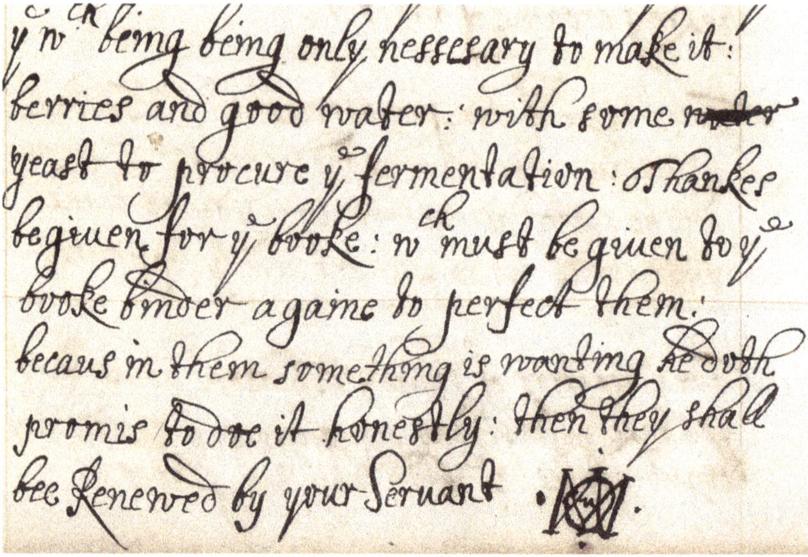
y^{e} wch being being only nessesary to make it:
berries and good water: with some
yeast to procure y^{e} fermentation: Thankes
be giuen for y^{e} booke: wch must be giuen to y^{e}
booke binder againe to perfect them:
becaus in them something is wanting he doth
promis to doe it honestly: then they shall
bee Renewed by your Seruant

Fig. 7.2 Signature Insignia

For ye matters of spaw Sr Theodore adviseth y^{t} you take ye first day six glasses: ye second day eight glases: ye third day ten glases::

[117r] Hee understands y^{t} a hundred ounces: wch is ye quantity y^{t} he would have you continue: and to drinke it if it may bee: ye whole quantity within an houre after you begin: and at your meale to drinke: either to drinke small ale and renish wine or renish wine and your spawe waters: alsoe if you observe after you have taken them three or foure dayes that these doe not pass: he adviseth to leave them of: but he would have it donne by some phisitian: that may Judge of y^{e} operation but this quantity I find: to much for my stomacke: for I could never give above eight beare glasses:

An exelent Glister for y^{e} Colicke as foloweth[:] Take of malowes: pelletories of y^{e} wall: mellilot and cam=momile floures of each a handfull: bay berries: and Juniper berries of each six drams: cumin seeds three drames: seana six drams: bruse all and boyle them in three [^pynts] of watter: untill: a pynte remayne [117v] then with the liquore strained add of diacatholicon and honey of roses of each an once: this is easily made and with small charge[752]: and I dare affirme y^{t} a better cannot be given:

Sr Theodor likes well of your purging after ye matters: but not with mercurius dulcis least[753] you should make it too fammiliar with your body: and soe when [you] have most need of it: it will miss of its effect and operation: but this is the purge that hee would have you take too or 3 times[:] Take of seana halfe an ounce: cream of tarter a dram and a halfe: anysseeds[754]: gallingall[755] roots and cina=mond: of each halfe a dram: Lavender flowres a quarter of a handfull: infuse all twelve houres in six ounces of white wine: with an ounce of tamarinds: then pass it gently through a cloth and add to it: too ounces of y^{e} sirope of y^{e} Juce of buckthorne berries [which] hath not been long knowne: but of very late time:

[118r] And is yet rare amongst us: but used generally by all ye able phisitians with us: and indeed purgeth color beter then any syrops what ever and more freely then any:

The Electuary of sassafras

Take of ye best Sassafras sliced thin: two ounces: faire watter three pyntes: boyle it to ye consumption of a third parte[756]: adding in ye later end of ye boyling halfe an ounce of sinamond[757]: then straine it and add too pound of fine sugar and boyle it to a very thick sirop: and when it is [taken] from ye fier: strow[758] amongst it: pouder of sinamond a dram: nutmege ten graines[,] ambergreece thirty too graines: musk three graines: leaves of gold ten graines: spirit of vitrioll four drops[;] keep it in a pot well covered for your use.

[118v]

Mr Puets[759] Leter to His Exelency ye Lord Marquis of Newcastle.

[752] I.e. it is inexpensive.

[753] Lest.

[754] Aniseeds.

[755] Galangal.

[756] I.e. until a third of the liquid evaporates.

[757] Cinammon.

[758] Strew.

[759] Unknown. Active mid-seventeenth century.

My Lord

I have heare inclosed ye mass of Pills, of which too[760] scruples, which is to graines, is to be taken, at a time, as you weere formerly directed,

I have sent a small quantity of ye balsum a shell[761] of which is put up[762] with ye other things, ye balsum is to bee taken, halfe a dram at a time, in three pills when you goe to bed, and as much in ye morning fasting, and soe continue 14 dayes, and if you have not too stools a day you are to take once or twice a weeke, too scruples of ye amber pills,

[119r-194v are blank]

[195r]

Of Weights./

Haber de Poyos[763] is .16. ounces to the Pound/
and 8. Drammes to ye ounce./
And this serves to way[764] all manner of things, But,–
Gold, Silver, & Bred./

Troyo weight is .12. ounces to the Pound,[765]/
And .20. weights to the ounce./
And this wayes Gold, Silver, and Bred./

Venice weight, y^t silke-men sell their, Gold, Silver, and Silke Lacy by, But is not alowed./

It should be as the Haver-de Poyos,[766] But wants a quar=ter of an ounce in an ounce./

760 Two.

761 Almost certainly the fruit of the balsam tree, which Gerard described as "hollowed like the palme of an hand, two inches long, halfe an inch thicke, covered with a thicke smooth rind, of the colour of a drie Oken leafe; wherein is contained a kernell (of the same length and thicknesse, apt to fill the said *shell* or rinde) of the substance of an Almond" (Gerard, *The Herball*, 1527; emphasis added).

762 To put up. To place in a receptacle for safe keeping or for future use (*OED*, v. 1a, a).

763 This is almost certainly a strangely spelt reference to the avoirdupois system of weights (see Transcription, n. 11).

764 Weigh.

765 For more on the systems of "Troyo weight" and "Venice weight", see Chapter 3, 33–34.

766 Avoirdupois.

[196r]

How to fatt Chickens./

Take white-bred, and soked in Milke, & wrought to a Past, and with it Cramm ye Chickens, and if you put suger into it rost/

Another way with ground Malt, ye Chickens to be in a Pen, darke, & warme/

Ale, & Malt [are] best for ye tight-Chickens And Milke for ye losse ones, Let it be made up in Litle gobbs[767] and ye Penns to be made so as they cannot come one at ye other, nor roome to turne them selves, and always is meat & ale or milke before them in severall vessalls, and these not to have anything given them but what they eat them selves/

Honey is beter then suger/

Sente[768] to my wife, frome the Lady Boswell[769] att the Hage, Our resident wife*

[767] Gob. A mass, lump, or heap (*OED*, n. 1, 1a).

[768] This sentence is in a different hand, probably William Cavendish's.

[769] Margaret Boswell (1595–1682).

CHAPTER 8

Appendices

A. Glossary of Terms

The standard spellings and definitions of key chymical, culinary, herbal, and medical terms in Pw V90 and Harley 6491	*Alternate spellings in Pw V90 and Harley 6491*
Acori veri. A rush-like plant, *Acorus Calamus*, widely distributed in the North Temperate zone, growing in water and wet places, with an aromatic odour, and having a thick creeping rootstock of a pungent aromatic flavour; called "sweet flag" in English (*OED*, "sweet" adj. A, C2).	
Adianthus. One of two types of fern, *Adiantum capillus-veneris*, or maidenhair spleenwort, *Asplenium trichomanes*, which were not clearly distinguished in early use; also called "maidenhair fern" (*OED*, n.).	adianth
Agaric. The bracket fungus *Fomitopsis officinalis*, which grows mainly on larch; a preparation of this, formerly used medicinally chiefly as a laxative and purgative (*OED*, n. A, 1).	agarick
Agnus castus. The plant *Vitex agnus-castus* or its various parts, originally used to reduce sexual potency or desire, and later mainly to treat menstrual and menopausal symptoms; also called "vitex", "chasteberry", or "chaste tree" (*OED*, n. 1).	

(continued)

J. Begley and B. Goldberg, *The Medical World of Margaret Cavendish*, Palgrave Studies in Medieval and Early Modern Medicine,
https://doi.org/10.1007/978-3-030-92927-5_8

(continued)

The standard spellings and definitions of key chymical, culinary, herbal, and medical terms in Pw V90 and Harley 6491	*Alternate spellings in Pw V90 and Harley 6491*
Agrimony. Any of various plants of the genus *Agrimonia*, comprising perennial herbaceous flowering plants native chiefly to temperate areas of the Northern Hemisphere; a medicinal preparation of this plant (*OED*, n. 1).	agremony, agromoine, agrimoney, egremony
Agro di cedro. An Italian name for the lemon or citron (lit. "sour cedar"), which could refer to a variety of citrus fruits; sometimes called *Citrus medica* (Dunglison, *Medical Lexicon*, 40 and 210).	agrodicedro
Ague. An acute or high fever, or a disease characterised by such a fever, often linked to malaria when the fever recurred serially (*OED*, n. 1).	
Album graecum. The dry, white excrement of a dog, hyena, or another bone-eating carnivore, esp. when used as an ingredient in medicinal preparations or in dyeing and tanning (*OED*, n.).	
Alembic. An early apparatus used for distilling, consisting of two connected vessels, a typically gourd-shaped cucurbit containing the substance to be distilled, and a receiver or flask in which the condensed product is collected; also called "limbeck" (*OED*, n. 1).	alembick
Alkanet. The southern European plant, *Alkanna tinctoria*; also called "dyer's bugloss" and "orcanet" (*OED*, n. 2).	alkanett
Alkermes. The dried bodies of scale insects (females of *Kermes ilicis* and *K. vermilio*), found in the Mediterranean region on the kermes oak (*Quercus coccifera*) and formerly used medicinally and as a source of red dye; a medicinal preparation containing these insects, esp. confection of alkermes. They were once thought to be the fruit of the tree upon which the insect lives (*OED*, n. 1).	alkermis, al-chermes, alchemes, kermis, kermes-berry
Aloes. An aromatic resin or wood; spec. the resin or decaying heartwood of any of several South-East Asian trees of the genus *Aquilaria*, burnt or used as incense; esp. that of *A. malaccensis*. More generally, any of various plants constituting the tropical genus *Aloe*, comprising succulent shrubs or trees, which include aloe vera and many other plants with medicinal uses. More particularly, a drug made from the concentrated or dried juice of plants of the genus *Aloe*, having a bitter taste and unpleasant odour, and used mainly as a purgative and laxative (*OED*, n. 1, 2, and 3a).	

(continued)

(continued)

The standard spellings and definitions of key chymical, culinary, herbal, and medical terms in Pw V90 and Harley 6491	*Alternate spellings in Pw V90 and Harley 6491*
Althea. Any plant of the Eurasian genus *Althaea*, comprising herbaceous perennial plants including the marsh mallow, *A. officinalis*; the hollyhock, *Alcea rosea*; and the widely cultivated hibiscus, *Hibiscus syriacus* (*OED*, n.).	
Ambergris. A wax-like substance having a brownish-grey colour and a sweet, earthy scent, formed as a natural secretion in the bile duct of sperm whales; also called "grey amber" (*OED*, n.).	ambergrees, ambergreece, amber greece, ambergreese
Angelica. An aromatic umbelliferous plant, *Angelica archangelica*, cultivated for its seeds and stems, which are used as a flavouring and in confectionery (*OED*, n. 1a).	angelic
Anima hepatis: see Salt of Steel.	
Aniseed. The seed of the anise, used as a carminative, and in the preparation of Oil of Anise, Spirit of Anise, Anise water, and Anisette (*OED*, n. 1a).	seeds of anis, seeds of annis, annis, anis seed, anisseed, anyseed, anysseed
Antimony. One of the elementary bodies (atomic number 51), a brittle metallic substance, of bright bluish-white colour and a flaky crystalline texture. Its metallic characteristics are less pronounced than those of metals generally, and it forms the fourth member of the natural series that includes nitrogen, phosphorus, arsenic, antimony, and bismuth. Historically, "antimony" was applied to a native trisulphide; also called "gray antimony", "stibnite", or, when calcined and powdered, "crude" or "black" antimony (*OED*, n. 1a).	antimoney, antimonie, antimonye
Aparine. The climbing plant *Galium Aparine* or goosegrass, which adheres by its minute, hooked bristles to hedges or shrubs, the clothes of passers-by, etc.; also called "cleavers" (*OED*, "cleavers", n.).	aparin
Apoplex. A malady, very sudden in its attack, which arrests the powers of sense and motion; it is usually caused by an effusion of blood or serum in the brain, and preceded by giddiness, partial loss of muscular power, etc.; also called "apoplexy" (*OED*, n. 1a).	appoplex
Apozem. A decoction or infusion (*OED*, n.).	apozeme
Aqua benedicta Rulandi. A purgative that, according to the medical writer Samuel Boulton, was originally "described by *Hartman* in his *Chymica Praxi*: It is made of *Antimony* and *Vitriol anæ*, and twice or thrice as much of Sal niter into *Crocus metallorum*, which being exquisitely sweetned, is given by infusion ℈1, or somewhat more, as the sickness requires in ℥3 of whitewine" ((1656), *Medicina magica tamen* physica (London), 85).	aqua benidictæ Roulandy

(continued)

(continued)

The standard spellings and definitions of key chymical, culinary, herbal, and medical terms in Pw V90 and Harley 6491	*Alternate spellings in Pw V90 and Harley 6491*
Aquafortis. The early scientific, and still the popular, name of the nitric acid of commerce (dilute HNO_3), a powerful solvent and corrosive (*OED*, n. 1).	aqua-fortis
Aqua Imperialis. A medicine that consisted of many ingredients including "the roots of Cypress, Orris, Florentine, [and] Calamus Aromaticus", which are infused in rose water and white wine (Culpeper, *Pharmacopoeia*, 66).	imperiall water
Aqua Mirabilis. Latin for "wonderful water". According to Culpeper, it consists of "Cloves, Galanga, Cubebs, Mace, Cardamoms, Nutmegs, [and] Ginger" as well as "Juice of Celandine", "spirits of Wine", and "white Wine" (Culpeper, *Pharmacopoeia*, 67).	
Aquapendente's Pills. These pills were a compound of aloes and rose water dried in the sun (see John L. Heilbron (2012), *Galileo* (Oxford), 154). The pills were named after William Harvey's teacher in Padua, Fabricius ab Aquapendente.	Aquapendentis Pills, Aquapendentis-Pills
Aqua-vitae: see Spirit of Wine.	aquavitæ, aqua vita
Aretnuß. An antiquated German term for "areca nut", which is the seed of the areca palm (*Areca catechu*) (see François Roux (1809), *Nouveau Dictionnaire* (Halle), 59).	
Aristolochia. A genus of shrubs, of which one common species is *Aristolochia clematitis*, the common birthwort. It was often divided into "round" and "long" kinds, which are now called *A. rotunda* (smearwort or English mercury) and *A. longa* (sarrasine), respectively (*OED*, n.).	aristolochy
Aromaticum rosatum. As the name indicates, "red rose" is the chief ingredient in the recipe for "Aromaticum Rosatum", which Culpeper maintained "strengthens the brain, heart, and stomach, and all such internal Members as help towards concoction, it helps digestion, consumes the watry excrements of the bowels, strengthens such as are pin'd away by reason of the violence of a disease, and restores such as are in a consumption" (Culpeper, *Pharmacopoeia*, 121).	
Ash-Key. The winged, two-celled seed or samara of the ash-tree (*OED*, "ash" n. 1, C, 2a).	ash key, ashen key
Astringent. A medicine meant to bind one's tissues together via constriction, so as, for example, to stop the flow of blood (*OED*, n. B).	

(continued)

(continued)

The standard spellings and definitions of key chymical, culinary, herbal, and medical terms in Pw V90 and Harley 6491	*Alternate spellings in Pw V90 and Harley 6491*
Balm. An aromatic substance, consisting of resin mixed with volatile oils, exuding naturally from various trees of the genus *Balsamodendron*, and much prized for its fragrance and medicinal properties; or, more generally, an aromatic ointment used for soothing pain or healing wounds (*OED*, n. 1, I, 1 and 5). Also see: Balsam, Opobalsamum.	baulm, balme
Balneo: see Mary's Bath.	
Balsam. An aromatic, resinous product or vegetable juice. "True balsam", or "balsam of Mecca", the earliest known sort, is also called "balm of Gilead". The discovery of America brought knowledge of many other natural balsams or oleo-resins, e.g. balsam of Acouchi, balsam of Copaiba, balsam of Peru, and balsam of Tolu, all used medicinally (*OED*, n. A, I, 1a and b). Also see: Balsam of Sulphur, Balm, Opobalsamum, Tolu Balsam.	balsum, balsume, balsom
Balsam of Peru: see Balsam.	balsum of Peru, balsum of Perue, balsom of Peru, balsume of Perue, balsum of Hyspanicum
Balsam of Sulphur. A medicinal compound consisting of sulphur dissolved in oil, turpentine, and/or wax (*OED*, "balsam", n. A, I, 2b). Also see: Balsam, Sulphur.	balsome of sulpher, balsum of sulfer
Barberry. The berries of the Barberry shrub, *Berberis vulgaris* (*OED*, n. 1 and 2).	barbery
Barm. The froth that forms on the top of fermenting malt liquors, which is used to leaven bread, and to cause fermentation in other liquors; yeast, leaven (*OED*, n. 2, 1a).	barme
Basilicum. The name of several ointments that were supposed to possess "sovereign" virtues; from "basilicus" meaning royal; also called "basilicon" (*OED*, n.).	bassilicum
Basilic Vein. The large vein of the arm starting from the elbow and discharging into the axillary vein. So-called for its supposed great importance; the right and left basilic veins were thought to be in direct communication with the liver and spleen respectively (*OED*, "basilic", adj. 2).	basilicke vaine
Bays. A berry or small fruit, esp. that of the laurel or bay-tree (*OED*, n. 1, 1a).	bayes
Bay-Salt. Salt, obtained in large crystals by slow evaporation; originally, from sea-water by the sun's heat (*OED*, n.).	bay salt
Bead Amber: see Yellow Amber.	bead

(continued)

(continued)

The standard spellings and definitions of key chymical, culinary, herbal, and medical terms in Pw V90 and Harley 6491	*Alternate spellings in Pw V90 and Harley 6491*
Benjamin. A dry and brittle resinous substance with a fragrant odour and slightly aromatic taste, obtained from the *Styrax benzoin*, a tree of Sumatra, Java, etc. It is used in the preparation of benzoic acid, and in medicine and perfumery. For scientific distinction, it is now termed "gum benzoin"; also called "benzoin" (*OED*, n. 1).	bennjomin, benjamine
Betony. A plant (*Stachys betonica*) of the *Labiate* order, having spiked, purple flowers and ovate crenate leaves, and to which medicinal and magical virtues were attributed (*OED*, n. a).	betoney, bettony, bettoney
Bezoar. In general, an anti-poison, but often, more specifically, a stony concretion found in the bowels of ruminants, frequently called "bezoar-stone", which has long been popular in both Western and Eastern medicine. Often used to make various medicinal waters (bezoar-water) (*OED*, n. 1, 2a, and 2c).	East bezoar, orientall-bezarstone, beozer
Birthwort: see Aristolochia.	
Biscuit Cakes. Baked, unleavened bread (*OED*, n. C, 3a).	bisqueet-cakes
Bismuth. One of the elementary bodies; a reddish-white metal, found native, and also in combination, in numerous ores; it is brittle and melts at a low temperature. Chemically, bismuth (Bi) is closely allied to antimony. Its chief use in the arts is as an alloy; the oxide and some salts are used in medicine (*OED*, n.).	bismutho, besmuthum
Blind-Head. A cover for an alembic or retort; a retort with such a cover (*OED*, n. 2).	blind head, blinde head
Bole Armoniac. A soft, friable, fatty earth, usually pale red, originating from Armenia. It is an astringent, and was used as an antidote and styptic (*OED*, "ammoniac", adj. A, 1b).	bole-armoneck, bole-armeneck, bolus-armonus, bolearmenak
Bolus. A medicine shaped into a ball, usually larger than a pill (*OED*, n. 1a).	
Borage. A genus of plants, giving its name to a family (*Boraginaceæ*), spec. the common British species (*Borago officinalis*), which was formerly esteemed as a cordial (*OED*, n.).	
Boxwood. The wood of the common box, *Buxus sempervirens*, characteristically light cream to yellow in colour, and used esp. by woodturners and wood engravers because of its fine grain and high density (*OED*, n. 1).	boxe-wood
Bragget. A drink made of honey and ale fermented together (*OED*, n.).	braccet
Brimstone: see Sulphur.	brimston

(continued)

(continued)

The standard spellings and definitions of key chymical, culinary, herbal, and medical terms in Pw V90 and Harley 6491	*Alternate spellings in Pw V90 and Harley 6491*
Broom. A shrub, *Sarothamnus* (or *Cytisus*) *Scoparius*; also the genus to which this belongs, and the allied genus *Genista*, which includes the white, giant, or Irish broom. These are commonly cultivated in gardens (*OED*, n. 1).	broome
Bruscus. Any of the low evergreen shrubs constituting the chiefly Mediterranean genus *Ruscus*; esp. butcher's broom, *R. aculeatus*, which has pale green flowers and is commonly grown in gardens; also called "ruscus" (*OED*, "ruscus", n.).	
Bryony. The English name of the plant-genus *Bryonia*; and spec. the common wild species (*B. dioica*); the common plant is white, while the black is an endogenous climbing plant, known as "lady's seal" (*Tamus communis*) (*OED*, n. 1 and 2).	briony
Buckthorn. The shrub *Rhamnus catharticus*; the berries of which yield sap-green and other pigments, and were formerly used as a powerful cathartic (*OED*, n.).	buckthorne
Bugle. Any of numerous herbaceous plants of the Old World genus *Ajuga*, many of which are low plants with an erect stem of blue or purple flowers, often used for ground cover. *Ajuga reptans* or "the common bugle" was thought to have the same virtues as "lapis prunella" (for which see Sal prunella) (*OED*, n. 1).	buge
Bugloss. Any of several hairy herbaceous plants of the family *Boraginaceae*, esp. those of the genera *Anchusa* (as the small annual *A. arvensis*) and *Echium* (in English "viper's bugloss") (*OED*, n.).	buglosse, buglasse, bueglas
Burgundy Pitch. The resinous juice of the Spruce-fir (*Abies excelsa*), sometimes called white pitch, though it is reddish-brown or whitish-yellow. Its name is derived from the fact that it is obtained from the neighbourhood of Neufchâtel, which was once Burgundian territory (*OED*, n.).	Burgondy pitch, Burgundy-pitch
Calcination. The action or process of calcining; reduction by fire to a "calx", powder, or friable substance; the subjecting of any infusible substance to a roasting heat. Extended to other processes producing similar results; or used as synonymous with oxidation in general (*OED*, n. 1a and b).	
Campane. A bell-shaped vessel forming the head or upper part of an alembic (*OED*, n. 2).	

(continued)

(continued)

The standard spellings and definitions of key chymical, culinary, herbal, and medical terms in Pw V90 and Harley 6491	*Alternate spellings in Pw V90 and Harley 6491*
Camphor. A whitish, translucent, crystalline volatile substance, belonging chemically to vegetable oils, and having a bitter aromatic taste and a strong characteristic smell: it is used in pharmacy, and was formerly reputed as an anaphrodisiac (*OED*, n. 1).	camfor, campher, cimford
Canary. A sweet, fortified white wine produced in the Canary Islands (*OED*, n. 3).	Canarie
Cantharides. The pharmacopoeial name of the dried beetle, *Cantharis vesicatoria* (or "Spanish Fly"). Used externally as a rubefacient and vesicant; internally as a diuretic and stimulant to the genito-urinary organs, etc. Formerly considered an aphrodisiac (*OED*, n. 2).	
Carabe: see Yellow Amber.	carebe
Caranna. A resin obtained from the tree *Bursera acuminata* (*OED*, n.).	caramna
Cardamom. The aromatic seed capsules or seeds (whole or ground) of the South Asian herbaceous plant *Elettaria cardamomum* (family *Zingiberaceae*), used as a spice and as a medicinal agent; also the plant itself. Also, in early use, the seed capsules or seeds of any of several other plants of the family *Zingiberaceae* (*OED*, n. a).	cardamam
Carduus benedictus. A thistle-like plant, now classified as *Cnicus benedictus*, in the family *Asteraceae*. It is also known in English as "St. Benedict's thistle", "blessed thistle", "holy thistle", or "spotted thistle" (*OED*, n.).	cardus-benedictus, cardus
Cassia. An inferior kind of cinnamon, esp. the bark obtained from *Cinnamomum Cassia*; thicker, coarser, less delicate in flavour, and cheaper than the true cinnamon; also called cassia lignea (*OED*, n. 1, 1a).	casia, cassice lign
Cassia fistula. The name given in the Middle Ages to the Pudding Pipe tree, a native of India, but cultivated in Northern Africa, the West Indies, etc., which produces the *cassia pods* containing a pulp used as a laxative; also a pod of this tree (*OED*, "cassia", n. 1, 4a and b).	cæssia fistula
Castor. A reddish-brown unctuous substance, having a strong smell and nauseous, bitter taste, obtained from two sacs in the inguinal region of the beaver; used in medicine and perfumery; also called "castoreum" (*OED*, n. 1, 2).	castorium

(continued)

(continued)

The standard spellings and definitions of key chymical, culinary, herbal, and medical terms in Pw V90 and Harley 6491	*Alternate spellings in Pw V90 and Harley 6491*
Catarrh. The profuse discharge from the nose and eyes that generally accompanies a cold, and which was formerly supposed to run down from the brain. It can also refer to the inflammation of a mucous membrane; usually restricted to that of the nose, throat, and bronchial tubes, causing an increased flow of mucus, and often attended with sneezing, cough, and fever; constituting a common cold (*OED*, n. 1 and 3).	catarr
Catholicon. An electuary; a universal remedy or prophylactic; panacea (*OED*, n. 1a). Also see: Electuary.	catholicum, cathalicum
Caudle. A warm drink of sweetened or spiced wine or ale thickened with gruel or other ingredients, given chiefly to invalids, expectant mothers, etc. (*OED*, n. a).	
Celandine. The name of two distinct plants, bearing yellow flowers, which were regarded by the old herbalists as species of the same plant, and identified with the "greater and lesser chelidonia". Common or greater celandine (*Chelidonium majus*) yields a thick yellow juice and was deemed a powerful remedy for weak sight; also called "swallow wort". Small or lesser celandine is now called "pilewort" or "figwort" (*Ranunculus Ficaria*), and was traditionally used to treat piles (*OED*, n. 1a and b).	chelendine, caladine
Centaury. Either of two European plants of the gentian family (*Gentianaceae*), *Centaurium erythraea* (more fully, common centaury, less centaury, lesser centaury, small centaury, smaller centaury), and yellow-wort, *Blackstonia perfoliata* (in full yellow centaury, more centaury, great centaury, greater centaury) (*OED*, n. 1a).	centory, century, centuary
Ceterach. A genus of ferns; the most common species, *Ceterach officinarum*, is known as "rustyback" (*OED*, n.).	cetrach
Chafing-Dish. A vessel to hold burning charcoal or other fuel, for heating anything placed upon it; a portable grate (*OED*, n.).	chafing-disshe
Chamaedrys. Any of the plants constituting the genus *Teucrium*, esp. common or wall germander, *T. chamaedrys*, an evergreen shrub with bright pink flowers, formerly used medicinally and now commonly grown as an ornamental (*OED*, "germander", n.).	chamedrys
Chamaepitys. A species of flowering plant of the family *Lamiaceae*, *Ajuga chamaepitys*, said to be named from its resinous smell; also known in English as "yellow bugle" or "ground-pine" (*OED*, "ground-pine", n. 1).	chamepythis, champythis

(continued)

(continued)

The standard spellings and definitions of key chymical, culinary, herbal, and medical terms in Pw V90 and Harley 6491	*Alternate spellings in Pw V90 and Harley 6491*
Chervil. A garden pot-herb (*Anthriscus cerefolium*), the young leaves of which are used to impart an aromatic flavour to soups, stews, salads, etc. (*OED*, n. 1).	cherveil
Chicory. The plant *Cichorium Intybus*, found wild in the south of England, and elsewhere in Europe and Asia, cultivated for its roots (*OED*, n. 1).	cichory, cicory, succory, shuckory, suckory
China-Root. The root-stock of *Smilax China*, which is similar to Sarsaparilla, and was quite popular in early modern medicine as a decoction (*OED*, n. 2, 1a).	china root
Clyster: see Glyster.	clystor
Colic. A name given to severe paroxysmal griping pains in the belly, due to various affections of the bowels or other parts; also, to the affections of which such pains are the characteristic symptom (*OED*, n. A).	colick, collick, collicke, cholike, cholicke, cholick
Coloquint. The bitter-apple (*Citrullus colocynthis*), which contains a light, spongy, and extremely bitter pulp, furnishing the well-known purgative drug. Also, the fruit of this plant, and the drug prepared from it; often called "colocynth" (*OED*, n.).	coloquinte
Coltsfoot. The common name of *Tussilago Farfara*, a common weed in waste or clayey ground, with large spreading cordate leaves that are downy beneath, and yellow flowers appearing in early spring before the leaves; formerly used to treat, among other things, asthma (*OED*, n. 1a).	coultsfoote
Comfrey. *Consolida maggiore* (or "common comfrey") now has the scientific name *Symphytum officinale*. It is a tall plant, common on the margins of streams and ditches, with rough leaves, and drooping clusters of yellowish-white or reddish-purple bell-shaped flowers. Also applied to other plants, chiefly as a rendering of Latin *Consolida* or *Symphytum*; formerly esteemed as a vulnerary (*OED*, n. 1a and 2).	comfre
Complexion. In the Galenic tradition, the blend of humours (black bile, blood, yellow bile, phlegm), and associated qualities (hot, cold, dry, wet), which (at least in part) determine the physical characteristics of a person; also called "temperament" (*OED*, n. I, 1a).	
Coney. A rabbit as hunted, bred, sold, or prepared for food (*OED*, n. 1, I, 2a).	connie
Confit. A sweetmeat made of fruit, root, etc., preserved with sugar; now usually a small, round, or oval mass of sugar enclosing a caraway seed, almond, etc.; also called "comfit" (*OED*, n. a).	comfitt

(continued)

(continued)

The standard spellings and definitions of key chymical, culinary, herbal, and medical terms in Pw V90 and Harley 6491	*Alternate spellings in Pw V90 and Harley 6491*
Conserve. A confectionary or medicinal preparation made from some part of a plant plus sugar; often mixed or beaten into a uniform mass (*OED*, n. 2a).	concerve
Consolida maggiore: see Comfrey.	cansolida majorie
Consumption. Originally, this referred to some abnormality or loss of humours, resulting in wasting of the body (extreme weight loss); such wasting (*OED*, n. 2a).	consumtion, consummtion
Contracture. A condition of persistent contraction and rigidity in the muscles or the joints (*OED*, n. 1).	
Contrayerva. The root-stock and rhizome of species of *Dorstenia* (*D. Contrayerva* and *D. braziliensis*) native to tropical America. The English term is imported from Spanish and means "counter herb". It is used as a stimulant, tonic, and was formerly reputed as an antidote to snake bites (*OED*, n.).	contra-yerva, contraverva
Cordial. A medicine, food, or beverage that is stimulating, "comforting", or invigorating to the heart (*OED*, n. B, a). Also see: Julep.	cordiall
Costmary. A perennial aromatic herb, *Tanacetum balsamita*, having clusters of small, yellowish, button-shaped flowers and silvery, oblong, serrated leaves, smelling of balsam and mint, probably native to the Mediterranean but widely cultivated in Europe from the Middle Ages as a medicinal and flavouring herb (*OED*, n.).	
Cowslip. The common name of *Primula veris*, a well-known wild plant found in pastures and on grassy banks, blossoming in spring, with drooping umbels of fragrant yellow flowers; also called paigle (*OED*, n. 1a).	cows lipe
Crab's Eyes: see Gascoigne's Powder.	crabs eyes, crabbs eyes
Cream of Tartar. The purified and crystallised bitartrate of potassium, used in medicine and for various technical purposes (*OED*, "cream", n. 2, 4a). Also see: Tartar.	creame of tarter, cræme of tartar, creame of tartar
Crocus martis. A name given to various yellow or red powders obtained from metals by calcination; in Pw V90 the crocus of iron (sesquioxide or peroxide of iron). It is sometimes referred to as "crocus of mars" because "mars" was the astrological symbol for iron (*OED*, "colcothar", n. a).	crocus of mars

(continued)

(continued)

The standard spellings and definitions of key chymical, culinary, herbal, and medical terms in Pw V90 and Harley 6491	*Alternate spellings in Pw V90 and Harley 6491*
Crocus metallorum. A fused sulphide of antimony used as an emetic; also known as "crocus of antimony, "liver of antimony", or "hepar-antimony" (*OED*, "crocus", n. 3a).	crocus metalorum, crocus-metalorum, crocus mettallorum, crocus mettalorum, croqus metalorumm
Crystalline Tartar: see Cream of Tartar.	christalline tarter
Cubebs. The berry of a climbing shrub Piper Cubeba or *Cubeba officinalis*, a native of Java and the adjacent islands; it resembles a grain of pepper, has a pungent spicy flavour, and is used in medicine and cookery (*OED*, n. a).	cubeb
Cucurbit. A vessel or retort, originally gourd-shaped, used in distillation and other chymical processes, or for keeping liquids, etc., in; forming the lower part of an alembic (*OED*, n. 1, 1).	cucurbite
Danewort: see Dwarf Elder.	dainewort
Dead Head. The residuum remaining after the distillation or sublimation of any substance; also called in Latin "caput mortuum" (*OED*, "caput mortuum", n. 2).	
Defluxion. A supposed flow of humours to a particular part of the body in certain diseases. Also, the flow or discharge accompanying a cold or inflammation; a running at the nose or eyes (*OED*, n. 2). Also see: Catarrh.	
Dephlegmed. Used to describe a spirit or acid that is freed from "phlegm" or watery matter; rectified (*OED*, adj.).	deflegmed
Diacatholicon. An old term for a laxative electuary, which was so-called for its manifold compositions, or, according to some, for its general usefulness: hence, a universal remedy (*OED*, n.).	diacathoholicon
Diamber. An old stomachic and cordial containing ambergris, musk, and other aromatics (*OED*, "diamber", n.).	
Diaphoretic. Having the property of inducing or promoting perspiration (a sudorific); a medicinal agent having this property (*OED*, adj. A; n. B).	diaphorettick, diphoretike
Diarrhodon abbatis. A cordial powder that apparently derives its name from the Abbot who invented it. Consisting of many ingredients, it was thought to strengthen the heart, stomach, and liver; assist in digestion; and prevent vomiting (Ephraim Chambers (1728), *Cyclopædia, or, An Universal Dictionary of Arts and Sciences*, Vol. 1 (London), 208).	diarodan abatis

(continued)

(continued)

The standard spellings and definitions of key chymical, culinary, herbal, and medical terms in Pw V90 and Harley 6491	*Alternate spellings in Pw V90 and Harley 6491*
Digestion. An operation of maturing or preparing a substance using gentle heat (*OED*, n. 5a).	
Diuretic. A substance having the property of promoting excretion or the evacuation of urine (*OED*, n. B).	diueretick
Dock-Root. The roots of the common dock (*R. obtusifolius*), well known as a popular antidote for nettle-stings (*OED*, "dock", n. 1, 1a and C1).	dockroote
Dragons. The plant *Dracunculus vulgaris*; also called "dragonwort" (*OED*, n.).	
Dragon's Blood. A bright red gum or resin from the fruit of a palm, *Calamus draco.* Formerly also applied to the inspissated juice of the dragon-tree, *Dracæna draco*, and to exudations from *Pterocarpus draco*, *Croton draco*, and other plants (*OED*, n.).	dragons bloud
Dropsy. An illness characterised by the accumulation of fluids in the connective tissues of the body; also linked with incredible thirst, and used metaphorically, e.g. as with greed (*OED*, n. A, 1a, 2, and 3). Also see: Hydroptic.	dropsie, dropsey
Dulcified. Sweetened, which, in an early modern medical context, might also mean to make less corrosive or acidic (*OED*, adj. 1).	
Dwarf Elder. A *Sambucus ebulus* or, in English, danewort (*OED*, "danewort", n. and "elder", n. 1, 2a and b).	dwarfe-elderne, dwarfe elder
Eglantine. A species of rose, *Rosa rubiginosa* (and some other species, as *R. micracantha*), having strong hooked prickles, pink single flowers, and small aromatic leaves; frequently cultivated in gardens; also called "sweet-briar" (*OED*, n. 1, 1).	eglintine
Elder. A low tree or shrub, *Sambucus nigra*, called the "Common Elder" or "Black-berried Elder" (*OED*, n. 1).	eldere
Elecampane. A perennial composite plant, Horse-heal (*Inula Helenium*), formerly used as a tonic and stimulant (*OED*, n. 1).	ilycampany, elecompane, alicampana
Electuary. A medicinal conserve or paste, consisting of a powder or other ingredient mixed with honey, preserve, or syrup of some kind (*OED*, n. 1a).	ellectuary
Elemi. A stimulant resin, obtained from various trees, such as *Canarium commune* (Manilla), *Icica icicariba* (Brazil), and *Elaphrium elemiferum* (Mexico) (*OED*, n.).	elemy
Elixir: see Philosophers' Stone.	elixer, elixier

(continued)

(continued)

The standard spellings and definitions of key chymical, culinary, herbal, and medical terms in Pw V90 and Harley 6491	*Alternate spellings in Pw V90 and Harley 6491*
Endive. The name of two species of Chicory (*Cichorium*). The first is *C. Intybus*, now called Wild Endive, Succory, or Chicory; it is indigenous to Europe, and common in a wild state in many parts of England. The second is *C. Endivia*, alleged by some writers to have been imported into Europe from China in the sixteenth century. Of this there are two varieties, the Batavian or broad-leaved, formerly also called Scariole, and that with a curled or frizzled leaf, which is commonly blanched for use as salad, etc. (*OED*, n. a and b). See also: Chicory.	
Epileptic. Of or pertaining to epilepsy; of the nature of epilepsy; also, being affected with epilepsy (*OED*, adj. A, 1 and 2).	epilepticke
Epsom Water. The water of a spring near Epsom, Surrey, which is rich in magnesium sulphate (*OED*, n. C).	Epsam (water)
Faeces. Sediment; dregs, lees, subsidence, refuse (*OED*, n. 1).	feces
Fat. A vessel of large size for liquids (*OED*, n. 1, 2a).	fatt
Female Fern. Any of various tall, graceful ferns; spec. any fern of the genus *Athyrium*, occurring worldwide in moist, shady habitats, esp., *Athyrium filix-femina*, which is often cultivated for its delicate lacy foliage; also called "lady fern" (*OED*, "lady", n. 2C, c). Also see: Fern.	femal fearne, female ferne
Fenugreek. A leguminous plant (*Trigonella Fœnum Græcum*) cultivated for its seeds, which are used by farriers (*OED*, n. 1).	fænu=greek, foenugreekseed, fonngreek
Filipendula. An obsolete term for the drop-wort (*Spiræa filipendula*) (*OED*, n.).	philipendula, phillipendula
Fistula. Originally: any abnormal tube-like passage between the skin and an organ or cavity within the body, typically exuding pus or other matter. Also, the condition of having such an abnormal passage or something resembling a fistula in the body, esp. a hidden source or conduit of moral corruption that is difficult to eradicate (*OED*, n. 3a and c).	
Five-Leaved Grass. The plant cinquefoil (*Potentilla reptans*) (*OED*, "five-leaf", n.).	five leave grasse
Flower. In chymistry, the pulverulent form of any substance, esp. as the result of condensation after sublimation (*OED*, n. 2c).	
Flux. An abnormally copious flowing of blood, excrement, etc. from the bowels or other organs; a morbid or excessive discharge, spec. an early name for dysentery (*OED*, n. I, 1a).	fluxe

(continued)

(continued)

The standard spellings and definitions of key chymical, culinary, herbal, and medical terms in Pw V90 and Harley 6491	*Alternate spellings in Pw V90 and Harley 6491*
Fomentation. The action or process of applying a heated therapeutic preparation, typically in the form of a moistened poultice or pad of material, to the surface of the body to relieve swelling, inflammation, or pain (*OED*, n. 1a).	
Fumitory. A plant of the genus *Fumaria* (or the related *Corydalis*), usually *F. officinalis* (*OED*, n. 1a).	fumetory, fumitorye, fumetera
Galangal. The aromatic rhizome of certain Asian plants of the genera *Alpinia* and *Kaempferia*, of the ginger family. The type most commonly used in cooking is *A. galanga*, sometimes known as "greater galangal", while "lesser galangal" refers to the more pungent *A. officinarum* (*OED*, n. 1a).	gallingall
Galbanum. A gum resin obtained from certain Persian species of *Ferula*, esp. from *F. galbaniflua* and *F. rubricaulis* (*OED*, n. 1).	
Gallipot. A small earthen glazed pot, esp. one used by apothecaries for ointments and medicines (*OED*, n. 1a).	gally-pott
Gascoigne's Powder. A powder made primarily from crab's-eyes, which are round concretions, found in the stomachs of crayfish and some other crustacea, consisting mainly of carbonate of lime; it has been used as an absorbent and antacid (*OED*, "crab's-eyes", n. 1).	Gascoigne powder, Gascons powder
Genitor. An alternative for "genitory", which originally referred to the male genitals, spec. the testicles, but later also the female genitals (*OED*, n.).	geniture
Gentian. Any plant belonging to the genus *Gentiana*, esp. *G. lutea*, which yields the gentian-root of pharmacopoeia; known in English as the "great yellow gentian" (*OED*, n. A, 1a).	gentiana
Germander: see Chamaedrys.	
Gillyflower. A clove, or several plants of the genus *Dianthus*, which have flowers with a spicy, sweet scent similar to that of cloves (*OED*, n. 1 and 2).	gillyflowres, gilly flowres, gilly-flowers
Glyster. A medicine injected into the rectum to empty or cleanse the bowels, afford nutrition, etc. This form of medicine was popular in the early modern period, esp. among the upper classes (*OED*, n. 1a).	glister
Gob. A mass, lump, or heap (*OED*, n. 1, 1a).	gobb
Grains. In full Grains of Paradise: the capsules of *Amomum Meleguetta* of Western Africa, used as a spice and in medicine; also called Guinea grains (*OED*, n. 1, I, 4a). Also see: Cardamom.	graines

(continued)

(continued)

The standard spellings and definitions of key chymical, culinary, herbal, and medical terms in Pw V90 and Harley 6491	*Alternate spellings in Pw V90 and Harley 6491*
Gravel. Aggregations of urinary crystals that can be recognised as masses by the naked eye (as distinguished from sand); the disease of which these are characteristic. Also popularly used to indicate pain or difficulty in passing urine with or without any deposit (*OED*, n. 4).	gravell
Green Sickness. A disease characterised by greenish discolouration (chlorosis), which was associated with virginity and unmarried women, frequently with reference to the abnormal appetite (pica) accompanying this disorder; also called "green jaundice" (*OED*, n. 1 and 2).	greene sicknesse
Green Vitriol. Crystalline ferrous sulphate, a pale blue-green salt formed by the action of sulphuric acid on iron or its compounds (*OED*, "green", n. C1, d.). Also see: Vitriol.	greene vitrioll
Gromwell. The common name for plants of the genus *Lithospermum*, characterised by hard stony seeds, which were formerly much used in medicine (*OED*, n. a).	
Groundsel. Any plant belonging to the genus *Senecio*, esp. *S. vulgaris* (common groundsel), a common European weed, which is given as food to cage-birds and was formerly largely used for medical purposes (*OED*, n. 1).	groundsell
Guaiacum: see Lignum vitae.	guaick
Gum Ammoniac. A gum-resin, of peculiar smell, and bitterish taste, the inspissated juice of an umbelliferous plant (*Dorema ammoniacum*) found wild from North Africa to India. Employed in medicine, and as a cement (*OED*, "ammoniac", adj. A, 2).	gumm amoniack
Haemorrhoidal Veins. Veins that are distributed to the rectum and adjacent parts (*OED*, "haemorrhoidal", n. 2).	hemerehoidale veines, hemrodveynes
Haemorrhoids. A disease characterised by tumours of the veins about the anus; also called "piles" (*OED*, n. 1).	hemrodes, hemoroides, hamorhoids, hemærhoides
Hartshorn. The horn or antler of a hart (i.e. a male deer) or various other horned mammals. Spirit of hartshorn, sometimes simply called hartshorn, is an aqueous solution of ammonia (whether obtained from harts' horns or otherwise), while the salt of hartshorn is the carbonate of ammonia (or "smelling salts"). "Hartshorn" also refers to one of two plants having leaves branched like a stag's horn, either Buck's-horn Plantain (*Plantago Coronopus*) or Swine's Cress (*Senebiera Coronopus*) (*OED*, n. 1, 2, 3).	harts horne, harts-horne, harteshorne, hartes horne, hartts-horne
Hart's Tongue. The common name of the fern *Scolopendrium vulgare*; also extended to other species of the genus (*OED*, n.).	harts tongue, hartes-tongue

(continued)

(continued)

The standard spellings and definitions of key chymical, culinary, herbal, and medical terms in Pw V90 and Harley 6491	*Alternate spellings in Pw V90 and Harley 6491*
Hectic Fever. A prolonged, debilitating fever causing wasting of the body; consumption (*OED*, n. 1, C1).	hectick feavor
Hellebore. Any of various medicinal plants valued chiefly for their purgative properties, mostly from the genera *Veratrum* and *Helleborus* (*OED*, n. 1a).	helebore, ellebore, ellebory
Hepar Antimony: see Crocus Metallorum.	hepar-antimones
Herb of Grace: see Rue.	
Hermes' Seal. An airtight closure of a vessel, esp. a glass vessel, by fusion, soldering, or welding; also called "hermetic seal" (*OED*, "hermetic", adj. 2b).	hermis seal
Hermodactyl. A bulbous root, probably of a species of *Colchicum*, formerly imported from the East (*OED*, n. 1).	
Hiera picra. A purgative drug composed of aloes and canella bark, sometimes mixed with honey and other ingredients (*OED*, n. 1a).	hiera pigra, hiera
Hippocras Bag. A conical cloth bag used to strain or filter liquids finely. Hippocras was a popular drink made from wine sweetened with sugar or honey and flavoured with spices and other ingredients, typically drunk as a restorative or a digestive (*OED*, "Hippocrates", n. C).	hipocras-bagg
Hogshead. A large cask, esp. for storing liquids; spec. one of a definite capacity; made from the staves of a bourbon cask with new oak ends (*OED*, n. 1).	
Horehound. A labiate herb, *Marrubium vulgare*; its aromatic bitter juice is much used as a remedy for coughs, etc. (*OED*, n. 1a).	
Hound's Tongue. Any of various plants constituting the genus *Cynoglossum*, esp. a European weed, *C. officinale*; also called "Dog's tongue" (*OED*, n.).	houndes tongue
Hyacinth. A reddish-orange gem, which is usually a variety of zircon; also applied to varieties of topaz and garnet (*OED*, n. 1b).	hiacynth
Hydroptic. Swollen with water or fluid; also called "hydropic" or "dropsical" (*OED*, adj.). Also see: Dropsy.	hydroptick
Hypericon: see St. John's Wort.	
Hypochondria. A common diagnosis within humoural medicine, which was thought to stem from an imbalance caused by the production of black bile and vapours by the liver and spleen, characterised by low spirits and pains of various kinds, usually dyspepsia or abdominal discomfort (*OED*, n.).	hypochondry

(continued)

(continued)

The standard spellings and definitions of key chymical, culinary, herbal, and medical terms in Pw V90 and Harley 6491	*Alternate spellings in Pw V90 and Harley 6491*
Hyssop. A small, bushy aromatic herb of the genus *Hyssopus*, spec. the common species *H. officinalis*, a native of Southern Europe, formerly much used medicinally, esp. in decoctions (*OED*, n. 1a).	hysope, hysop
Imperial Pills. Pills that consist of many ingredients including aloes and sena along with "Nutmegs, Cloves, Spicknard, Mastick", and Syrup of Violets (Culpeper, *Pharmacopiea*, 141).	imperiall pills
Impostume. A purulent swelling or cyst in any part of the body; an abscess (*OED*, n. 1).	empostume
Iron Scales. A coating of ferric oxide that forms on iron and steel during hot working and can be separated as thin flakes (*OED*, "iron", n. 1, C2).	scales of iron
Isinglass. A firm, whitish, semi-transparent substance obtained from the sounds or air-bladders of some freshwater fishes, esp. the sturgeon; used in cookery for making jellies, etc., for clarifying liquors, in the manufacture of glue, and for other purposes (*OED*, n. 1).	isinge-glasse
Issues. An incision or other superficial wound made to drain an accumulation of fluids, very often with a small object (such as a pea) used to keep the wound open (*OED*, n. 4a).	isues
Jacinth: see Hyacinth.	jacinthe
Janders. A morbid condition caused by obstruction of the bile, and characterised by yellowness of the conjunctiva, skin, fluids, and tissues, and by constipation, loss of appetite, and weakness; also called "jaundice" (*OED*, "jaundice", n. 1a).	
Julep. A sweet drink; often, a liquid sweetened with syrup or sugar and used as a vehicle for medicine; sometimes, a medicated drink used as a demulcent, "comforting", or gently stimulating mixture (*OED*, n. 1a). Also see: Cordial.	juleb, julip
Juniper. A genus of coniferous, evergreen shrubs and trees, and, in a medical context, usually *Juniperus communis*, which has a pungent taste, yielding a volatile oil (oil of juniper), and is used as a stimulant and diuretic, and also in the manufacturing of gin (*OED*, n. 1a).	junipery
Keel. A tub or vat for holding liquor (*OED*, n. 4).	keele
Lapis prunella: see Sal prunella.	lap: prunella
Laudanum. A name for various preparations in which opium was the main ingredient (*OED*, n. 1).	labdanum, laudenam

(continued)

(continued)

The standard spellings and definitions of key chymical, culinary, herbal, and medical terms in Pw V90 and Harley 6491	*Alternate spellings in Pw V90 and Harley 6491*
Lenitive. A soothing medicine that is gently laxative (*OED*, adj. A, 1).	
Life-Honey. Fresh honey which runs from the honeycomb, or which may be extracted after the virgin honey has been removed (*OED*, n.).	life hony
Lights. The lungs (*OED*, n. a).	
Lign-Aloes. A short form of the Latin "lignum aloes" (i.e. the wood of aloes). A bitter drug derived from aloes; an aromatic wood obtained from a Mexican tree of the genus *Bursera*; also called "xyloaloes" (*OED*, "lign-aloes", n. a and c). Also see: Aloes.	lignum aloes
Lignum nephriticum: see Nephritic Wood.	lignum nefreticum
Lignum rhodium. An obsolete name for Candle-wood, *Amyris balsamifera* (*OED*, n. 1, C).	lignum-rhodium, lignum roodium
Lignum vitae. A genus of trees and shrubs (*N.O. Zygophyllacæ*); a tree of this genus, esp. *Guaiacum officinale* and *G. sanctum*; also called "guaiacum" (*OED*, n. 1).	lignum vite
Lily of the Valley. A beautiful spring flower, *Convallaria majalis*. The name "lily of the valley" represents the Vulgate *lilium convallium*, a literal translation from the Hebrew of Cant. ii.I that apparently stems from the sixteenth-century German herbalists (*OED*, "lily", n. 2a).	lilly of ye valleys
Limbeck: see Alembic.	limbeke, lymbeck
Litharge. Protoxide of lead (PbO) prepared by exposing melted lead to a current of air (*OED*, n. 1a).	
Liverwort. In early use, the plant *Marchantia polymorpha*, which has lobed, liver-shaped leaves, and was thought to be effective against liver disorders. In later use, any plant of the division Marchantiophyta, which consists of small, non-vascular, moss-like plants chiefly of moist habitats, and which are united in having single-celled rhizoids and lacking stomata. It may also refer to Common agrimony, *Agrimonia eupatoria*, formerly used medicinally in the treatment of liver disorders (*OED*, n. 1 and 2).	liver-woort
Lixivium. Water containing alkaline salts leached from wood ashes; lye; (more widely) any solution obtained by leaching and esp. used for washing or soaking (*OED*, n. 1).	lixium
Lohoch. A linctus; that is, a medicine meant to be licked (*OED*, n.).	loohoc

(continued)

(continued)

The standard spellings and definitions of key chymical, culinary, herbal, and medical terms in Pw V90 and Harley 6491	*Alternate spellings in Pw V90 and Harley 6491*
Lye. Alkalised water, primarily that made by the lixiviation of vegetable ashes, but also applied to any strong alkaline solution, esp. one used for washing (*OED*, n. 1, 1a).	ly
Mace. An aromatic spice consisting of the fleshy aril or covering surrounding the seed in the fruit of the nutmeg tree, *Myristica fragrans*, dried and used (chiefly in powdered form) to flavour savoury dishes, sauces, etc. (the kernel of the seed being the source of nutmeg) (*OED*, n. 1, 1).	
Magistery. Not to be confused with "magistralis", this term, when applied to pearl and coral, indicates that these substances have been heated and made into a "calx" or "philosophick calcination" that causes the "corrosion" of the "Spirit of Sulphur" (Pierre Thibaut (1668), *The Art of Chymistry* (London), 176).	magisteres
Magistralis. "Of the master", suggesting a well-established medicine. In an early modern medical context, the English "magistral" often referred to a sovereign, supremely effective remedy (*OED*, "magistral", adj. A, 2a).	
Malaga. A white or red fortified wine originally from Spain or the Canary Islands. The word is probably derived from the name of the town where the wine was thought to have originated: Málaga (*OED*, n. 1).	Malego, Malegoe, Maligo
Mallow. A common European plant of waysides and waste places, *Malva sylvestris*, with showy, deeply notched, rose-purple flowers, widespread in Britain (*OED*, n. 1).	mallowe, malowe, common mallow
Malmsey. A strong, sweet wine, originally the product of the district of Monemvasia (Napoli di Malvasia) in the Peloponnese, later also from other parts of the Mediterranean, the Azores, the Canaries, Madeira, and elsewhere (*OED*, n. 1).	
Manna. A dried, sweet exudate or gum produced by various plants when cut, damaged, or punctured; esp. one rich in mannitol exuded from the branches of the manna ash, which was used medicinally as a mild laxative (*OED*, n. 1, II, 3a).	mana
Manus Christi. A kind of restorative confection or cordial drink, with refined sugar boiled in water, usually with roses (*OED*, n.).	manus cristie, manus-cristie

(continued)

(continued)

The standard spellings and definitions of key chymical, culinary, herbal, and medical terms in Pw V90 and Harley 6491	*Alternate spellings in Pw V90 and Harley 6491*
Marjoram. Any of various plants of the genus *Origanum*, comprising aromatic herbs and low shrubs whose leaves are much used in cooking; esp. *O. vulgare* (in English "wild marjoram" or "common marjoram"). "Sweet marjoram" is the less pungent *O. majorana* (*OED*, n. 1).	majoerem, margerum, marjorum
Marshmallow. A shrubby plant, *Althaea officinalis*, native to Eurasia and North Africa, which has ovate leaves, pale pink flowers, and a mucilaginous root (*OED*, n. A, 1a).	march-mallowe, march mallowe
Mary's Bath. A heated bath used to warm things gently or keep them warm (*OED*, "bath", n. 1, IV, 14).	Marys bathe, balneo Mariæ, Maria Prophetissa bath, Balneum Mariæ
Mastic: see Pills of Mastic.	mastick, mastich, mæstick
Matrix. The womb; the uterus of a mammal (*OED*, n. I, 1).	
Maidenhair: see Adianthus.	maiden-haire, mayden haire
May-butter. Unsalted butter preserved in the month of May and sometimes used medicinally (*OED*, n. a).	may butter
Melilot. Any of various Eurasian plants constituting the genus *Melilotus*; esp. the yellow-flowered *M. officinalis*, the dried flowers of which were formerly much used in making plasters, poultices, etc., and the white-flowered *M. alba* (*OED*, n.).	mellilot, mellilott, melilott
Mercurius dulcis. Mercurous chloride or "protochloride" of mercury; a preparation much used in medicine in the form of a white powder with a yellow tinge, becoming grey on exposure to light; also found native as horn-quicksilver in crystals; frequently called "calomel" (*OED*, "calomel", n.).	
Mercury. The herb *Mercurialis annua*, which now chiefly appears as a garden weed, but was formerly grown for its supposed medicinal qualities and used esp. in enemas (*OED*, n. IV, 10a).	
Mercury. The metal mercury as used medicinally, usually as a preparation of an inorganic salt such as mercurous chloride (calomel) or mercuric chloride (corrosive sublimate). Mercurial preparations were used extensively in western medicine from the late fifteenth to the early twentieth century, notably for the treatment of syphilis (*OED*, III, 7b).	mercurye
Mesaraic. Of, or relating to, a mesentery, esp. that of the small intestine (*OED*, "mesenteric", adj.). Also see: Mesentery.	meseraick

(continued)

(continued)

The standard spellings and definitions of key chymical, culinary, herbal, and medical terms in Pw V90 and Harley 6491	*Alternate spellings in Pw V90 and Harley 6491*
Mesentery. Originally: the folded sheet of peritoneum in which the jejunum and ileum are suspended from the dorsal abdominal wall. Later also any of several other folds of peritoneum serving a similar function for other organs; the embryonic precursor of these structures, a double layer of splanchnic mesoderm attached to both the dorsal and ventral walls of the body, which also temporarily supports the organs of the chest (*OED*, n. 1).	
Metheglin. A spiced or medicated variety of mead, originally popular in Wales (*OED*, n.).	matheglen
Milt. The spleen (*OED*, n. 1a); also, fish semen (*OED*, n. 2).	
Mithridate. Any of various medicinal preparations, usually in the form of an electuary compounded of many ingredients, believed to be a universal antidote to poison or a panacea (*OED*, n. 1a). It was also called "treacle", a term that stems from the Greek "θηριακή" (or the Latin "theriac"), an adjective that means "pertaining to animals" in reference to the belief that it could counteract dangerous bites.	mythridate, mithridat, metridate, mithredate, mithradate
Mugwort. A tall perennial plant, *Artemisia vulgaris*, which has small flower heads in long, leafy panicles and aromatic, pinnatifid leaves, and is found widely on waste ground, roadside verges, etc. (*OED*, n. 1).	
Mummia. A substance that purportedly comes from mummified human flesh. Belief in the medicinal powers of the bituminous liquid that could be extracted from the bodies of ancient Egyptian mummies apparently arose because of its resemblance to pissasphalt. Later, similar powers were ascribed to mummified flesh itself, which was often used in the form of a powder (*OED*, "mummy", n. 1, I, 1a).	
Murrain. Infectious disease, plague, pestilence (*OED*, n. A, 2a).	murraine
Muscadel. Wine made from muscat or similar grapes; esp. sweet white wine that is sometimes fortified (*OED*, "muscat", n. 1).	muscadell
Muscadine: see Muscadel.	muscadin, muskedine, muskadine
Muscovado Sugar. Raw or unrefined sugar obtained from the juice of the sugar cane by evaporating it and draining off the molasses (*OED*, n.).	muscuvadoe suger

(continued)

(continued)

The standard spellings and definitions of key chymical, culinary, herbal, and medical terms in Pw V90 and Harley 6491	*Alternate spellings in Pw V90 and Harley 6491*
Musk. A reddish-brown substance with a strong, persistent odour secreted by a gland of the male musk deer, esp. *Moschus moschiferus*, and highly prized in perfumery. Also any of various substances secreted by other mammals, esp. for scent marking (*OED*, n. 1a).	muske
Myrobalan. The astringent plumlike fruit of any of various tropical trees of the genus *Terminalia*, esp. *T. bellirica* and *T. chebula*, formerly used medicinally but now chiefly in dyeing and tanning (*OED*, n. 1).	mirabolonn
Myrrh. A bitter, aromatic gum resin exuded by various Arabian and African trees of the genus *Commiphora*, esp. *C. abyssinica* and *C. myrrha*, which was formerly important esp. in perfumery and as an ingredient of incense. Also in pharmacology: a tincture made from this, used medicinally as an astringent and expectorant (*OED*, n. 1, 1a).	mirrh
Naples Biscuit. A kind of biscuit flavoured with rose water; also called "Naples cake" (*OED*, n. 4).	Naples biskett
Neat's Tongue. The tongue of a cow or ox, esp. as an article of food (*OED*, n.).	neatstounge
Nenuphar. A waterlily, esp. the white waterlily, *Nymphaea alba*, or the yellow water lily, *Nuphar luteum*. Also, the roots, leaves, flowers, or seeds of these plants, used in medicinal preparations (*OED*, n. 1a).	
Nephritic Wood. *Lignum nephriticum*, a tropical wood used, in the form of an aqueous infusion, to treat diseases or pain of the kidneys (esp. kidney stones) (*OED*, n.).	nephretick-wood, nefratick wood
Opobalsamum. A fragrant oleoresin obtained from the tree *Commiphora opobalsamum* of Arabia and north-east Africa, formerly used for incense and in medicine and perfumery; also called "Mecca balsam" and "balm of Gilead" (*OED*, "Mecca", n. I, 1). Also see: Balm, Balsam.	opabalsamum
Orpine. Any of various stonecrops, but usually the common Eurasian stonecrop *Hylotelephium telephium* (*OED*, n. 1). Also see: Stonecrop.	orpin
Oxymel. A medicinal drink or syrup made of vinegar and honey, sometimes with other ingredients (*OED*, n.).	oximell
Oxymel Julianizan. One of the two principal kinds of oxymel (along with squill). According to Culpeper, Oxymel Julianizan includes, among other ingredients, "Bark of Caper roots, the roots of Orris, Fennel, Parsley, Bruscus, Chicory, Sparagus, [and] Cypress" (Culpeper, *Pharmacopoeia*, 112).	

(continued)

(continued)

The standard spellings and definitions of key chymical, culinary, herbal, and medical terms in Pw V90 and Harley 6491	*Alternate spellings in Pw V90 and Harley 6491*
Pap. Semi-liquid food, such as that considered suitable for babies or invalids, usually made from bread, meal, etc., moistened with water or milk (*OED*, n. 2, 1a).	pappe, papp
Paroxysm. An episode of increased acuteness or severity of a disease, esp. one recurring periodically in the course of the disease; a sudden recurrence or attack, e.g. of coughing (*OED*, n. 1).	paroxysme
Parting glass. A glass vessel used in parting gold and silver with acid (*OED*, n. "parting", C2).	parting glasse
Pastille. A small, flat, usually round sweet, often coated with sugar and sometimes medicated; a lozenge (*OED*, n. 1a).	pastill, pastiel, pastoele
Pasty. A cake or pie containing medicinal ingredients (*OED*, n. b).	pastey
Pearmain. A variety of pear; also any of several varieties of apple that have a somewhat conical shape and a skin often flushed or streaked on the side exposed to the sun (*OED*, n. 1 and 2).	pearemanes
Peck. A unit of capacity for dry goods equal to a quarter of a bushel, now equivalent (in Britain) to two imperial gallons (approx. 9.09 litres) or (in the U.S.) to eight dry quarts (approx. 8.81 litres) (*OED*, n. 1, 2a).	peeck
Pectoral. Of a medicine, food, or drink: good for digestive or (esp.) respiratory complaints (*OED*, adj. 2).	pectorall
Pellicle. A small or thin skin; a fine sheet or layer covering a surface or (occasionally) enclosing a cavity; a membrane, a cuticle, a film (*OED*, n.).	pellicula
Pellitory. Either of two plants of the genus *Parietaria*, *P. officinalis* and *P. judaica*, which tend to grow upon or at the foot of walls; also simply called "Pellitory of the Wall" (*OED*, n. 1).	pelletory, pelitory, pellitory of the walle, pelitory of the wall, peletory of the wall, pelletories of ye wall
Penide. A stick of boiled sugar, resembling barley sugar, formerly used esp. as a remedy for colds, or in treating other medical conditions (*OED*, n.).	penny dice
Pennyroyal. A small-leaved, creeping mint, *Mentha pulegium*, native to damp heaths and similar places in Europe and western Asia; also the fresh or dried leaves of this plant, formerly used medicinally and in cooking (*OED*, n. 1).	peneyroyall, peney royall, pennyroyall, penny-royall
Peony. Any of various herbaceous or shrubby plants of the genus *Paeonia*, which were formerly valued chiefly for the supposed medicinal properties of the root, flowers, and seeds (*OED*, n. 1a).	peoina

(continued)

(continued)

The standard spellings and definitions of key chymical, culinary, herbal, and medical terms in Pw V90 and Harley 6491	*Alternate spellings in Pw V90 and Harley 6491*
Periwinkle. Any of several plants of the Old-World genus *Vinca*, cultivated as ornamentals and locally for their medicinal properties (*OED*, n. 1, 1a).	peruincke
Pestilential. Carrying, producing, or tending to produce pestilence or epidemic disease, esp. bubonic plague. Also of the nature of, or symptomatic of, pestilence or epidemic disease (*OED*, adj. 1a and 1b).	pestilentiall
Pewter. A grey alloy of tin, originally with about 20% lead (and sometimes other elements) but now with about 10% antimony and a small quantity of copper, used chiefly for ornaments and (esp. formerly) utensils (*OED*, n. A, 1a).	putor
Philosophers' Egg. An egg-shaped glass vessel with a long neck used for heating substances over a sustained period (*OED*, "philosopher", n. C3c).	filosophers egge
Philosophers' Stone. A mythical solid substance, supposed to change any metal into gold or silver and (according to some) to cure all wounds and diseases and prolong life indefinitely (*OED*, n. 1a).	philosopher stone, philosophers stone
Phthisic. Wasting disease, esp. one involving the lungs; spec. tuberculosis; from the Latin *phthisicus*, and, earlier, the Greek for "consumption" (Φθισις) (*OED*, n. 1a).	tisike
Piles: see Haemorrhoids.	pills
Pills of Mastic. Pills made from mastic, an aromatic gum or resin which exudes from the bark of the lentisk or mastic tree, *Pistacia lentiscus*, used chiefly in making varnishes and formerly in medicine (*OED*, "mastic", n. 1a and C1a).	pills of mastick, pils mastichina, pill mastick, pilulea masticum
Pills of Pistacia: see Pills of Mastic.	pills of pistashes
Pilulae. Little balls of medicines meant to purge the head, and, perhaps, to make the ingredients more palatable (Culpeper, *Pharmacopoeia*, 139–140).	pilule, pilulea
Pilulae cochiae minores. According to Culpeper, it consists of, among other things, aloes, scammony, syrup of wormwood, and purging thorn, which are made into a mass (Culpeper, *Pharmacopoeia*, 140); also called "Pilulae cochiae the lesser".	pilulæ cochiæ minores, pilule cocci, cochias minores
Pilulae rudii. This medicine, according to Culpeper's recipe, involved ingredients including agrick, scammony, and the roots of black hellebore, and could "clenseth both head and body of choller, flegm, and melancholly" (Culpeper, *Pharmacopoeia*, 142).	rudius

(continued)

(continued)

The standard spellings and definitions of key chymical, culinary, herbal, and medical terms in Pw V90 and Harley 6491	*Alternate spellings in Pw V90 and Harley 6491*
Pimpernel. Any of several plants of the genus *Pimpinella* (the anises); esp. Burnet saxifrage, *Pimpinella saxifraga* (*OED*, n. I, 1b).	pimpernell
Pipkin. A small (usually earthenware) pot or pan. Also a drinking vessel; a bottle (*OED*, n. 1).	
Pippin. A seed or pip of various fleshy fruits, or, more particularly, a kind of sweet apple (*OED*, n. 1a and 3a).	
Pismire. An ant (*OED*, n. 1).	
Pistacia lentiscus: see Pills of Mastic.	lentisque
Plantain. Any of various low-growing plants constituting the genus *Plantago*, with dense, cylindrical spikes of inconspicuous flowers and leaves in a basal rosette usually close to the ground (*OED*, n. 1, 1).	plantine, plantin, plantaine
Pleurisy. An abscess of the ribs or inner surface of the chest; pain in the chest or the side, often stabbing and exacerbated by inspiration or coughing; any disease resulting in such pain (*OED*, n. 1).	pleurice
Polypody. The fern *Polypodium vulgare*, which is found growing on moist rocks, old walls, and trees in many temperate regions; also any of various ferns belonging to the genus *Polypodium* (*OED*, n. 1, 1).	polipode, polipody
Pomander. A mixture of aromatic substances usually made into a ball, and carried in a small box or bag, or suspended by a chain from the neck or waist, esp. as protection against infection or unpleasant smells (*OED*, n. 2a).	
Pomatum. A scented grease or ointment applied to the skin as a perfume, cosmetic, or salve; also called "pomade" (*OED*, n. 1).	
Possett. A drink made from hot milk curdled with ale, wine, or another liquor, flavoured with sugar, herbs, spices, etc., and often drunk for medicinal purposes (*OED*, n. 1).	posset, poset, poset drink, posset-drink, posset drinke, posett drinke
Pottle. A pot, tankard, or similar container. Also a unit of capacity used chiefly for liquids, equal to half a gallon (approx. 2.3 litres) (*OED*, n. 1, 1a and 1b).	potle
Poultice. A moist, often heated, mass of a substance with a soft, pasty consistency, applied to the skin, usually by means of a bandage or dressing, in order to promote healing, reduce swelling, relieve pain, etc.; also called a "fomentation" or a "cataplasm" (*OED*, n. 1). Also see: Fomentation.	pultess, pultis

(continued)

(continued)

The standard spellings and definitions of key chymical, culinary, herbal, and medical terms in Pw V90 and Harley 6491	*Alternate spellings in Pw V90 and Harley 6491*
Precipitate. Any of various solid mercury compounds formed by methods involving precipitation or separation from solution; formerly frequently attributive, designating pharmaceutical preparations containing such compounds. More generally, any precipitated substance; a solid, often powdery or flocculent, which separates or is deposited from a solution as a result of chemical action, cooling, etc. (*OED*, n. 1 and 2a).	
Pulmonaria. Originally the plant lungwort, usually *Pulmonaria officinalis*, but later any other plants of the genus *Pulmonaria* (*OED*, n.).	pulmoaria
Purples. Any dark red or purplish lesion of the skin (*OED*, n. B, 3a).	
Purslane. A low-growing succulent plant, *Portulaca oleracea*, which is widely cultivated (chiefly in warmer regions) as a salad vegetable or herb but is also widespread as a weed (*OED*, n. 1).	percelaine
Quartan Ague. A fever that recurs (by inclusive reckoning) every fourth day (i.e. at intervals of approximately seventy-two hours) (*OED*, "quartan", n. A, 1).	quarteran ague, quartern
Quicksilver: see Mercury.	quickesilver
Raddle. A red, natural, earthy material or clay, which is rich in iron oxide; a pigment made from such a material; also called "red ochre" (*OED*, "ochre", n. A, 1a).	
Recipient. A (glass) container typically used for holding liquids (*OED*, n. A, 1a).	recippent, recipent
Rectified. Purified or refined, esp. by repeated or continuous distillation; redistilled (*OED*, adj. 2).	rectifi'd
Red Cicer. A chickpea. (Now used only as the botanical name of the genus) (*OED*, "Cicer", n.).	
Red Grocers' Corinths. This could refer to two different fruits. The first is "raisins of Corinth", the raisin or dried fruit prepared from a dwarf seedless variety of grape, grown in the Levant; much used in cookery and confectionery. The second definition, which is more likely in the context of Pw V90, is "grocers' currants", the small round berry of certain species of *Ribes* (*R. nigrum*, *R. rubrum*) called Black and Red Currants (*OED*, "currant", 1a and 2a).	red grocers corinths
Red Lead. A red oxide of lead obtained from litharge by exposing it to hot air, much used as a pigment; also called "minium" (*OED*, "lead", n. 1, 2).	

(continued)

(continued)

The standard spellings and definitions of key chymical, culinary, herbal, and medical terms in Pw V90 and Harley 6491	*Alternate spellings in Pw V90 and Harley 6491*
Red Sugar. Sugar made with any of various ingredients that produce a reddish colour; also brown or unrefined sugar (*OED*, "red", n. C1, c, i).	red suger
Reins. The kidney and/or loins (*OED*, n. I, 1 and 2b). The turn of phrase "running of the reins" referred to the flowing or discharge of fluid from the body, esp. gonorrhoea (*OED*, n. III, 23a).	raines
Retort. A (usually glass) container with a bulb and a long downward-curving neck, used in distilling liquids and in other chemical operations (*OED*, n. 1, 1).	retorta
Rhodium. A rose-scented wood derived from the rootstock and lower part of the stem of either of two species of bindweed, *Convolvulus floridus* and *C. scoparius*, native to the Canary Islands; also called "rhodium wood" (*OED*, n. 1, 1).	rodium
Ripen. Of an abscess, pustule, etc. that has come to a head; that is ready to discharge its contents or to be drained. Also, that is ready for curative or definitive treatment (*OED*, "ripe", adj. A, 7c).	
Rue. Any of various southern European dwarf shrubs constituting the genus *Ruta*, esp. *R. graveolens*, which were formerly much used for medicinal purposes; also called "common rue", "garden rue", or "wall rue" (*OED*, n. 2, 1a).	
Sack. A general name for a class of white wines formerly imported from Spain and the Canaries (*OED*, n. 3 a).	sacke
Sal ammoniac. A hard, white, opaque crystalline salt, supposed to have been originally prepared from the dung of camels near the temple of Jupiter Ammon in Egypt; chemically ammonium chloride (NH4Cl), formerly called muriate of ammonia; used in tinning iron, in pharmacy, and for the manufacture of ammonium alum for the dyer; also called "armoniac", which is a corruption that is perhaps due to an association with the Greek ἁρμονία (fastening or joining), based on the use of sal ammoniac in the joining of metals (*OED*, "ammoniac", adj. A, 1a).	armoniack
Sal-Nitre: see Saltpetre.	sall niter
Sal-prunella. A preparation made by fusing potassium nitrate with sulphur, originally used medicinally to treat diseases of the throat; also called "lapis prunella" or "prunella salt" (*OED*, "prunella", n. 2, 2).	

(continued)

(continued)

The standard spellings and definitions of key chymical, culinary, herbal, and medical terms in Pw V90 and Harley 6491	*Alternate spellings in Pw V90 and Harley 6491*
Sallow. A plant of the genus Salix, a willow; chiefly, in narrower sense, as distinguished from "osier" and "willow", applied to several species of Salix of a low-growing or shrubby habit. Also one of the shoots of a willow (*OED*, n. 1).	
Salt of Steel. An oxide of iron, called by various names since its popularisation by Paracelsus and his followers in the sixteenth century, including "crocus martis", "green vitriol", and "sal martis". The last was Paracelsus's chief name for it, following in the alchemical tradition (where the astrological sign of Mars is related to iron) (see Paracelsus (1573), *Philippi Aureoli Theophrasti Paracelsi Bombast eremitae, summi inter Germanos medici & philosophi. Chirurgia magna, in duos tomos digesta* (Strasbourg), 206–207 and 242). There have been numerous recipes for this substance, including the recipe for what Mayerne calls in Pw V90 "anima hepatis". Also see: Crocus Martis, Green Vitriol.	salt of steele
Saltpetre. A supposed volatile substance or chemical principle related to or present in saltpetre, said to exist in the air or in plants and to give rise to various physico-chemical or vital phenomena (*OED*, "nitre", n. 3).	salt-peter, salt-peeter, salt peeter
Samphire. The plant *Crithmum maritimum* (growing on rocks by the sea), the aromatic, saline, fleshy leaves of which are used in pickles; also called "rock samphire" (*OED*, n. 1a).	samphere, sampere
Sanders. An antiquated term for "sandalwood", a scented wood obtained from several species of *Santalum*. White sandalwood is obtained from *S. album*, a tree resembling the myrtle, found on the Malabar coast; citron or yellow sandalwood is from *S. Freycinetianum*, found in the South Sea Islands; red sandalwood is the deep red heartwood of a South Asian leguminous tree, *Pterocarpus santalinus* (*OED*, "sandalwood", n. 1).	sandar
Sanicle. The umbelliferous plant *Sanicula europæa* (more fully "wood sanicle"). Also, in an extended sense, any plant of the genus *Sanicula*, as *S. marilandica*, the black snakeroot (*OED*, n. 1).	
Sarsaparilla. The dried roots of plants of the various species of *Smilaceæ*, esp. Jamaica sarsaparilla, *Smilax officinalis*. Also a medicinal preparation of the root; the early mentions often relate to the supposed efficacy of the drug in the treatment of syphilis (*OED*, n. 1b).	salsa-perilla

(continued)

(continued)

The standard spellings and definitions of key chymical, culinary, herbal, and medical terms in Pw V90 and Harley 6491	*Alternate spellings in Pw V90 and Harley 6491*
Satyrion. Any of various European orchids having tuberous roots or rhizomes; esp. the early purple orchid, *Orchis mascula*. Also, a tuber or rhizome of any of these plants, formerly used medicinally either to increase or to suppress sexual desire (depending on the species or the condition of the root). Occasionally also any of several other herbaceous plants, esp. cuckoo pint, *Arum maculatum*, and dragon arum, *Dracunculus vulgaris* (*OED*, n. 1).	sayircon
Saxifrage. Any plant of the genus *Saxifraga*, esp. *S. granulata* (White Meadow Saxifrage) (*OED*, n. 1a).	saxafage, saxafrag, sasafras, sassafras, sassaphrass
Scabious. Any of the herbaceous plants of the genus *Scabiosa*, formerly believed to be efficacious for the cure of certain skin-diseases (*OED*, n. 1a).	scabiose, scabions, scabiosa, scabius
Scammony. A gum-resin obtained from the tuberous roots of *Convolvulus Scammonia*, used in medicine as a strong purgative; also, the dried tuberous root from which the drug is prepared (*OED*, n. 1a).	scamony
Scorbute. French for "scurvy" (*OED*, n.). Also see: Scorbutgrass.	scurbruteary
Scorbutgrass. A cruciferous plant, *Cochlearia officinalis*, believed to possess anti-scorbutic properties; also called "common scurvygrass" or "spoonwort" (*OED*, n. 1a).	scorbutgrasse
Scordium. A name for the Water-Germander, *Teucrium Scordium*, a plant formerly in use in medicine as a sudorific, an antidote for poisons, etc. (*OED*, n.).	scordi
Scorzonera. A plant of the modern genus *Scorzonera*, esp. *S. Hispanica* (or black salsify), much cultivated in Europe for its root which is used as a vegetable and somewhat resembles the parsnip; also called "viper's grass" (*OED*, n.).	scorzonira, scorzanery
Scrophularia. A genus of monopetalous plants (the figworts), typical of the *N.O. Scrophulariaceæ*; a plant of this genus (*OED*, n.).	
Searce. A sieve or strainer (*OED*, n.).	searse
Semicupium. A half-bath, one that comes up only to one's hips; in French, demi-bain (Dunglison, *Medical Lexicon*, 288).	semi cupium
Sena. A shrub of the genus *Cassia*, native in tropical regions; also the dried leaflets of various species of *Cassia*, used as a cathartic and emetic (*OED*, n. 1a and 2a).	senna, seana, sennea, cena, cenna
Sciatica. Pain in the hip; a disease causing such pain (*OED*, n. 1a).	

(continued)

(continued)

The standard spellings and definitions of key chymical, culinary, herbal, and medical terms in Pw V90 and Harley 6491	*Alternate spellings in Pw V90 and Harley 6491*
Snake-Root. The root or rhizome of one or other of several American plants reputed to possess properties antidotal to snake-poison, esp. the dried root of *Polygala senega* and *Aristolochia serpentaria* used largely in medicine; the medicinal preparation obtained from this (*OED*, n. 1a). Also see: Sanicle.	snake-roote
Sol. In alchemy, a term referring to gold, which was associated with the sun (*OED*, n. 1, 2a).	
Solutive. A laxative or purgative medicine (*OED*, n. B, 1).	
Sorrel. Certain small perennial plants belonging to the genus *Rumex*, characterised by a sour taste, and cultivated for culinary purposes; esp. the common wild species, *R. acetosa* (*OED*, n. 1, 1a).	sorrell
Southernwood. A hardy southern European shrub, *Artemisia abrotanum*, formerly cultivated for medicinal purposes; also called "common southernwood" and "garden southernwood" (*OED*, n.).	southerwood
Species dianthus. A medicinal powder that, according to Culpeper, consists of numerous herbs and other ingredients, including roses, violets, ginger, cinnamon, cloves, and nutmeg (Culpeper, *Pharmacopoeia*, 123).	dianthos
Speedwell. One or other species of *Veronica*, an extensive genus of small herbaceous plants with leafy stems and small blue flowers; a plant belonging to this genus (*OED*, n. 1).	
Spermaceti. A fatty substance, which, in a purified state, has the form of a soft, white, scaly mass, found in the head (and some other parts) of the sperm whale and various other whales and dolphins. It is used in various medicinal preparations, and in the manufacturing of candles; it is sometimes confused with "ambergris" (*OED*, n. 1).	spermacetj
Spina cervina. See Buckthorn.	
Spirit. An essence, distilled extract, or alcoholic solution of a specified substance (*OED*, "spirit", n. VI, 23a).	spirite
Spirit of Clary. A sweet liquor consisting of a mixture of wine, clarified honey, and various spices, such as pepper and ginger (*OED*, "clary", n. 1).	
Spirit of Salt. An extracted essence of salt, which usually refers to hydrochloric acid; also known as "muriatic acid", an aqueous solution of hydrogen chloride (Dunglison, *Medical Lexicon*, 609).	

(continued)

(continued)

The standard spellings and definitions of key chymical, culinary, herbal, and medical terms in Pw V90 and Harley 6491	*Alternate spellings in Pw V90 and Harley 6491*
Spirit of Wine. A term that alchemists applied to ardent spirits or unrectified alcohol; also called "aqua vitae" (*OED*, "aqua-vitae", n. 1).	sperit off wine, spirit of wyne
Splenetic. Of, or relating to, the spleen (*OED*, adj. A, 1).	spleeniteque, spleneticke
Spotted Fever. Any of various febrile diseases characterised by the appearance of (esp. petechial or purpuric) spots on the skin, as meningococcal meningitis and typhus (*OED*, "spotted", n. C3).	spotted feaver
Squill. The bulbous-rooted sea-shore plant *Scilla* (or *Urginea*) *maritima*; the sea-onion; also, any other species of the genus *Scilla*. In the names of various preparations made from these bulbs (*OED*, n. 2a and 1b).	scilitic, scills
Stag's Heart. Stags, deer, and other ruminants have bones in their heart, called "heterotopic bones", which provide them with strength or rigidity; also known as "*os cordis*" (Alfred Sherwood Romer and Thomas S. Parsons (1986), *The Vertebrate Body*, Vol. 7 (Philadelphia), 199).	staggs-heart
Stitch. A sharp, sudden, local pain, like that produced by the thrust of a pointed weapon (*OED*, n. 2a).	stich
St. John's Wort. Any plant of the genus *Hypericum*, esp. *H. perforatum*, the flowers and foliage of which are used in herbal medicines, originally for a wide variety of purposes including the treatment of wounds and of pulmonary, urinary, and nervous diseases (*OED*, n. 1). Also see: Hypericon.	St John-worth, St John-wort, St John wort, S^{t}. Johns-wort, S^{t}. Johnwort
Stœchas. French lavender, *Lavandula stœchas* (*OED*, n.).	stoecas
Stomachic. Of, or pertaining to, the stomach (*OED*, adj. A, 1a).	stomachick
Stonecrop. The common name of *Sedum acre*, a herb that grows in masses on rocks, old walls, etc.; also applied to other species of *Sedum*, esp. those of similar growth, and of allied genera, as the North American *Penthorum* (*OED*, n. a).	stone cropp
Storax. A gum-resin described by ancient writers. In early modern use applied to the resin of the tree *Styrax officinalis*; in later commercial and pharmaceutical use to the balsam of the tree *Liquidambar orientale* (*OED*, "storax", n. 1).	styrax
Sublimate. A substance produced by sublimation; a solid, typically crystalline deposit of a substance that has sublimed. Also, a refined or concentrated product; the essence or distillation of something (*OED*, n. 2a and 3).	

(continued)

(continued)

The standard spellings and definitions of key chymical, culinary, herbal, and medical terms in Pw V90 and Harley 6491	*Alternate spellings in Pw V90 and Harley 6491*
Succus. Originally a hard tumour or swelling; an abnormally hard area of tissue, esp. in a glandular organ. In later use a cancer with a hard texture resulting from the presence of dense fibrous connective tissue in the stroma; more commonly called a "scirrhus" (*OED*, "scirrhus", n. 1a).	succes
Suet. The solid fat around the loins and kidneys of certain animals, esp. the ox and sheep, which, chopped up, is used in cooking, and, when rendered down, forms tallow (*OED*, n. a).	cuiet, suiet, cyuet
Sugar of Saturn. Lead acetate; also called "sugar of lead" or "English sugar" (*OED*, n. 3a).	suger of saturne
Sulphur. A greenish-yellow non-metallic substance, found abundantly in volcanic regions, occurring free in nature as a brittle crystalline solid, and widely distributed in combination with metals and other substances. In a refined state, used medicinally as a laxative, a resolvent, a sudorific, and an ingredient of various ointments, esp. for skin diseases. For some chymists, sulphur was one of the ultimate elements of all material substances (*OED*, n. 1a, 1b, and 2a). Also see: Balsam of Sulphur, Brimstone, Flowers of Sulphur, and Sulphur of Hartmannus.	sulpher, sulfer, sulfor
Sulphur of Hartmannus. A form of sulphurated potash popularised by Johannes Hartmann (1568–1631), which, when used internally, acted as a narcotic (see Hartmannus (1633), *Praxis Chymiatrica*, Chap. II, Sec. 14, 44–45). Also see: Sulphur.	sulpher of Hartmannus, sulpher narcotick of Hartmannus
Sweet Calamus. Some eastern aromatic plant or plants (supposed by some to be *Andropogon Schoenanthus*, the Sweet-scented Lemon Grass of Malabar). Also applied by some English herbalists to the native Sweet Flag or Sweet Rush (*Acorus Calamus*) (*OED*, n. 2).	sweete calamus
Sweet Maudlin. One of two aromatic plants, costmary (*Tanacetum balsamita*) or sweet yarrow (*Achillea ageratum*); also simply called "maudlin" (*OED*, "maudlin", n. 2). Also see: Costmary.	sweet mandelyn, sweete maudlin
Syrup of Diacodium. A syrup prepared from poppy-heads, used chiefly as an opiate (*OED*, "diacodium", n.).	syrope of diacodium
Syrup of Meconium. A syrup made from the thickened juice of the opium poppy, used as a sedative or analgesic (*OED*, "meconium", n. 1).	syrup ad melancoliamto, magistralis ad melancholiam

(continued)

(continued)

The standard spellings and definitions of key chymical, culinary, herbal, and medical terms in Pw V90 and Harley 6491	*Alternate spellings in Pw V90 and Harley 6491*
Tacamahaca. An aromatic resin, used for incense, and formerly extensively in medicine. Originally, that yielded by a Mexican tree, *Bursera* (*Elaphrium*) *tomentosa*; extended in the West Indies and South America to similar resins obtained from other species of *Bursera* and the allied genus *Protium*, and subsequently to resins imported from Madagascar, Bourbon, and the East Indies; chiefly the product of species of *Calophyllum* (*OED*, n. 1).	takamaka
Tallow. A substance consisting of somewhat hard animal fat (esp. that obtained from the parts about the kidneys of ruminating animals), separated by melting and clarifying from the membranes, etc., naturally mixed with it (*OED*, n. 2).	
Tamarind. The fruit of the tree *Tamarindus indica*, a brown pod containing one to twelve seeds embedded in a soft brown or reddish-black acid pulp, valued for its medicinal qualities, and also used in cookery as a relish (*OED*, n. 1).	tammarin
Tamarisk. A plant of the genus *Tamarix*, esp. *T. gallica*, the common tamarisk, a shrub or small tree growing in sandy places in Southern Europe and Western Asia; a decoction or other preparation of the leaves of this plant, formerly used in medicine (*OED*, n. 1a and 1b).	tamariscus
Tartar. Bitartrate of potash (acid potassium tartrate), present in grape juice, deposited in a crude form in the process of fermentation, and adhering to the sides of wine-casks in the form of a hard crust (*OED*, n. 1, 1a). Also see: Cream of Tartar and Oil of Tartar.	tarter, tarrter
Tartarous. Of the nature of, consisting of, or containing tartar (*OED*, adj.).	tartarus
Terra Lemnia. An astringent bole of fatty consistency and reddish colour, obtained from Lemnos in the Aegean Sea; also called "terra sigillata" or "lemnian earth" (*OED*, n.).	terra-Lemna
Tertian Ague. Of a fever or ague: characterised by the occurrence of a paroxysm every third (i.e. every alternate) day (*OED*, "tertian", adj. A, 1).	
Thunderbolt-Stone. A term applied to various stones, fossils, or mineral concretions, formerly supposed to be thunderbolts; also called "thunder-stone" (*OED*, "thunderbolt", n. 3).	thunder-boalt stone
Tilia. A tree of the genus *Tilia* (family *Tiliaceæ*), esp. *T. europæa*, a common ornamental tree having heart-shaped leaves and many small, fragrant, yellowish flowers; also called "the linden" (*OED*, "lime", n. 3, 1).	tylett, tilliett

(continued)

(continued)

The standard spellings and definitions of key chymical, culinary, herbal, and medical terms in Pw V90 and Harley 6491	*Alternate spellings in Pw V90 and Harley 6491*
Tincture. The (supposed) essential principle of any substance obtained in solution; also the extraction of this essential principle (*OED*, n. 7a).	
Tisane. A wholesome or medicinal drink. Originally an infusion made with barley water, and subsequently a medicinal tea or infusion made from herbs (*OED*, n. 1).	ptysan, ptisan
Tolu Balsam. A balsam obtained by incision from the bark of the tolu-tree, *Myrospermum* (*Myroxylon*) *toluiferum*, a leguminous tree of tropical South America; used in medicine and perfumery (*OED*, "tolu", n.). Also see: Balsam.	balsum de tolu
Tormentil. A low-growing herb, *Potentilla tormentilla* (*Tormentilla repens*), having strongly astringent roots; in use from early times in medicine, and in tanning (*OED*, n. a).	tormentill
Tragea. A mixture of diverse powders with sugar and strong wine (Humphrey Llwyd (1585), *Treasury of Healthe Conteynyng Many Profitable Medycines Gathered out of Hypocrates, Galen and Auycen, by one Petrus Hyspanus* (London), unpaginated).	
Treacle: see Mithridate.	
Troche. A flat, round tablet or lozenge, made of a powdered medicinal substance that is worked into a paste with mucilage or the like, and dried (*OED*, n. 2).	
Turbith. A cathartic drug made from the root of East Indian jalap, *Ipomœa Turpethum*, an Indian and Australian plant; also the plant itself, or its root (*OED*, n. 1a).	turpeth
Turpentine. A term applied originally to the semifluid resin of the terebinth tree, *Pistacia Terebinthus*. Also a volatile oil, contained in the wood, bark, leaves, and other parts of coniferous trees, and usually prepared by distilling crude turpentine; also called "oil of turpentine" (*OED*, n. 1a and 3).	terpentine
Tussilago. A genus of composite plants, including the coltsfoot, *T. Farfara* (*OED*, n.). Also see: Coltsfoot.	tussilage
Ureter. Either of the fibro-muscular tubes or vessels, which convey the urine from the pelvis of the kidney to the bladder; a urinary duct (*OED*, n. a).	uritories
Urinal. In chemistry, an oblong glass vessel, used for making solutions (*OED*, n. 2).	urinall

(continued)

(continued)

The standard spellings and definitions of key chymical, culinary, herbal, and medical terms in Pw V90 and Harley 6491	*Alternate spellings in Pw V90 and Harley 6491*
Valerian. One or other of the various species of herbaceous plants belonging to the widely-distributed genus *Valeriana*, many of which have been used medicinally as stimulants or antispasmodics (*OED*, n. 1).	valeriana
Vehiculum. A substance, esp. a liquid, which acts as a means for the distribution and application, or action, of another substance dissolved in or mixed with it (*OED*, n. 1).	
Verdigris. A green or greenish-blue substance obtained artificially by the action of dilute acetic acid on thin plates of copper (or a green rust naturally forming on copper and brass), and much used as a pigment, in dyeing, the arts, and medicine (*OED*, n. a).	verdegrice
Vermillion. Cinnabar or red crystalline mercuric sulphide, esp. in later use that obtained artificially, much valued on account of its brilliant scarlet colour, and largely used as a pigment or in the manufacture of red sealing-wax; also, any red earth resembling this and similarly used as a pigment; occasionally confused with "red lead" (*OED*, n. A, 1a).	vermillon
Vitriol. One or other of various native or artificial sulphates of metals used medicinally; "oil of vitriol" is concentrated sulphuric acid; "spirit of vitriol" is a distilled essence of vitriol (*OED*, n. 1a, 4a, and 4b).	vitrioll
Vitriolic. Of or belonging to vitriol; having the nature or qualities of vitriol; impregnated with vitriol (*OED*, adj. A, 1a).	vitriollick, vitriollique, vitriolique
Wallflower. A plant of the genus *Cheiranthus*, esp. *C. Cheiri*, growing wild on old walls, on rocks, in quarries, etc., and cultivated in gardens for its fragrant flowers (*OED*, n. 1a).	wall flower
Water Betony. A figwort (*Scrophularia aquatica*) (*OED*, "betony", n. b).	water bettoney, water-bettony
White Amber: see Ambergris.	w^{ht} Amber
White Lead. A mixture of lead carbonate and hydrated lead oxide; largely used as a white paint, formerly also in medicine for ointments, etc.; also called "ceruse" (*OED*, "ceruse", n. 1a).	
White Tartar: see Cream of Tartar.	
Wig. A kind of bun or small cake made of fine flour (*OED*, n. 1).	wigg

(continued)

(continued)

The standard spellings and definitions of key chymical, culinary, herbal, and medical terms in Pw V90 and Harley 6491	*Alternate spellings in Pw V90 and Harley 6491*
Wild Carrot. An umbelliferous plant, *Daucus carota*, having a large, tapering root, which, in cultivation, is bright red, fleshy, sweet, and edible; also called "Bird's nest", "Bishop's lace", and "Queen Anne's lace" (*OED*, "carrot", n. 1).	wilde carrot
Wood sorrel. The common name of *Oxalis Acetosella*, a low-growing woodland plant having delicate trifoliate, leaves and small, white flowers streaked with purple, appearing in spring (*OED*, n. a).	woodsorell, wood-sorrell
Wormwood. The plant *Artemisia Absinthium*, proverbial for its bitter taste. The leaves and tops are used in medicine as a tonic and vermifuge, and for making vermouth and absinthe; formerly also to protect clothes and bedding from moths and fleas, and in brewing ale. It yields a dark green oil. The ashes of wormwood were often used to make salt of wormwood, an impure carbonate of potash (*OED*, n. 1a and 1c).	wormewood, worme-wood, worme wood
Xyloaloes: see Lign-Aloes.	zyloaloes
Yellow Amber. A hard, translucent fossilised resin, typically yellow, orange, or brown, used for jewellery and ornaments since ancient times and also in perfumery and medicine; "oil of amber" is a yellow liquid obtained by distilling amber (*OED*, "amber", n. 2 A, I, 1a and 1b).	
Zedoary. Any of several species of the genus *Curcuma*, esp. *C. zedoaria*, which are native to South Asia and have aromatic, tuberous rhizomes. Also, the rhizome of such a plant, used in cooking and (formerly) as a medicinal agent (*OED*, n.).	zedoar

B. Introduction to Harley 6491

Harley 6491 principally consists of "A short treatise of the nature of minerals given to my right honourable Lord My Lord Marquis of Newcastle by Doctor Davissone at his Lordships desire in Paris the 20 of July 1645". The "Doctor Davissone" here is of course William Davisson, who provided several pieces of advice in Pw V90. His treatise begins by describing various metals under headings such as "Of Iron", "Of Gold", "Of Silver", and "Of Tinne", and proceeds to explain how these metals can be used in various chymical preparations. Under "Of Iron", for instance, Davisson lays out "The preparation of aperitive Iron or Steele" and "The preparation of astringent Iron". At the end of the treatise, he also includes "A general review and observations upon the practicke of Vegetals animals and minerals", in which he divides all natural bodies into "volatile" or "fixe" based upon whether or not they flee (volatile) or remain present (fixed) when put over a fire. More generally, Davisson's treatise emphasises the power of fire, and he claims that his observations are not founded upon "the authority or speculation of vaine authors, but upon true experience maistresse of all things".[1]

While Davisson's treatise demands further study in its own right, this is not the aim of the present appendix. Rather, at the end of the treatise, there are a set of hitherto unnoticed recipes given to the Cavendishes. These are recorded in the hands of William (ff. 29r–33v; Appendix 8C, 268–274), an amanuensis (ff. 34r–37r; Appendix 8C, 274–280), and, we believe, Margaret (ff. 38r–51r; Appendix 8C, 280–290). If more than half of Harley 6491 is in Margaret's hand, as we believe could be the case, this would not only substantiate our earlier claims about her contributions to the household and the collection of recipes, but would also offer the most substantial known example of her hand other than her love letters to William in British Library Add. 70499, ff. 259r–297r. In terms of their dates, the recipes in Harley must have been written after Davisson's treatise and gifted to William in July 1645. Since Pw V90 was composed between 1647 and 1654, it would seem that the Cavendishes collected recipes in Harley 6491 immediately after their marriage in 1645 until they began to gather them more systematically in Pw V90.

Overall, Harley 6491 contains roughly 24–25 recipes, of which only about 16 are medical. Further, unlike Pw V90, there are no advice letters from physicians, nor extensive pieces of physic, though there are some interesting discussions of the use and dosage of various medicaments on the part of both the recipe collector and several physicians, including some observations about their efficacy and the modifications needed to increase it. The manuscript also contains extensive discussions of how to make an assortment of medicaments, which are more detailed in terms of procedure than what is found in Pw V90 and many recipe collections from the time. The organisation of Harley 6491 is also more haphazard than Pw V90, and it is not organised chronologically. In fact, there are no

[1] Harley 6491, ff. 1r–2v.

dated letters or temporal markers whatsoever, which makes dating the recipes in Harley 6491, beyond their attachment to Davisson's treatise, quite difficult.

Not only is Harley 6491 organised rather arbitrarily, but it also contains revisions of a sort that do not appear in Pw V90. To give one instance, a recipe for "A perfumed Balsom for the lipps" is provided, but, on the next page, "observations about the lipp salve found out by experiance" are offered, in which the hand that we think is Margaret's revises the initial recipe.[2] Similarly, at the end of "A rare Medisin to cure a Cancer in a Breast", it is written (in a lighter ink that also seems to be in Margaret's hand but to have been added later) that "it has not Cured a Cancer upon Triall".[3] William is also more self-conscious in Harley than in Pw V90, reflecting at one moment that "Iff I had nott a huge opinionn off Itt I woulde never have takenn this payns to have writtenn Itt oute with my owne hande", while there are a number of occasions where the individual transcribing the recipe reminds himself (or herself) of ingredients that were discussed in previous recipes.[4] In addition, quite a few recipes—especially those that we suppose to be in Margaret's hand—are completely devoid of medical utility. These include recipes for foodstuffs such as a "greate cake on of the best I ever eate" and "oring Pudding", along with cosmetics such as a "Pomatum for the Lipps".[5] Since, as we suggested in Chapter 2, Pw V90 (unlike Harley 6491) was a "starter collection" designed to be passed down from one generation to the next, it is far cleaner and the Cavendishes were more selective in which recipes they included.

On a broader social and intellectual level, Harley 6491 continues to underscore the role that religious individuals of various confessional stripes played in the invention and dissemination of medical knowledge. While figures such as George Morley, the future Bishop of Winchester, and "an nunn, att Antwerpe" feature in Pw V90, the first recipe in Harley 6491 is from "a Friar In Franse".[6] As noted in the biography of the

[2] Appendix 8C, 283–284.

[3] Appendix 8C, 288.

[4] Appendix 8C, 273.

[5] Appendix 8C, 285–286 and 281.

[6] While it is impossible to identify this friar, on chymistry and religious orders at this time, see Andrew Campbell, Lorenza Gianfrancesco, and Neil Tarrant (2018), Alchemy

Antwerp nun in Appendix 8D.IV, Margaret associated medical knowledge with "Monastical lives", and this recipe is further evidence for why she would have done so.[7]

Harley 6491, moreover, contains several interesting, if cryptic, references to Parisian chymists and natural philosophers with whom the Cavendishes may well have interacted. In the recipe for the universal powder of gold from the Friar, for instance, there is a reference to an "Alcumiste att Paris" who supposed that if a lump of mercury and gold "with the sole heate off your hande, shall become a runninge mercury, the Philosopher Stone, was Made".[8] Having transcribed the recipe of the Friar, William proceeded to evaluate it briefly, clarifying, in the process, that the chymist in question was Davisson. While he was unable to verify this particular claim about mercury and the heat of the hand facilitating the creation of the philosophers' stone, William nonetheless supposed that the powder of gold "Is the beste medecinn Inn the Worlde".[9] It is noteworthy here that mercury and gold appear in a number of recipes in Pw V90, including in one from Mayerne (Transcription, 128) and one from Davisson (Transcription, 130). Indeed, Davisson writes in Pw V90 that to use the powder of gold and mercury is to "Conforme to reason & Experience".[10] In Harley 6491, William also calls the powder of gold and mercury a "Universall Medsin", which is a concept that (as discussed in Chapter 6) Margaret argues against in *Observations*. Of course, Pw V90 is filled with a variety of medicines that were meant to be universal cures—including Dudley's elaborate recipe and the various treacles and mithridates referenced throughout—but the actual phrase "universal medicine" does not appear in Pw V90.

Several recipes that appear in Pw V90 are also included in Harley 6491 (such as "the Countess off Kents pouder") while a number overlap

and the Mendicant Orders of Late Medieval and Early Modern Europe. *Ambix* 65.3: 201–209 and Martha Baldwin (1993), Alchemy and the Society of Jesus in the Seventeenth Century: Strange Bedfellows? *Ambix* 40: 41–64. On friars in particular, see Roger French and Andrew Cunningham (1996), *Before Science: The Invention of the Friars' Natural Philosophy* (Aldershot).

[7] See Appendix 8D.IV.

[8] Appendix 8C, 269.

[9] Appendix 8C, 269.

[10] Transcription, 130.

(including the two recipes for fattening chickens).[11] With that being said, Harley 6491 includes significantly fewer named recipes attributed to medical professionals, with only three of its recipes being attributed to physicians (Drs "Moore", "Davisonn", and "Waldron"), and none to apothecaries. Interestingly, like in Pw V90, there is a single instance of "probatum est", with a recipe for the pox being approved by a mysterious "Greate personn thatt I knewe".[12] In the end, even though the same individuals had a hand in producing Pw V90 and Harley 6491, the latter is a very different kind of recipe collection. Like many manuscript receipt books from the period (including Pw V90), Harley 6491 is the product of the movement of various recipes from a scratch copy to a presentation volume.[13] Yet it is significantly lacking in indications of curation. In addition to being considerably shorter and less well-organised than Pw V90, it is full of deletions, revisions, and additions, and cosmetic and culinary recipes also make up a far greater percentage of its overall entries.

Before presenting the transcription itself, it is finally worth highlighting one particularly interesting recipe in Harley 6491: "Daffyes Elixier salutis given mee by Docter Waldron".[14] Since the recipes in Harley 6491 were almost certainly recorded in the mid-1640s, it contains what is in all probability the earliest example of Daffy's elixir, which was a popular patent medicine that was marketed and sold by Anthony Daffy (1620s?–1690s?) from the 1680s until his death, and continued to be made and used well into the twentieth century.[15] Peter Elmer has helped us to identify the "Docter Waldron" who provided the recipe to the Cavendishes as Edward Waldron, an English physician who received his medical degree in Padua in March 1641.[16] Waldron lived and worked in Mansfield, Nottinghamshire, and, given his proximity to William Cavendish, whose

[11] Transcription, 225 and Appendix 8C, 289.

[12] Appendix 8C, 274.

[13] On the various procedures involved in early modern recipe collecting, see Leong, *Recipes and Everyday Knowledge*, 71–98.

[14] Thanks to Jonathan Barry, Peter Elmer, and Patrick Wallis for helping to untangle the historical knots here.

[15] David Boyd Haycock and Patrick Wallis (2005), Quackery and Commerce in Seventeenth-Century London: The Proprietary Medicine Business of Anthony Daffy. *Medical History*, Supplement 25: 1–216.

[16] Peter Elmer, March 23, 2021, personal communication. Also see the Will of Edward Waldron of Mansfield (1662), The National Archives, PROB 11/309/148, Kew.

first title was Viscount of Mansfield, it makes perfect sense that William would have been in contact with him. How exactly Waldron procured this recipe from a “Daffy” almost forty years before Anthony Daffy began to promote the recipe, however, requires some untangling. It is significant here that David Boyd Haycock and Patrick Wallis have argued that Anthony Daffy was likely not the original author of the recipe, but that it was rather invented by Thomas Daffy (1616/17–1680), a clergyman who “may have produced his cure to make an income after being ejected from his living in Harby [...] in 1648”.[17] Harley 6491 thus provides compelling evidence for this supposition. There is indeed some circumstantial evidence that Waldron could have been in contact with Thomas Daffy. In particular, Daffy may have also been the minister of Harby, Leicestershire, from 1647, which is relatively close to Mansfield. As all of this indicates, Harley 6491 provides a significant piece of the puzzle that is the history of Daffy’s famous elixir.

What follows is lightly annotated, and only provides explanations for those terms that do not appear in the transcription of Pw V90.

C. Transcription of Cavendish Recipes in Harley 6491

[29r]

The Universall Medsin, or the pouder off Golde, Itts preperation & howe Itt is made,—Givenn Mee, by a Friar Inn Franse,

---

Take common quickesilver, & without any other preparation, putt Itt in to a glass, whereoff the head is made like ann urinall, & with a longe necke. putt into a due heate Untill 3: parts off foure bee turnde, Into a red pouder. & thenn Itt is prepared,.—

Take Golde, & cleanse Itt with Antimonye, meltinge Itt firste, with 3: times Itt’s waight off Antimonye; agayne melte Itt, with 2: times Itts wayght, & 3:^{ly} with once Itts wayght, & beinge thus cleansed Itt Is prepared.—

To your mercury prepared, putt the tenth parte off Itts waight off Golde prepared & Inn the same heate as before, Itt will all, Soone come

17 Haycock and Wallis, Quackery and Commerce, 3.

to a red pouder, Which maye then bee used, iff you please; Butt iff you will render Itt more perfect,—

Reduce this laste pouder agayne, to a runninge mercurye, Itt is redused to quicksilver, by puttinge Itt Inn a greate heate thatt Itt flye nott a waye in a Smoke, & thatt Smoke beinge taken Inn a Glass Imediatlye turnes to a runninge quicksilver,—& to Itt putt off your prepared Golde a fixt parte, off the waight off the mercurye, & as before Reduce Itt to a pouder.—

[29v] Ande agayne reduce Itt to mercury, & with more Golde agayne reduse Itt to pouder, & the offtner you reiterate this worke, The better your pouder will bee; & att Everye time hee will take more Golde thenn before; Till the quantetye off the Golde made mercurye, shall bee so much Inn respecte off the mercurye Simple, as beinge colde you shall see the Golde, made more. Settle Itt selfe like a lumpe, frome which you maye poure the mercurye putt to Itt att firste, & thatt lumpe with the sole heate off your hande, Shall become, a runninge mercurye.–

& thenn beinge braughte to a red pouder, Itt is in Itts perfectionn.—

The dose maye bee 4: gr. or Inn a violente accident as off poyson &. wher the danger off nott workinge Spedely[18] maye bee mortall, the dose maye bee 6: gr. – I have never knowne Itt doe harme to anye personn. nott to Childrenn, woemenn with Childe, persons Inn Consumptions.

A rare Alcumiste[19] att Paris had ann opinion & was verye confidente off Itt thatt this was the trewe waye to make the greate Elixer, & Swore thatt iff thatt lumpe, with the sole heate off your hande, shall become a runninge mercury, The Philosopher Stone, was Made,.—

Docter Davisonn a moste Exselente kimiste[20] sayde absolutlye, this was the waye to the Philosophers Stone,—howe soever I amm shure Itt Is the beste medecinn Inn the Worlde.

[30r]

The Use off the pouder off Golde, & The operation off Itt;—

[18] Speedily.

[19] Alchemist.

[20] Chemist.

The dose maye bee 3: or 4: gr: & I have founde Itt to worke the gentler, when the quantetye is greater, because thenn with less resistance Itt workes Itts Effect.

For the manner off takinge Itt, any way is good, Itt maye bee made into a pill, with the pulpe off one raison off the Sun, as theye call themm, which will bee much less thenn a pease,[21] Butt I have used to take Itt, in simple water, Inn a spoone, & so washe Itt doune[22];—

The maner off workinge is by evaculation[23] all the wayes, thatt nature cann purge Itt Selfe; Thatt Is by Stoole, vomit, Sweat, urine, & by Salivation, Iff the state off the patiente requier Itt.—

Inn a bodye thatt is nott very foule, Itt never vomitts, much less causes a Spittinge; which Is ann oparation offensive, butt off good Effecte[;] Itt lasts nott above 2: or 3: dayes, I have knowen Itt worke thatt waye, with 2: woemenn only, butt never with any mann. Butt iff thatt waye off workinge bee troublesume, the only waye Is to take ann other dose off the Same pouder.

Inn a body that Is cleane & Sound Itts operation Is Imperceptible, eyther by urine, or by a gentle Sweat like Bezoar, But Inn a verye foule body, Itt workes offte by vomitt, butt after Itt has donne Itts vomitinge, the same dose workes commonly By Stoole.—

[30v] Ande after that by Sweat; butt iff a mann have naturally any difficulty Inn vomitinge, I never knewe Itt worke, with himm that waye; for as a freinde to nature Itt workes thatt waye, which nature findes moste Easeye;—

Butt iff Itt chance to trouble you, & worke churleshly[24] that waye, beyonde what you are willinge to suffer, a litle off the Countes off Kents pouder[25] stops Itt Inn a Momente.

Ande Itts operationn beinge stopt, Itt does no harme; For lack off the Countess off Kents pouder, take 6: gr. off Bezoar; &c.

Itts operationn by urine Is nott Inn the quantetye off Itt, butt Inn Sendinge thatt ~~a~~ waye the Cause off the disease, the colour, & substance

[21] Piece.

[22] Down.

[23] Evacuation.

[24] Churlishly: that is, roughly.

[25] See Transcription, 188.

will bee changde, I have harde[26] the maker off Itt saye; thatt hee never knewe Itt fayle off Itts cuer what soever the disease was, iff Itt wraught by urine;—

Take Itt fastinge unless ther bee nesesety to the contrary, & faste after Itt, the longer the better, you need no brothes, nor Possetts, &. milke Is observ[^e]de nott to bee a freinde to Itts operationn. nothinge Is nesesary, butt only to keep warme after Itt, you May sleep after Itt,—

[31r] I never knew itt doe harme to any body in what disease soever. yett I have knowne Itt taken by childran off 20: monthes olde for a cough, & by weemen with childe, whom withoute nesesety for all that, I woulde not advise to take Itt;

Itt is proper for all venerian[27] diseases, fluxes, feavors, consummtions, &c.—Iff the disease bee fluxes, or feavors, &c.—Itt cures ordinarely with once takinge. Butt iff Itt bee an Inveterate disease this Is the use off Itt.—The firste dose findinge the body foule, does butt voyde the Excrements.-take ann other dose, the next daye, & that will finde oute the root off the disease, & you shall perceve Itt to worke Eyther by urine, or stoole, more or less, for Summ Space off time, & when you perceave Itts operationn to cease, take ann other dose, & when that has donne, take ann other, tell[28] you finde your selfe perfectly well; I have knowne a mann finde Itt worke a month together,[29] by 2: or 3: stooles a day, & yet Inn all thatt time, to growe sensibly stronger, &c.—

Itt is a certine & Imediate cure off burninge feavors, & off fluxes, bloody, or other, wether they bee joined together, or no. I have knowne Itt cure the palsey in a personn off 70: yeares olde & the falinge[30] sicknes Inn a childe, that had Itt frome the wombe off her Mother; The childe had ordinarely 3: or 4: fitts a weeke, & att summ times off the Moone,[31] 3: or 4: fitts in a night,

[26] Heard.

[27] Venereal.

[28] Till.

[29] Without intermission, continuously, uninterruptedly (*OED*, adv. A, 5).

[30] Falling.

[31] I.e. Some times in the lunar year. On ideas about lunar effects in medicine, see Anna Marie Roos (2000), Luminaries in Medicine: Richard Mead, James Gibbs, and Solar and Lunar Effects on the Human Body in Early Modern England. *Bulletin of the History of Medicine* 74.3: 433–457 and Louise Hill Curth (2005), Astrological Medicine and the Popular Press in Early Modern England. *Culture and Cosmos* 9.1: 73–93.

[31v] Ande frome the first takinge off the pouder never had any fitt, & Inn all tooke only 8: gr. att 4: times, I have knowne Itt doe rare thinges for the stone, & Inn all kinde off defluxions, & Itt Is admirable for the Plage[32]; butt thenn I woulde double the dose because the disease Is churlish, & Inn those payns off the stomach whereof Phititians[33] ofte, finde neyther cause, nor cure, I have knowne Itts operation, to bee wonderfull;—

The same pouder mixt with a litle pomatum or may butter & aplied Externaly, shall Eate a-way dead fleshe, & yett cure a green wounde, & takes awaye without payne, those thinges that are used to bee[34] Eaten away by corrosives, with Exsedinge torture,

I have knowne Itt cure the small poxes withoute strikinge themm[35] out by workinge a waye the matter which Caused themm,

The maker off Itt had a rare Experimente off Itt in a Childe, which with ann other off Like Age, had takenn poyson; A Phititian tooke his choyse, off the 2: childerenn, his died, thatt which tooke the pouder lived, & Inn 24: howers, was perfectly well, Hee tolde mee as a cheef matter, that Itt did rare thinges for weemenn, stoppinge wher ther was to much, & provokinge, wher ther was to litle, I have nott knowne Itts Effect for the Goute, Butt I have a greate opinonn off Itt.

[32r] This Friar that gave mee this Medcin tolde mee, Thatt a Gentlemann att London was givenn over by all the beste Phetitians[36] theare, affter greate Expense, & time, havinge the greate Poxe[37] so Extreamly, & thatt this pouder withinn a fewe dayes perfectly curde himm, to the astonishmente off Doctor Moore,[38] his former Phitiann thatt thenn mett himm lustelye[39] walkinge Inn the Streets, wher uponn the same Docter

[32] Plague.

[33] Physicians.

[34] I.e. usually.

[35] I.e. by attacking the cause of smallpox rather than physically removing individual pustules.

[36] Physicians.

[37] "Great pox" is syphilis.

[38] Possibly John Moore, who was a Licentiate of the College, admitted c. 1618; died 1641 (see William Munk (1878), *The Roll of the Royal College of Physicians of London* (London), 174–175).

[39] Lustily. With vigour or energy (*OED*, adv. 2).

baught summ off Itt, att 20:^{s} agr.[40] att leaste, & whilste hee had the pouder hee did very greate Cures.

The same Friar tolde mee hee beinge Governor to a younge Gentle-mann Inn Italey, the Gentle-mann fell sick off a desperate fevor, & all the Phetitians ther gave himm over, so the Friar gave himm summ off this pouder, & the nexte day hee was quit off his feavor, to the astonishmente off all the Phetitians Ande Itt wraughte by urine moste,[41] the Friar ~~th~~ sayenge Itt was the strangeste black, & mudeye water, thatt Ever hee sawe, iff Itt worke by urinn Itt never fayles the Cure, as the Friar sayes

Iff I had nott a huge opinionn off Itt I woulde never have takenn this payns to have writtenn Itt oute with my owne hande

W: NEWCASTLE:—

[32v is blank]

[33r]

S^{r}. Thomas Cadaman Phetitian to the Queen, his newe & moste rare secrett & receyte to cuer the Poxe, by the flewe de pores, & nott du boushe,[42]

Take halfe a pounde off white Sublimate finly powderde, disolve Itt Inn fowre Gallans,[43] off rayne water, or cleer river water, so this will keepe alwayes uncorupted, Espetialye Inn glass vessells.—one gallon off this putt to as much river, or rayne water, as will fill a litle Bathinge Tubb, Is Enough to seasonn all, one Bath will serve thrise, & Itt is suffitiente, to sitt Inn Itt in the morninge, thre quarters off ann hower,[44] or one hower,

[40] 20 shillings a grain.

[41] I.e. mostly.

[42] This is a strange mixture of French and English, but the meaning is clear: "by the flow of the pores and not of the mouth". On medical conceptions of pores at this time, see Mieneke te Hennepe (2012), Of the Fisherman's Net and Skin Pores. Reframing Conceptions of the Skin in Medicine 1572–1714. *Blood, Sweat and Tears: The Changing Concepts of Physiology from Antiquity into Early Modern Europe*, eds. Herman F.J. Horstmanshoff, Helen King, and Claus Zittle (Leiden), 523–548.

[43] Four gallons.

[44] An hour.

ackordinge to the forses,[45] off the Infirme, warme as maye bee verye well, & with pleasure Indured, The Tubb, beinge Coverde, to the throte off the Bather, with the heade oute, with a warme sheet, & blankett, att the goinge oute, the body, muste bee dride, with warme napkins, & putt to a warme bed, for three howers after, on[46] hower takinge a draught off warme broth, made off veale, mutton, & a well fleshte chicken; to the quantety off a pinte, & so much agenn one hower before hee goes oute off Bedd; which muste bee all the nourishmente for 12: dayes, whilste the Bath Is Continude, Exsepte 3: or 4: times a daye, a good draughte, [33v] Off the Decoction, off Salsa-perilla,[47] aramotisde[48] with a blade off mace, & swetned with white Sugercandye,—Itt Is nesesarye to purge 3: dayes before Consequenter,[49] with a Benigne Medcinn, & once affter, & to swett gentlye Inn Bedd, for a boute one hower, affter the goinge oute off the Bath,.—

Probatumm Este, by a
greate personn thatt I
knewe,

[34r]

Mes Grace Johnsons[50] Plague Water

[45] Forces.

[46] One, which is often spelt as "on" in what follows.

[47] Sarsaparilla.

[48] Aromatise. To render aromatic or fragrant; to impart a spicy flavour to; to season with spice (*OED*, v. 1).

[49] Latin for "consequently", which means, in this context, "in following time or order; consecutively, subsequently (*OED*, adv. 1a).

[50] We have been unable to identify "Grace Johnson".

Take Dragons,[51] Rosemary, Wormewood, Sage, Scor=dium, Mugwort,[52] Scabius, Balme, Cardus, Tormentill[53] roots & leaves, Angelica Rootes & leaves, Marygold flowers and leaves, Century[54] Tops, Bettony flowers & leaves, Pimpernell[,][55] woodsorell, Egremony,[56] Rue, of every one of^{e} these, half a pound; of^{e} Liquorish 4: Ounces; Alicampana[57] rootes, 2: Ounces. Wash the hearbes & shake them drye in a cleane Sheete or sheets, shread them, and slice the rootes, put them all into 3 Gallons of the best white wyne; & let them stand 2 nights close covered, stirr them morning & Evening; then take some of the hearbs quickly out of the wyne .& fill up an ordinary still betimes in the morning, let them distill 12: houres at the Least w:th a reasonable quick fire; Distill the hearbs & rest of ye wyne in a Lymbeck, as yo^{u} do other strong waters, till all the strength of the hearbs & wyne bee gone.[58] For when it begins to bee soure either in Still or Lymbeck then give over: So it is good towards the latter end to set lesser Glasses, that yo^{u} may the better match ye Change.[59]/

Yo^{u} may if yo^{u} please reserve 3: or 4: Spoonfulls of the first running of the Lymbeck. It is good to rub palsey Lymbs. or ones temples & Nose in

[51] According to Culpeper, dragons (or dragonwort) can "cleanse mightily, and take away proud, or dead flesh", and, "in Oyntments, they take away Scurf, Morphew, and Sunburning" (Culpeper, *Pharmacopoeia*, 6).

[52] Mugwort was thought to be "an herb apropriated to the foemine sex, it brings down the terms, brings away both birth and after birth, easeth pains in the matrix" (Culpeper, *Pharmacopoeia*, 14).

[53] Culpeper claims that tormentil was "exceeding good in pestilences, provokes sweat, staies vomiting, cheers the heart, expels poyson" (Culpeper, *Pharmacopoeia*, 10).

[54] Centuary.

[55] Pimpernel was deemed to "draw thorns and splinters out of the flesh, amend the sight, clense ulcers, help infirmities of the liver and reins" (Culpeper, *Pharmacopoeia*, 14).

[56] Agrimony.

[57] Elecampane.

[58] I.e. the "strength" is no longer in the herbs but in the decoction.

[59] The idea here is that, when the decoction is heated and begins to ferment, there may be some runoff, so one should adapt to the changing circumstances and set some smaller containers around the limbeck to catch this liquid.

a sound.[60] Or that & Capons grease are good for the Stomach in a cold. or such like, as any ach &c./

[34v] When yo^{u} have all the waters; put the first of the Still & Last of the Lymbeck togeather & the first of the Lymbeck & the Last of the Still; Keepe the second of the Still for Children, never desire to give it very strong to any body. And when yo^{u} have or see reason to give it very often, as in feavers, great heat, or Sick; Let it not bee of the strongest but middling w:^{ch} yo^{u} may order in the mingleing as yo:^{u} please./

We hold that of the Still, w:^{ch} has most hearbs to bee rather the better./

If yo^{u} give it every 2: or 3: houres, either in small Pox or Measles, when yo^{u} feare they are goeing in, or that ye Party bee very sick yo^{u} may safely doe it & feare not./

It will quench thirst in feavers, the smaller of it,[61] better then any Julips or Beare.[62] And though it give a small heat at very first, yet they shall soon find the benefitt of it./

Tis good for almost anything but the Spleen, w:^{ch} almost noe strong waters are good for./

A Pynt of plaine Succory water, 2: Spoonfulls of wood=sorrell .& 2 of Gillyflowers, w:th as much of this Cordiall water as will make it a little strong like small white=wyne is an Excelent Julip./

[35r]

Cosmeticum[63] of Silver

Take of the purest refined Silver garbles[64] i ounce[,][65] put it in a parting glasse, & poure upon it of strong & well rectified Aqua=fortis 2: Ounces .& set it in warme sand till all y^{e} silver bee disolved, then take it out of the sand and let it cool. Then take of bay salt such as is well purified six ounces .& dissolve it in so much spring water as will dissolve it, and

60 Sound. A swoon or fainting-fit. Usually with prepositions in or into. Very common c. 1530–1650 (*OED*, n. 4, 1a).

61 Refuse, such as husks, dust, or soil, that is separated out from a commodity (*OED*, n. 2).

62 Beer.

63 Latin for "cosmetic".

64 To remove husks, dust, soil, or other refuse from (spices, dyestuffs, etc.); to sift or cleanse of refuse (*OED*, v. 1).

65 1 ounce.

being dissolved filter that solućon[66] through browne paper. Put this salt water in a glasse bason, poure ye aforesaid solućon of the silver into it, and all ye silver will precipitate in a white pouder. Let it rest so for half a day. Then put to it so much cleane and well filtred spring water, as ye rest of the Liquor is,[67] that standeth over the silver powder. And so let it rest another half day, or till next day. Then poure of the Liquor gently that none of the powder come over w:th it and poure on very cleane & well filtred spring water as much as almost filleth the bason and stir the powder very well w:th a glasse stick and all yo:^{r} Liquor will turn white like milk, and afterwards cleer it self againe, & the powder settle in the bottome. Thus stir it againe & againe every half houre for that day; and so let it rest till ye next. And then [^poure] of ye water gently from ye powder & poure an other and stirr it againe as before And the next day poure off ye water againe, and poure on fresh & stir it as before—

[35v] And this proceeding continue till ye water dip'd into w:th yo:^{r} finger & tasted, is quite insipid w:th out any tartnes. Then poure off that water, and dry y^{e} powder in a warme place, and all the powder will become of a deep blew[68] colour. Then take half a pound of salt Armoniack, and of bay salt purified, decrepitated,[69] melted, as much .& sublime y^{e} salt Armoniack from the bay salt. This sublimed salt Armoni=ack, mix againe w:th y^{e} like quantity of fresh bay salt, done as before & sublime it againe, & thus do thrice. Then take a quart of the best dephlegmed spirit of wine, such as will burne all away. And put one ounce of this sublimed salt Armoniack into it, & put them in a glasse body w:th a blind head, & set it in a balneum vaporosum.[70] And when ye salt Armoniack is dissolved in the spirit of wyne. Then take of the blind head, & put on an Alembick or head w:th a snout & set it in Balneum Mariæ, and apply a fitt receiver, and close all the chinks[71] very well w:th bladders[72] and distill in a gentle heate all that will come over .& the salt Armoniack that remaneth in the

66 Solution.

67 I.e. add as much water as there is liquor.

68 Blue.

69 To calcine or roast (a salt or mineral) until it no longer crackles in the fire (*OED*, v. 1).

70 Latin for a vapour or steam bath.

71 A fissure caused by splitting; a cleft, rift, or crack; a crevice, gap (*OED*, n. 2, 1a).

72 The substance of a sheep's or ox's bladder used for airtight coverings (*OED*, n. 4).

bottome undissolved, mix againe w:th y^{e} spirit w:^{ch} yo^{u} have distill'd over and digest it in balneo=vaporoso being coverd w:th a blind head as before. And when y^{e} salt Armoniack is dissolved, take off the blind head and put on a head w:th a snout, & put it in balneum Mariæ applying a fitt receiver, and losing the Chinks, distill as before./

[36r] This proceeding repeat so often till all the salt Armoniack beecome over[73] w:th the spirit of wyne And in like manner do w:th all y^{e} remaineing sublimed salt armoniack, till all by degrees bee fetched over w:th y^{e} spirit of wyne then take yor blew colour'd silver powder, & grind it finely, & then put it into a parting glass, and power[74] upon it of that spirit of wyne im=pregnated[75] w:th that salt Armoniack (w:^{ch} was taught even now how to make it) so much as that it reach an inch or two over it, and shake it well about. And then set it in a warme place well stop'd, and shake it now & then and when yo^{u} see the spirit of wyne well tinged w:th y^{e} blew colour. Then take it from ye warme place and let it rest for an houre or 2: Then poure off gently ye tinged spirit and poure on fresh and do as before./ and so goe on till all the blew tincture bee extracted, and noe spirit will tinge any more. Then take ye silver pouder out of the parting glass, and put it in a glass bason, and put on spring water well filtred, and stir it now and then w:th a glass stick, for one day, the next day poure on fresh filtred spring water. and do as the day before. The next day poure off the water & drye the powder gently, & by degrees. And being drye scrape it out of yo:^{r} bason w:th an Ivory knife & grind it, & keep it for use/

[36v]

Cosmeticum e Bismutho.[76]

[73] I.e. covers.

[74] Pour.

[75] To fill (a substance or portion of matter) *with* some active principle, element, or ingredient, diffused through it or mixed intimately with it; to imbue, saturate (*OED*, v. 3a).

[76] Latin for "bismuth".

Take of the best Besmuthum[77] (commonly called tinglass) one ounce[,] beat it into powder .& put it in a middle sized parting glass. Then poure upon it of the sharpest Rhenish viniger, so much as to reach 2: Inches over. shake it about & when y^{e} vinigar is black, poure an other, & do as before, and so continue till the vinigar changes noe more. Then poure it off & poure on half an Ounce of good Aquafortis, and it will make an Ebullition, w:^{ch} ceasing poure on half an Ounce of ~~good~~ fresh Aqua=fortis, and shake it, & so do twice, till yo^{u} have poured on two ounces of Aquafortis by half an ounce at a tyme, & when the Ebullition ceaseth cover y^{e} glass w:th a peice of paper & let it rest in a warme place overnight. The next day filter it through Cotton wooll stop'd into the nozell of yo:^{r} Glass funnell, into another parting glasse or any Glass bottell .& being well filtred, from all it's feces, put the solućon into a Large glass bason .& poure into it one pynt of well dephlegmed spirit of wine & stir it w:th a glasse stick, & there wilbe an Ebullition w:^{ch} ceasing a white powder will precipitate or sink to ye bottom. Then for the more exact precipitating of all, and also the more suppling[78] and dulcifing of the powder, pour into it two ounces of the best Oyle of Tartar, made of pure white salt of Tartar [37r] par deliquium[79] (as they call it) that is by liquation[80] in the cold most air, and stir all well about w:th a Glass stick .& there wilbe againe an Ebullition, w:^{ch} ceaseing, let it rest coverd in a warme place, over night. The next day pour of all the Liquor gently from the white powder. Then pour on, of well filtred & cleare spring water about the quanti=ty of a quart, & stir up y^{e} powder w:th a glass stick divers tymes & then let it rest, & so do every six or twelve hours, for 3: or 4: dayes, each tyme poureing on fresh spring water, & stirring up y^{e} powder. Then (the powder being well settled in the bottom.) trye w:th yor finger dip'd in the water w:^{ch} standeth over the powder how it tasteth .& finding all the tartnes gone .& ye water taste like it self, pour it off

[77] Bismuth.

[78] The action of making something more supple, soft, or flexible (*OED*, n. 1).

[79] "Per deliquium" is Latin for "by defect". The suggestion is that the white tartar comes from old, unclean wine barrels, and, when left out in the cold (as if neglected) in these wine barrels, it becomes purified.

[80] The process of making or of becoming liquid; the condition or capacity of being melted (*OED*, n. 1).

gently from the powder [^&] keep ye vessell w:^th^ the powder cover'd, in a warme air, but set it not in the Sun nor neare any hot place for th^n^ it would drye too violently. and afterwards cast a yellow hew. but being dryed & Leisurely it wilbe very white. then scrape it off from ye bason, & grind it upon a marble very fine .& so keep it for yo:^r^ use/

[37v is blank]

[38r]

Lac Virginis[81]

Take fore or five ounces of Poudered Litharge of silver of [*sic*] or Gold or red Lead or white Lead or burnd Lead and too[82] or three pounds of distild vinegar[;] putt them into a glass pipkin and sett them over a gentle fire[;] the Pott beeing close covered take of the cover and stir them offten with a wooden spattula and then cover the Pott againe[;] continew soe doe=ing till thare remaine about on pint of the vinegar which philter[83] warme into a cleane Glass[;] Then take a large hand full of Kitchin salt and too pints of spring watter in another vesell[;] lett them boyle togather to the consumtion of halfe and Then philter it into another glasse[;] keepe the Glassis close stopt. when you would euse[84] them take of each of these Liquors a like quantitie[,] mix them togather and thay will make a white Liquor very safe and cooleing in which moysten duble Linin Cloths and aply them to the inflamed part and lett them lie thare till thay bee drie then moysten them againe and aply them untill all the heate bee drawne out which will bee in a short time

[38v is blank]

[39r]

[81] Latin for "Milk of the Virgins". According to Culpeper, this concoction "takes away Pimples, redness, freckles and sunburning" (Culpeper, *Pharmacopoeia*, 74).

[82] Two, which is often spelt as "too" in what follows.

[83] Filter.

[84] Use, which is often spelt as "euse" in what follows.

Pomatum for the Lipps or Chops[85] in the hands or nose

Take of the greate black grapes[,] squeese out the Juce of them with ons[86] hands[,] straine[87] it thorrow[88] a lin=in Cloth[;] take a pound of fresh Butter with out salt and a quarter and a halfe a pound of new yalow[89] wax[;] melt them togather then putt to them 3 pints of the Juce of grapes on has redy and lett all boyle Gently togather keepeing it stiring continew=aly and when it has boyled halfe an houer take a quince Aple and with an aple scoope scoope out all the inward part of it and putt it in to the Licquor keepeing it still Boyleing and stiring it as beefore[;] when it has boyled 3 quarters of an houer and that on finds it well mingle=d and thick Pouer it out into an Earthen Pott ~~stire~~ stiring it till it bee cold[;] this will keepe 2 or 3 yeares[;] tis the best thing for lipps that ever I did euse[;] rub it well in to the lips over night and rub it of with a cloth next morneing and all the loose skinn will com of[90] and the lipps [will be] as whole and well as can bee

[39v]

To make Wiggs

Take three Pound and a halfe of the finest wheate flower but the Halfe Pound must bee to work them up[91][;] take a Pint of Ale yest and a fill of sack[92][;] lett the sack [^and] Ale yest worke togather 2 Houers beefore you beegin to make them[93][;] then straine it beefore you Putt the yest to the sack[;] putt a graine of Muske and an other of Amber greece [^in] to the sack and a nutt megg beaten[;] make all this in to a Past[94] like white bread with a pint of New milk and keepe it warme for an Houer to rise[;]

[85] Chaps. A painful fissure or crack in the skin, descending to the flesh: chiefly caused by exposure of hands, lips, etc., to frost or cold wind (*OED*, n. 1, 1b).

[86] One's.

[87] Strain.

[88] Through.

[89] Yellow.

[90] Come off.

[91] I.e. half a pound of flower is necessary to trigger the yeast to start working.

[92] I.e. a quantity sufficient to fill the sack.

[93] I.e. the wigs.

[94] Paste.

then worke in to it a Pound of sweete Butter and a pound of Carraway comfitts[;] then make them in to wiggs with Papers under [^the] bottams and bake them very quick

[40r]

To Make serup of Althea

Take of Marsh mallow rootes 2 ouneces[,] rootes of Grass Asparagus[,] licoris[,][95] rason of the sunn stond[96] of Each halfe an ounce[;] the Leaves of Marsh mallows[,] common Mallow[,] Pelitory[,] Pimpernell[,] saxifrage[,] Plantaine[,] black and white Maden Haire of Each one hand-full[;] red cicers one ounce[;] the fore greate and lesser cold seeds of Each three drams[;] Infuse all these a whole day in 6 Pints of spring watter then boyle them untill 2 pints [^bee] boyled away then straine out the Liquor and ad to it 3 pounds and a halfe of white sugar and boyle them to a syrup

[40v]

To Make Angell water

Take a Peck of Damaske roses[,] small Lavindor and sweete Margerum[97] of Each 3 handfulls[;] sweete Bassill and sweete Maudlin of each on handful[;] the tender tops of bay leaves[,] the herb mæstick[,] the tops of rose mary of each one handful[;] cost=mary and Eglintine of each on hand=full[;] take of Pouders the best sweete bag Pouder from the Perfumers of Benjamin and storax[,] sweete calamus[,][98] ciprus[99] rootes and cloves of eace one once ~~in~~ and of Laudenam 2 ounces[;] all grosely Poudered[;] Cutt the herbs small but not the roses[;] mingle the Herbs and roses well with ons hands[;] putt them in to a ~~dob~~ deepe Earthen Pott or Pitcher well glased[100][;] affter sum is in strew sum of the Pouders

[95] Liquorice.

[96] Stoned. To take the stones out of (fruit) (*OED*, v. 7).

[97] Marjoram.

[98] According to Culpeper, sweet calamus "binds, strengthens, stops fluxes of the belly, and immoderate flowing of the terms in women" (Culpeper, *Pharmacopoeia*, 3).

[99] Cypress.

[100] Glazed.

mingl=ed togather soe on a Layer of herbs and roses and a strewer[101] of Pouders [41r] all the herbs and Pouders bee in the Pott or Pitcher[;] then Pouer upon the herbs and Pouders a quart of white Muskadine or Malegoe[102] Sack and a Pint and a halfe of damaske rose water[;] lett all stand steepeing in the Pott close covered 2 days[;] then still[103] it of in [^a] cold still with a gentle fire[;] thare will com of 2 waters a stronger and a weaker[;] hange 8 graines of muske in [^a] bagg in the bottles and as much amber greece if you like it[;] the cakes that com out of the still when it is dun is very sweete to lay amongest Linin

[41v]

A Perfumed Balsom for the lipps

Take 3 ounces of the Pomatum men=tioned a little beefore in this booke while it boyles upon the fire or after it is cold[;] it matters not[;] it must bee made to boyle Gently and beecause it nether smells well nor lookes well Put in to it drops of these oyles a peece[104][:] Lignum Roodium[,][105] ~~Benjaber~~ Benjamine and storax cloves 3 graines a peece[;] of Ambergreese and Muske halfe a dram[;] of Naturall Balsom of Peru[;] mingle all well and with it a drame of the rind of the roote of Alkanett[106] sprinkled beefore with a little good sperit off wine keepe=ing it stiring upon the fire till it looke very red[;] then straine it thorow[107] a cleane linin cloth very fine and then stir it till it bee cold[;] cover it close and and: [*sic*] keepe it to euse but the Alkanett must bee Putt in beefore the Parfumes and allsoe the Ballsom of Peru

[42r]

observations about the lipp salve found out by experiance

101 I.e. a sprinkle.

102 Malaga.

103 To trickle down or fall in minute drops (*OED*, v. 2 1a).

104 Apiece: that is, severally.

105 Lignum Rhodium.

106 Culpeper notes that alkanet is "binding, good for old Ulcers" (Culpeper, *Pharmacopoeia*, 3).

107 Through; often spelt "thorow" in what follows.

Beefore it [^be] right as it should bee it must boyle as long againe[108] as the receipt directs[;] insted of a dram of Alkanett thare must bee too onces[;] it ads much to the vertue of the healeing Part as to the Perfumes[;] halfe a ounce off Naturall Balsam of Peru[,] 10 drops of oyle of cloves[,] 20 drops of oyle of ~~Bêiam~~ Benjamine[,] 20 of oyle of Rodium[,][109] 20 of oyle of Lemon Peele[,] 20 of oyle of Or=ing[110] Peeles[,] a knives Poynt of Licquid Storax[,] 3 Graines of Amber Greece[,] and 3 of Muske which 6 graines must bee incorporated togather with as much as a walnut of very fine hard sugar upon Marble and ad the oyles and Balsum to it and worke and incorporate all togather thorrowly by litle and litle and Put it in to the Licquar by a litle at a time while it is hot that it may disperce and thorrowly all over ~~it~~ the licquor that it may not gather togather in a lumpe[;] never leave stiring it while it can bee stird

[42v]

The Recept of Daffyes[111] Elixier salutis[112] given mee by Docter Waldron[113]

Take senna three ounces and a halfe[,] sassaphrass[,] Lignum vite[,][114] corriander seedes[,] Liquoris[,] sweete fenill[115] seeds[,] Anisseeds of Each too ounces[;] carraway seeds one ounce[,] Rhubarb on ounce and a halfe[,] Elecompane rootes six ounces[,] Cæssia fistula too ounces[,] and a halfe Rasins of the son stond one Pound

Cutt the rootes[,] Pouder the seeds and Lignum vite very well[,] Put all into a glass body with a blind head[;] Pouer upon them fore quarts of the best Brandy[;] Lute it carefully and sett it in a moder=rate heate for

[108] I.e. twice as long.

[109] Culpeper maintained that rhodium "Encreaseth Milk in Nurses" (Culpeper, *Pharmacopoeia*, 3).

[110] Orange; often spelt as "oring" in what follows.

[111] Thomas Daffy (1616/17–1680).

[112] Elixir of health.

[113] Edward Waldron (active mid-seventeenth century).

[114] According to Culpeper, lignum vitae "Dries, attenuates, causeth sweat, resisteth putrifaction, is admirable good for the French pocks, as also for Ulcers, Scabs and Leprosie, it is used in diet drinks" (Culpeper, *Pharmacopoeia*, 3).

[115] Fennel.

24 houers letting it stand soe a month[;] then Pouer it of from the feces and keepe it for euse[;] it maybee a good Purge but it is not right Elixier salutis[116]

[43r]

A Most Exelent Medsin for the cure of Man or Beast bit by a Madd doge that never failes if given with in nine days after the biteing

Take 4 ounces of Mithradate or london Treacle[,] 4 ounces of scraped Pewter very finely scraped small[,] 4 ounces of Herb of Grace or Rue slipt of the stalkes[,] 4 ounces of Garlike brused, boyle them all in 3 Pints of Ale till a 3 Part bee boyled away[;] straine it thorow a fine cloth and that which stayeth in the cloth lay to the wound if thare bee on[;] keepe the Licquer in a botle close stopt and as thare is oc=ation[117] give 7 or 9 spoonefulls every morneing fasting for 6 or 7 morneings togather[;] give it blood warme 7 or 9 spoonefulls to a man at a time[,] 9 for a beast and 5 for a child or a doge[;] first lett blood if age will Permitt[;] if thare bee many to give it to on must make as many quantities for one qua=ntitie will serve to cure but one[;] tis best to make it as on has ocation to euse it

[43v is blank]

[44r]

How to make Mrs Conwalisis[118] greate cake on of the best I ever eate

Take seven Pound of very fine flower[;] lett it bee soe dryd that it ~~is~~ may bee blood warme[;] Put in the Midle of it a quart of very good thick creame[,] 2 Pound of Butter which must bee melted in the creame[,] the yolkes of 16 eggs[,] the whites of 2 of them[,] a Pint of good Ale

[116] I.e. the sediment is a good purgative, but it is not the elixir itself.

[117] Occasion.

[118] This could be a reference to the renowned epistoler Jane Cornwallis (1580/1–1659), who was an associate of Charles II and Henrietta Maria, though Jane mostly went by "Lady Bacon" (taking the name of her second husband, Nathaniel Bacon) from 1626 onwards. A very similar, though unfortunately unattributed, recipe can be found in Kenelm Digby (1669), *The Closet of the Eminently Learned Sir Kenelme Digbie* (London), 258–260.

yeast[,] almost halfe a Pint of Rose water[,] 8 spoonefulls of sack[;] all these blood warme must bee strained into the flower[;] stir them round with ons hand and cover them with sum of the flower soe lett it stand att the fire to rise[;] then Putt in haveing Pict[119] and washt them thorowly in Hott water 10 Pound of curans[120][;] 2 Pound of Rasins of the sonn Pict[,] stond and cutt small[;] halfe a Pound of sugar[;] [^halfe] an ounce of mace[;] and a few cloves beaten and serced[121][;] a litle salt[;] the spice and su=gar must att first bee mingled with the flower[;] halfe a Pound of Preserved citron and oring cut in Bits Putt in with the curans[;] when all is thorrowly well incorporate Putt it in a tin ~~hop~~ hope[122] well buttered fitt for it[;] on may Putt 6 Grains of Amber Greece in the sugar iff on[123] will and on may Ice it with a Pound of Duble refined sugær with amber greece[,] with white of egg and rose water beefore it coms out of [the] oven if on like it[;] take care it bee thorowly well Baked or tis spoyld

[45v is blank]

[45r]

To make oring Pudding

Take 6 larg good oringis[,] Pare of the Rinds Prity thick[,] boyle them in severall waters to take out the biter=nes[;] when thay are Boyled tender Beate them very small in a marble morter[;] take the yolkes of 12 eggs and the whites of 2 of them[,] beate them very well[,] mix them with the Pill[124] and a litle creame[,] a litle oring flower water and as much Juce of oring as will make it sharpe to the tast[,] a spoone full or 2 of grated bread and sum grated Naples biskett[;] sweeten it with sugar Pritty sweete then stir all well togather[;] have redy as much fine Puft Past[125] as will cover the dish bottam and brim[,] the dish well buttered as on would for a Pie[;] then Pouer the Puding in to it and cover it with a shee[^te] of the same Past over it and bake it

[119] Picked.

[120] Currants.

[121] Searced.

[122] Hoop.

[123] What follows in this recipe is written vertically in the margin.

[124] Peel.

[125] In contradistinction to the above "Past" (which refers to "paste"), this is a reference to "puff pastry".

[45v]

To make the Bitter drinke for the wormes or recovery of a weake stomack

Take [^of] the Topps centuary[,] the flowers of camomill of Each a Pugle[;] The Rootes of Gentian cutt thinn halfe a scruple[,] cardus seeds a Dram[;] steepe these over night in a Pinte of Watter and boyle it in the morneing till it coms to 4 ounces which is the quantitie for one dose[;] on may make it to Purge with ading a dram of senna to it or Rubarb or other Purgeing drugg on would have

[46r]

To make a French Almond Tart

Take Halfe a Pound of Almonds[;] blanch them and beate them small with 2 or 3 spoonefulls of oring flouer water and Rose water to keepe from oyleing[;] then Put to them a quarter of a Pound of Biskett and beate that with them[126][;] Put halfe a Pound of beaten sugar and halfe a Pound Butter and beate that with it[;] ad 2 Aples Pared and cut small to them and beaten with it[;] beate 6 yolkes and 3 whites of Eggs very well an[d] Putt to them and an ounce of citron cutt small[;] beate them with it[;] the Juce of a Large Lemon beate all togather till [it is] well incorporated for Halfe an houer[127]

then Put it in to Puft Past in a Paty Pan[128] and Bar it over with Past[,] the Pan well buttered[;] it takes 3 quarters of an houer bake=ing and when it is baked stick[129] it with litle long bitts of citron cutt on purpose

[46v is blank]

[47r]

[126] Long space in original.

[127] Long space in original.

[128] Pattypan. A small tin or pan, frequently fluted, in which patties are baked (*OED*, n. 1a).

[129] What follows in this recipe is written vertically in the margin.

A rare Medisin to cure a Cancer in a Breast or any other thing

Take the Hornes that grow on the inside of a stond Horses[130] Leggs finely beaten and sifted on good bigg Thimble full[;] Mace finely beaten too Thimbles full[;] mix these too well togather[;] Putt them in to a quart of old Ale[;] stir it well togather[;] lett the Party drinke halfe a Pint of it in the morneing and as much att night dureing the cure[;] when one quan=titie is drunke make more[;] Rap up the Breast warme in a Red Cloth[131] but bee sure to take from it all Playsters and salves what soever it has not Cured a Cancer upon Triall[132]

[47v is blank]

[48r]

A most Exelent Playster for a cough or consumtion

Take Bees wax[,] Burgondy Pitch and Rosen[133] of Each 2 ounces[;] of course[134] Turpentine an ounce and a halfe[;] of oyle of Mace on ounce[;] Melt all these togæther till thay bee cold and Putt it in a Pott for euse[;] the Playster must bee of Lether cutt in the shape of ~~of~~ a hart[135] all most as large as too hands[;] it must bee laid upon the brest or chest with the Poynt downewards when it is ~~sped~~ spred to lay on[;] it must bee Prict[136] full of holes and strewed with grated nutmegg[;] when it dus[137] most good it itches much[;] when on Playster is dried away lay on another as long as on finds need of it[;] tis said to have dun greate cures upon consumptive coughs

[48v is blank]

[49r]

130 Stone horse. An uncastrated or entire horse; a stallion (*OED*, n. a).

131 Long space in original.

132 This last line is in the same hand but appears to have been added later.

133 Risen.

134 Coarse.

135 Heart.

136 Pricked.

137 Does.

A very good Receite for the Wormes

Take 2 Peny worth[138] of Tumerick[,] 2 Peny worth of Long Pepper[,] on peny worth of Graines[,][139] on Peny worth of Treacle[,] Worme wood leaves dried and Poudered[,] a good spoonefull swines Grease the bignes of a Turky Egg[;] stir and incorporæte them very well togather in a morter and ~~bred~~ spred it upon a cloth but it must bee quilted[140] beetweene 2 cloths or it will Role on a heape[;] it will lie on 9 days

[49v is blank]

[50r]

A recept to fatt chickins my Lady Coventrys[141] way

Take chickins about 6 weekes old and Putt them in to a Penn with Partis-tions Putting 2 and 2 togather[;] sett the Penn in a warme Place neer a fire[;] make a Past of Barly meale wett with stronge Beefe brauth[142] with out any salt and boyle in it [a] good ~~stir~~ store of sheepes suitt[143][;] shred ~~way~~ very small and a good quan=tiytie of coursest sugar to make it sweete[;] when it is boyled that all the sugar and suitt is melted Pouer it Hott in to a Bowle with Barly meale and worke it to a Past[;] when it is cold lay it beefore them[;] if Boyle ~~braught~~ brauth unsalted bee not allways to bee had boyle the ingredients in New milke makeing it fresh every day keepeing them very cleane and giveing the^{m} Lappings[144] of beare or Ale to drinke and in 3 weekes thay may bee very fatt

138 Twopenny. Of the value of, amounting to, or costing twopence (*OED*, adj. 1a).

139 Grains of Paradise.

140 To pad, line, or cover with material (*OED*, v. 1, 1a).

141 Elizabeth Coventry (1583–1653), who married Thomas, 1st Baron of Coventry (1578–1640), in 1610. Elizabeth and Thomas were, like the Cavendishes, part of Royalist circles. Thomas Coventry indeed worked for Walter Montagu, one of Queen Henrietta Maria's favourites, who was likely responsible for the publication of the *Queens Closet Opened* (see Richard Cust (2013), Coventry, Thomas, first Baron Coventry (1578–1640). *ODNB*). For her part, Elizabeth was profoundly interested in household management and particularly medicine, having contributed numerous recipes to *Natura Exenterata* (283–286).

142 Broth.

143 Suet.

144 Lap. To take up liquid with the tongue (*OED*, v. 5 1).

[50v is blank]

[51r]

A Medisin to Cure the Dropsey seldam failes

Take Dainewort which is dwarfe Elder[,] Wormewood[,] Barbery Barke[,] the inward rine of Greene Broome Each of these 3 handfulls[;] an ounce of Anis seeds and as much Licoris[;] bruse the seeds and slice the Licor=is[;] Putt all into 3 Galons of that Ale or beare on drinkes ~~constatly~~ constantly of but it must bee as stronge as 9 shilings beare[;] it must not bee drunke too new nor too stale soe [it] is best to fund[145] up in stor stands with spigots[146] att the bottom and the ingredients devided into severæll of those stands that on may Just bee redy to drinke when the first is drawne out[;] keepe the Tops close covered[;] on of these stands will bee redy in 2 days time[;] on must drinke no other drinke att all[;] it cures both dropsey and scurvie when thay have bin at a greate hight

D. Biographies

The biographies that follow are modelled on Noel Malcolm's bibliographical register in his magisterial edition of Thomas Hobbes's correspondences, but, rather than only being organised alphabetically, they are categorised according to the role of individuals in Pw V90 and their professional standing (whether as "Patients", "Royalists and Courtiers", "Physicians and Apothecaries", or "Servants and Untitled Characters").[147] While some of the figures could be placed in multiple sections, such as Fludd, who was both a physician and courtier, we have categorised them according to their main role or professional standing (i.e. Fludd is among the physicians). Within each section, the biographies are alphabetised, and emphasis is placed, where possible, on a given figure's medical contributions, however slight, and their relationship to the Cavendishes. In terms of inclusion, biographies are only provided for patients or contributors who are directly associated with recipes, and we

[145] Fund. A source of some material good; a stock or supply of material things (*OED*, n. I 1a).

[146] Small wooden pegs or pins used to stop the vent-holes of barrels or casks (*OED*, n. 1a).

[147] Noel Malcolm, ed. (1994), *The Clarendon Edition of the Works of Thomas Hobbes, Vol. 6: The Correspondence, Vol. I: 1622–1659* (Oxford), 777–920.

thus exclude ancients (e.g. Paul of Aegina) and early modern figures who are named in passing (e.g. Inigo Jones). We have not produced biographies for individuals who feature in Harley 6491 (see above, Appendix 8C). About a dozen of those individuals who are included in the biographies that follow lack *Oxford Dictionary of National Biography* entries, and, indeed, in some cases, such as that of Rust, we know of nothing that has been published about them. Although many of the individuals in Pw V90 do have *ODNB* entries, upon which we have usually drawn, every effort has been made to supplement them with additional printed (and manuscript) material, much of which has been gleaned from the classic biographies of Aubrey, Anthony Wood, and, in the case of physicians, William Munk. There are six figures about whom we can determine very little, but we include their names (or titles) and what we do know about them in Section IV for the sake of completeness; it is our hope that this might encourage future investigations into their identities.

When it comes to some of the lesser-known physicians and apothecaries, Peter Elmer of the "Medical Practitioners in Early Modern Britain" project at the University of Exeter has been extremely generous. He has not only helped us to identify individuals, but also directed us towards relevant archives and assisted us in filling in bibliographical lacunae, especially (though not exclusively) in the cases of Rust and Boucheret. Jonathan Barry has also greatly assisted with the biographies. While the biographical accounts of obscure figures such as Rust and Boucheret will probably prove more useful to historians than those of better-known ones, outlining the biographies of the latter helps to give a full sense of the Cavendishes' medical circles. Indeed, even in the case of some well-known individuals (such as Morley), this is the first time that their medical preoccupations have been recognised, whereas when it comes to Margaret, the biography as a whole should offer a fresh angle on her social and intellectual world that does not primarily focus on her relationship to fellow philosophers and women writers. Finally, organising the biographies according to the role of a given individual in the Cavendishes' medical networks should help to demonstrate that university-trained physicians were but one group among many that contributed to the dynamism of the seventeenth-century medical economy.

I. Patients

Henry Cavendish (1630–1691)

Henry Cavendish was the second surviving son of William Cavendish (1593–1676) and his first wife, Elizabeth Bassett Howard (1599–1643). Tutored at home along with his elder brother, Charles (c. 1626–1659), by Mark Anthony Benoist (a long-standing family retainer), Henry remained in England with his father until the battle of Marston Moor in 1644, at which point he accompanied him to the Continent.[148] Henry is the only Cavendish family member besides William and Margaret to feature as a patient in Pw V90, which makes sense given that his older sisters, Jane (1621–1669) and Elizabeth (1629–1663), remained in England, and Charles was healthy. In the autumn of 1647, however, Henry returned to England to assist with the recovery of the family's estates.[149] He remained in the country and married Frances (1629/30–1695), daughter of William Pierrepont (1607/8–1678), in May 1653. A few years later, after the death of his brother, he was made Viscount Mansfield.[150]

While Henry took less interest in poetry and philosophy than his father, he led an active political life, being elected as the representative for Northumberland in the Cavalier Parliament in April 1660, and also serving as Master of the Robes from May 1660 to March 1663, including at Charles II's coronation. Evidently successful in this role, Henry was promoted to Gentleman of the Bedchamber, and held this position until the end of Charles II's reign. He was also made Earl of Ogle in 1665, and, after William's death in December 1676, 2nd Duke of Newcastle. As joint lieutenant (1670–1677), Governor of Berwick (1675–1686), Lieutenant of Northumberland (1677–1688), and Lieutenant of Nottinghamshire (1677–1688), Henry was an outspoken opponent of the three bills that would have excluded the King's brother and presumptive heir, James, Duke of York, from the thrones of England, Scotland, and Ireland on

[148] For Benoist's correspondences with William, Frances, and Henry Cavendish, see Nottingham University Library, Pw 1/12–24.

[149] For this background, see Peter R. Seddon (2004), Cavendish, Henry, Second Duke of Newcastle upon Tyne (1630–1691). *ODNB*.

[150] Henry evidently had a close relationship with his father-in-law, with whom he stayed in regular contact at least for the first fifteen years of his marriage to Frances (see Nottingham University Library, Pw 1/371–384).

account of his Catholicism. Henry's support for the crown was unwavering; he went so far as to argue in the House of Lords that the bills were "not fitting to bee debated it is to Alter the succession", and he refused the oaths of allegiance to William III and Mary II.[151]

Even as he represented Northumberland, Henry remained at the family seat in Nottinghamshire, where he faced continual troubles throughout the 1680s. For one, riots erupted when the Nottinghamshire Whigs were defeated in the first elections under the new charter in September 1682. Henry, who also served as the local record keeper, collected evidence that could be used against the rioters and their Whig supporters and took advantage of news about the Rye House plot to attack the leading Nottinghamshire Whigs and search their homes. But the Whigs were nonetheless successful (for the first time) in the parliamentary election of 1685. In October 1686, Henry was appointed lieutenant of the three ridings of Yorkshire—which meant that he was responsible for guarding virtually all of the North East—and he was ordered to raise a regiment to secure Newcastle upon Tyne.[152] But he soon discovered (in November 1686) that a petition in favour of a freely elected Parliament was in circulation, and he left York in anger, which gave Thomas Osborne, Earl of Danby and Duke of Leeds, an opportunity to take control.[153] Henry had believed Danby's and the Earl of Devonshire's false claims of loyalty, and, when he finally recognised his errors and decided to take action, the militias ignored his orders and he found himself imprisoned in Welbeck Abbey. This episode left a stain on Henry's political career.

In 1680, Henry's only son died, the stress of which, combined with the struggle to find dowries for his five daughters, led him to split with his wife.[154] He died a decade later at Welbeck on 26 July 1691. His estates were left to Lady Margaret Cavendish (1661–1716), his third surviving and favourite daughter, who had married John Holles, 4th Earl of Clare,

[151] Esmond S. de Beer (1943), The House of Lords in the Parliament of 1680. *Historical Research* 20.59: 22–37 (34).

[152] David H. Hosford (1976), *Nottingham, Nobles, and the North: Aspects of the Revolution of 1688* (Hamden, CT), 82–83.

[153] On this episode, see Andrew Browning (1951), *Thomas Osborne, Earl of Danby and Duke of Leeds, 1632–1712*, Vol. 1 (Glasgow), 247–283.

[154] John Reresby (1936), *Memoirs of Sir John Reresby*, ed. Andrew Browning (Glasgow), 437 and 472.

in February 1690, upon the condition that any sons or daughters from the marriage would take the Cavendish name.

Margaret Cavendish [née Lucas] (1623?–1673)
Margaret Cavendish, the youngest daughter of Thomas Lucas (c. 1573–1625) and Elizabeth (d. 1647), was born at St. John's Abbey, Essex, in or around 1623. Having been home-educated, Margaret left for Oxford with her sister, Catherine Lucas, Lady Pye, in the autumn of 1642 to join the court. She then became a maid of honour to Queen Henrietta Maria (in 1643), and, the next year, moved to Paris to wait on the exiled Queen. In 1645, Margaret met William, and, despite the Queen's disapproval of their courtship, they were soon married. The couple remained in Paris for the next few years, but, after a brief sojourn in Rotterdam, settled at Rubenshuis in Antwerp in 1648. In August of that year, Margaret's brother, Sir Charles Lucas (1613–1648), was executed for his role in the siege of Colchester.[155]

Accompanied by her brother-in-law, Sir Charles Cavendish (1595?–1654), Margaret left Antwerp for England in November 1651 to appeal to the Committee for Compounding with Delinquents—which confiscated the estates of Royalists who fought against Parliament in the Civil Wars—in hopes of regaining some of William's sequestered property.[156] During the eighteen-month trial, Margaret lived in Covent Garden, and it was here that she penned her first two works: *Poems, and Fancies* and *Philosophicall Fancies* (published in London in 1653). While she commenced a lifelong friendship with Walter Charleton at this time, her stay in London was generally a lonely one.[157] What is more, despite her best efforts, she was ultimately deemed an illegitimate claimant for the estates, since she had married William after he had fought against parliamentary troops. This left the Cavendishes in some financial hardship. To make matters worse, Charles fell gravely ill during their stay and decided to remain in England when Margaret returned to Antwerp, passing away a few months later (in February 1654). Charles's death was a great blow to Margaret, since, as an adept mathematician with a strong interest in

[155] Cavendish, *Life of William*, 61–62.

[156] For more on Cavendish's time in England, see Whitaker, *Mad Madge*, 137–163, and, on the petition in particular, see Mary A. Green, ed. (1889–1892), *Calendar of the Proceedings of the Committee for Compounding* (London), 2021–2022.

[157] See Semler, The Magnetic Attraction of Margaret Cavendish.

natural philosophy, he had acted as a friend, tutor, and inspiration for her early output. In acknowledgement of his support, Margaret had dedicated to Charles both of the works that she published before his passing.[158]

Upon returning to Antwerp in May 1653, Cavendish continued to welcome visitors, help run their relatively modest household, and (no doubt) refine her ideas. But she only published three books in the nine years following her return to Antwerp: *The Worlds Olio* (1655), *Philosophical and Physical Opinions* (1655), and *Natures Pictures* (1656). One reason for her diminished output during these years is that Cavendish preferred to have her books printed in England, having already become frustrated by the number of typographical errors that even printers who were native English speakers introduced during the publication process. She was also daunted by the prospect of travellers shuttling precious manuscripts to London while being too far away to oversee the stages of production.

At the Restoration, William returned to England with Charles II, and Margaret soon followed. William secured his estates in September 1660, and, thenceforth, Margaret spent most of her time between Welbeck Abbey and Bolsover Castle. She took an increasingly active role in managing the estates both directly and through overseeing Francis Topp. It was also at this point that she pored over recent philosophical and medical publications, the contents of which she would have discussed, but not always fully digested, on the Continent. This is indicated in part by the fact that she began to cite and respond to specific passages from the likes of René Descartes, Jan Baptiste van Helmont, and William Harvey at some length in works such as her 1664 *Philosophical Letters*, but also by the fact that documents such as a bill from 9 July 1664 show that Benoist was paid 39.40.0 for her books.[159] This reclusive stint from 1662 to 1668 was the most productive of Margaret's life, granting her the time to publish, republish, or have translated no less than thirteen volumes including plays, orations, books of poetry, histories, and philosophical tomes, which she then had distributed to virtually all of the Oxford and Cambridge colleges and to many of her acquaintances in England and

[158] See *Poems, and Fancies*, A2r–A2v; *Philosophicall Fancies*, B4r–B5r; and *The World's Olio*, A3r.

[159] Nottingham University Library, Pw 1 592/2.

abroad.[160] At the invitation of Charleton, Cavendish was also the first woman to attend a meeting of the Royal Society of London (on 30 May 1667).[161]

While Cavendish became less productive after 1668 (only publishing two further redactions), she did not retreat from intellectual life entirely. Rather, Margaret and William continued to patronise poets and playwrights such as Richard Flecknoe and Thomas Shadwell. The latter not only bestowed manuscript drafts of his plays upon William and Margaret, which they marked up before publication, but also dedicated many of his major works to them, including *The Humorists* (1671).[162] Margaret died unexpectedly in 1673 and was buried in Westminster Abbey. A few years after her death (in 1678), *A Collection of Letters and Poems* appeared, which gathered epistles and epitaphs from many of Margaret's and William's most eminent acquaintances. This included controversial figures such as Hobbes and the philosopher and divine Joseph Glanvill but also more traditional ones such as the Dean of Christ Church and Bishop of Oxford John Fell and the Lady Margaret Professor and Archdeacon of Oxford Thomas Barlow. The collection was carefully constructed to showcase the extent of Margaret's contributions to intellectual life in England and beyond.

William Cavendish (bap. 1593–1676)

Born at Handsworth Manor, Yorkshire, and baptised in 1593, William Cavendish was the first surviving child of the MP Sir Charles Cavendish (1553–1617), and his first wife, Catherine (1570–1629).[163] Tutored at home (Welbeck Abbey), William entered St. John's College, Cambridge, in 1608 as a fellow-commoner.[164] After a brief stint in Cambridge, he

[160] See William Poole (2015), Margaret Cavendish's Books in New College, and Around Oxford. *New College Notes* 6: 1–7.

[161] See Samuel I. Mintz (1952), The Duchess of Newcastle's Visit to the Royal Society. *The Journal of English and Germanic Philology* 51.2: 168–176.

[162] See Nottingham University Library, Pw V 33 and Pw V 34. On Cavendish and Shadwell, see Begley, *Margaret Cavendish, The Last Natural Philosopher*, 242–268.

[163] The fullest study of William Cavendish, which has provided a great deal of material for the present biography, remains Geoffrey Trease (1979), *Portrait of a Cavalier: William Cavendish, First Duke of Newcastle* (London).

[164] John Venn (2011), *Alumni Cantabrigienses: A Biographical List of All Known Students, Graduates and Holders of Office at the University of Cambridge, From the Earliest Times to 1900*, Vol. 1, Part 2 (Cambridge), 311.

moved to the Royal Mews, where he was trained in *manège* under the renowned French riding instructor Pierre Antoine Bourdin, Seigneur de Saint-Antoine. In 1610, he attended Henry at his investiture as Prince of Wales, and, on the eve of the ceremony, was knighted. Shortly thereafter (in 1614), he was elected MP for Nottinghamshire. Upon his father's death in April 1617, William inherited the Cavendish estates, and proceeded to oversee the remarkable decorative scheme of Bolsover Castle (with the interior of the anteroom interestingly representing the four humours). It was also during these early years that William began to cultivate a friendship with and patronise the great playwright Ben Jonson.[165]

In 1618, William married the heiress Elizabeth Bassett, who was the widow of Henry Howard, the third son of the Earl of Suffolk. With Elizabeth, he had four children who survived infancy: two daughters (Jane and Elizabeth) and two sons (Charles and Henry). In subsequent years, William collected several titles: he was made Viscount Mansfield in October 1620, Lord-Lieutenant of Nottinghamshire in July 1626, and Earl of Newcastle as well as Baron Cavendish of Bolsover in March 1628. Upon the death of his cousin, William Cavendish, 2nd Earl of Devonshire (1590–1628), in June of 1628, William assumed the role of Lord-Lieutenant of Derbyshire, and he was also passed the barony of Ogle when his mother died in April 1629.

All the while, William remained a committed rider, and had the great architect John Smythson design his famous riding school at Welbeck Abbey. He also hosted James I and Charles I at his estates on various occasions. Most notably, he held a lavish banquet for Charles I in May 1633, when the latter was *en route* to his Scottish coronation. For this event, Jonson composed *The King's Entertainment at Welbeck*, and, in July

[165] For the works that Jonson prepared for William, see British Library, Harley 4955. On the relationship between William and Jonson more generally, see Nick Rowe (1994), 'My Best Patron': William Cavendish and Jonson's Caroline Drama. *The Seventeenth Century* 9.2: 197–212 and Timothy Raylor (2000), Newcastle's Ghosts: Robert Payne, Ben Jonson, and the "Cavendish Circle". *Literary Circles and Cultural Communities in Renaissance England*, eds. Claude J. Summers and Ted-Larry Pebworth (Columbia), 92–114.

1634, William hosted another banquet for the King and Queen Henrietta Maria for which Jonson composed *Love's Welcome at Bolsover*.[166] Having long expected to receive a position at court in part as a reward for his extravagant parties, William was finally appointed Gentleman of the Bedchamber, governor to the Prince of Wales, and Charles's equestrian trainer in 1638. He kept up his artistic and intellectual interests at this time, composing a handful of distinctly Jonsonesque plays, including *Witts Triumvirate, or The Philosopher* (1635) and *The Varitie* (1639–1641).[167] William also became a great admirer of the art of the Antwerp school, which drove him to present Charles with an Alexander Keirincx landscape and to sit for Anthony Van Dyck. Not only did William patronise many playwrights, poets, musicians, and painters, but (along with his brother Charles) he also supported several famous mathematicians and natural philosophers, including Hobbes, Walter Warner, and Robert Payne (the last of whom was his chaplain from 1632 to 1638).[168]

By 1642, it was clear that Britain was on the cusp of civil war, and William was appointed governor of Newcastle upon Tyne and given jurisdiction over Northumberland, Durham, Cumberland, and Westmorland; he started to train troops on his land. In December of that year, he assisted Thomas Glemham (1594–1649) and the Yorkshire Royalists against the Parliamentarians, and was made commander-in-chief of the North. The following year, he published *A Declaration ... in Answer to Six Groundlesse Aspersions Cast upon him by the Lord Fairefax*, in which he defended the rights of kings and argued that it was legitimate for Charles I to employ Catholics. This work also served as a self-defence, for William had accepted so many northern recusants into his forces that Parliamentarians such as John Hutchinson asserted that his force was "a papistical army led by an atheistical general".[169] In April 1643, while William was in the throes of battle, his wife, Elizabeth, died. He fought on anyway, taking Rotherham and then Sheffield that spring, before securing Howley Hall in June and Gainsborough in July. But, despite his best efforts,

[166] Edward Hyde (1702–1704), *History of the Rebellion*, Vol. 1 (London), 104. For a fuller account, see Lynn Hulse (1995), 'The King's Entertainment' by the Duke of Newcastle. *Viator* 26: 355–405.

[167] For *Witts Triumvirate*, see British Library, Add. 45865.

[168] See Timothy Raylor (2001), Hobbes, Payne, and *A Short Tract on First Principles*. *The Historical Journal* 44.1: 29–58.

[169] Lucy Hutchinson (1846), *Memoirs of the Life of Colonel Hutchinson* (London), 164.

William was stopped in his tracks by Hutchinson in August, who refused to give up Nottingham. Although Newcastle was made a marquess in October 1643, his luck on the battlefield only worsened the following year when he lost the famous Battle of Marston Moor in what would prove to be the worst Royalist defeat of the Civil Wars.[170]

Immediately after this vanquishment, William sailed from Scarborough to Hamburg, where he remained for six months, before proceeding to Paris (via Rotterdam and Brussels). He arrived in Paris in April 1645, at which point, as mentioned, he met Margaret, whose brother, Sir Charles, had been Lieutenant-General of the Horse under William's command. William and Margaret married in December of that year, and remained in Paris for some time, where they enjoyed the company of Hobbes, Descartes, Pierre Gassendi, Marin Mersenne, and many other high-profile philosophers, poets, and politicians. It was also at this point that the Cavendishes collected much of the medical advice and many of the recipes in Pw V90. By late 1648, however, William and Margaret decided to migrate to Antwerp, and it was there that they hosted Charles II during his three-day stay the following year. William also established a riding school at Rubenshuis, which attracted visitors from far and wide. Ten years later, in 1658, he published an influential and superbly engraved treatise on horsemanship, *La Methode et Invention Nouvelle de dresser les Chevaux*, which went through multiple editions in French, English, and German over the course of the seventeenth century. Around the same time, William penned another important (if very different) treatise: his well-known "Advice to Charles II", styled after Niccolò Machiavelli's *The Prince*.[171]

At the Restoration, William sailed from The Hague to London, and sat in the House of Lords in June and July 1660. All of the honours, lands, and tenements that had been stripped from him in October 1642 were restored with royal assent, and he gained a new title in March 1665: Duke of Newcastle upon Tyne. Compared to the pre-Interregnum years,

[170] See Elspeth Graham (2016), 'An After-Game of Reputation': Systems of Representation, William Cavendish and the Battle of Marston Moor. *Authority, Authorship and Aristocratic Identity*, eds. Peter Edwards and Elspeth Graham (Leiden), 83–110.

[171] See Conal Condren (1993), Casuistry to Newcastle: *The Prince* in the World of the Book. *Political Discourse in Early Modern Britain*, eds. Nicholas Phillipson and Quentin Skinner (Cambridge), 164–186. The "Advice to Charles II" is now catalogued as Bodleian Library, Clarendon 109.

William's life was relatively quiet following the Restoration. Although he continued to support artists and writers, he spent most of his time training horses at Welbeck. He also worked on plays including *Sir Martin Mar-All* (a collaboration with John Dryden, published in 1667) and two plays that were posthumously published in 1677: *The Humorous Lovers* and *The Triumphant Widow, or, The Medley of Humours*. Having developed what was probably Parkinson's, William died at Welbeck on Christmas Day 1676, and was buried alongside Margaret in Westminster Abbey.

Thomas Glemham (1595–1649)

The only non-Cavendish-household member for whom medical advice exists in Pw V90 is Thomas Glemham, who was treated by Richard Farrar for a cough and ague. Glemham was the son of Sir Henry Glemham (d. 1632), and his wife, Anne Sackville (b. 1595?), from whom he inherited land in Norfolk, Suffolk, Lincolnshire, Kent, and Nottinghamshire.[172] While Glemham matriculated at Trinity College, Oxford, in 1610, he did not take his B.A. and instead pursued a military career, fighting for various armies in Europe from 1610 to 1617. In September 1617, Glemham was knighted by James I, and he proceeded to serve as an MP for Reigate from 1621 to 1622 and for Aldeburgh from 1625 to 1626. A year later (in 1627), Glemham captained the Duke of Buckingham's expedition to La Rochelle, and, upon returning to England in 1629, he served as a deputy lieutenant for Suffolk.

Owing to his military experience, Glemham was commissioned as a lieutenant colonel during the Second Bishops' War from 1639–1640.[173] He was also regarded as a Royalist asset when the Civil Wars broke out in 1642, and, as governor of York and second-in-command to the Earl of Cumberland, was put in charge of a Yorkshire gentry's cavalry.[174] In this role, Glemham was ultimately unable to break through a Parliamentarian blockade in Hull, and he required reinforcement from William Cavendish, who marched south with approximately 8000 men in December 1642.[175] Glemham remained a colonel of foot and governor until November

[172] On Glemham's early life, see Andrew J. Hopper (2008), Glemham, Sir Thomas (1595–1649). *ODNB*.

[173] See Anthony Wood (1691), *Athenae Oxonienses*, Vol. 1 (London), 733.

[174] See David Lloyd (1668), *Memoires of* [...] *Excellent Personages* (London), 551.

[175] See Peter R. Newman (1981), *Royalist Officers in England and Wales, 1642–1660: A Biographical Dictionary* (New York), 158.

1643, when William appointed him Colonel-General of Northumberland in expectation of war with the Scots. When the Scottish invasion occurred in January 1644, however, Glemham retreated to Newcastle upon Tyne, where William's troops again provided him with support. After the Royalist defeat at the battle of Marston Moor, Glemham held the city for two weeks before negotiating terms of surrender on 16 July 1644. As colonel-general of the Royalist northern army's remnants, Glemham reassembled 3000 men in Cumberland the next month only to surrender this town too in June 1645.

Despite his military defeats, not all was lost for Glemham, since he was made Doctor of Law at the University of Oxford and appointed Governor of Oxford in October 1645. In this position, he greatly improved the defences of the city, and he managed to resist the surrender of Oxford for a month in 1646 (though he eventually succumbed in June at Charles I's command). In January 1648, upon release from a brief imprisonment, Glemham immediately fled to France. William had repeatedly come to Glemham's defence during the early years of the Civil Wars, and Pw V90 indicates that such camaraderie was not confined to the battlefield, since William procured a course of treatment from Farrar on Glemham's behalf in 1649. This treatment was, however, ultimately ineffective, and Glemham died later that year in exile, though it was arranged for his body to be returned to England, where it was buried in Little Glemham, Suffolk, in early 1650.

II. Royalists and Courtiers

Robert Atkyns (bap. 1621–1710)

The "Robert Atkins" who offers a "medicin for a Cough" is Robert Atkyns, a judge and politician from Hertfordshire, and the son of Sir Edward Atkyns (1587–1669), a baron of the exchequer, and Ursula (d. 1644).[176] Atkyns became a pensioner at Sidney Sussex College, Cambridge, in May 1637, entered Lincoln's Inn in October 1638, and was called to the Bar in 1645.[177] He stands out in Pw V90 because he is one of the few individuals who was not a straightforward Royalist: he was elected to the Commons to represent Carmarthen in 1656, and sat in

[176] Transcription, 198.

[177] Janelle Greenberg (2004), Atkyns, Sir Robert (bap. 1621, d. 1710). *ODNB*.

Richard Cromwell's Parliament as a member for Evesham, Worcestershire, in 1659.

While Atkyns's recipe was collected in Pw V90 before he had a title, he eventually shifted allegiances and was among the sixty-eight men whom Charles II made Knights of the Bath at the Restoration. With the backing of Charles II, Atkyns became Recorder of Bristol in 1661, and subsequently solicitor-general to the Queen.[178] His legal exploits, moreover, earned him a reputation as a learned man, and he was made serjeant-at-law and appointed a Justice of the Common Pleas in 1672. Despite these accolades, Atkyns remained a strong proponent of the independence of Parliament, arguing that "If the king would do the work of a parliament, we were like to be without a parliament".[179] He sat in the Cavalier Parliament for East Looe in Cornwall, and opposed Charles II's policy of censoring books without parliamentary approval; he also supported the people's right to petition peacefully.

Atkyns's convictions regarding parliamentary independence were front and centre in the major case that he presided over during the 1670s: *Barnardiston v. Soame*. Sir Samuel Barnardiston, an opponent of the court, had been elected to the Commons in Suffolk, but the county sheriff, Sir William Soame, sent in a double return (whereby multiple members are recorded as elected) on the grounds that many of Barnardiston's supporters were not forty-shilling freeholders—a requirement to vote. Atkyns was one of two dissenters in the exchequer who argued that common law courts should indeed compensate Barnardiston for the damages that he had incurred. As he spelt out in his reflections on the trial, even worse than the private wrong was the "design to model the parliament to the humour of the court".[180] Atkyns's position and argument were highly significant because, after the revolution of 1688, the case figured in the enactment of statutes 7 & 8 Wm. III c. 7, which made it illegal for a sheriff to make a double return in elections to the Commons.[181]

With the exclusion crisis, Atkyns's relationship with the Royal court was once again strained, and, under some pressure from the government,

[178] Much of what follows is derived from Michael Landon (1970), *The Triumph of the Lawyers: Their Role in English Politics, 1678–1689* (University, Alabama), 48–50.

[179] William Cobbett (1809), *Cobbett's Parliamentary History of England*, Vol. 5 (London), 309.

[180] Robert Atkyns (1734), *Parliamentary and Political Tracts* (London), 137.

[181] William H. Watson (1827), *A Practical Treatise on the Law Relating to the Office and Duty of Sheriff* (London), 305–306.

he resigned from the bench in 1680 and retired to Lower Swell, Gloucestershire. Atkyns's opposition to the Stuart monarchy continued under James II, whose Catholicising tendencies he strongly condemned, especially because, in order to place fellow Catholics in influential political positions, James had overturned statues that required religious conformity, most notably the Test Act.[182] In defiance of this move, Atkyns again deployed a version of the argument that, since the King and Parliament alike made statutes, such statutes could only be suspended with the support of both, and thus "None but the Law-Maker, can Dispence with the Law" (with the King only having a share in the legislature).[183] Atkyns's case relied on the radical ancient constitutionalist position, which he deemed to have been adopted by the government of the Saxons, rather than mediaeval divine right theory. He defended his standpoint in consecutive volumes, most notably *The Power, Jurisdiction and Privilege of Parliament* (1689).[184] In this work, and various others, Atkyns also expressed his wholehearted support for William and Mary, whose goodwill he, in turn, enjoyed during the final years of his life, as reflected in his appointment to the positions of Chief Baron of the Exchequer and Lord Speaker of the House of Lords after their coronation in 1689. Atkyns's view of kingship was also enshrined in the 1689 Bill of Rights, with its opening clause that abolished the suspending power. Having resigned from his professional duties and retreated for good to Sapperton Hall, Gloucestershire, in 1694, Atkyns died some sixteen years later, in February 1710.

Thomas Aylesbury (1579/80–1658)

While there are two Thomas Aylesburys whose dates are almost identical, the one who provided a "receit for the Colick in the stomacke" is the affiliate of the so-called Northumberland circle, centred around Henry

[182] Christopher W. Brooks (2019), Religion and Law in Early Modern England. *Law, Lawyers and Litigants in Early Modern England: Essays in Memory of Christopher W. Brooks*, eds. Michael Lobban, Joanne Begiato, and Adrian Green (Cambridge), 327–365 (363–364).

[183] Robert Atkyns (1689), *An Enquiry into the Power of Dispensing with Penal Statutes* (London), 20.

[184] See Corinne C. Weston and Janelle R. Greenberg (1981), *Subjects and Sovereigns: The Grand Controversy Over Legal Sovereignty in Stuart England* (Cambridge) and James C. Holt (2008), The Ancient Constitution of Medieval England. *The Roots of Liberty: Magna Carta, Ancient Constitution and the Anglo-American Tradition of the Rule of Law*, ed. Ellis Sandoz (Indianapolis), 22–56.

Percy, 9th Earl of Northumberland, rather than the Church of England divine. There are three reasons for drawing this conclusion. First, and most importantly, he is referred to as "Sir Thomas Aylesburie" in Pw V90 and only this one appears to have been knighted. Second, the Cavendishes (especially Charles) were in regular contact with Northumberland's affiliates.[185] Third, there is no evidence that the clergyman had any medical interests. The Aylesbury in question, then, was part of the Warwickshire branch of the ancient Aylesbury family. He was a Queen's Scholar at Westminster School in 1596 and matriculated at Christ Church, Oxford, in November 1588, taking his B.A. in 1602 and his M.A. in 1605. Aylesbury went on to pursue mathematical studies in Oxford until 1611, at which point he married Anne (1589–1661), daughter of Francis Denman, rector of West Retford, Nottinghamshire. He also became secretary to Charles Howard, Earl of Nottingham and Lord Admiral of England, around 1611.[186]

Despite lacking an institutional affiliation, Aylesbury cultivated relationships with leading mathematicians in the ensuing years. Not only was he an associate of Percy, whom he visited frequently during the latter's fifteen-year imprisonment in the Tower of London (from 1606 to 1621), but he was also friends with and a patron of the great mathematician and natural philosopher Thomas Harriot. The latter bequeathed to Aylesbury, as well as Robert Sidney, 1st Earl of Leicester, his papers, and, along with Warner, Aylesbury's patronage was instrumental to the eventual publication of Harriot's *Artis analyticae praxis* (1631). This is an important work in the history of mathematics, not least because it is the first algebraic work in the Latin tradition expressed in purely symbolic notation, helping to grant algebra the same epistemic status as Euclidean geometry with respect to formal proofs.[187] Towards the end of his life, Warner also gave to the great mathematician John Pell, who was in regular correspondence with Charles Cavendish, an account of the experiments on Snell's

[185] See Begley, *Margaret Cavendish, The Last Natural Philosopher*, 61–81.

[186] On Aylesbury's early life, see Colin Alsbury (2008), Aylesbury, Thomas (bap. 1597, d. 1660/61). *ODNB*.

[187] Muriel Seltman and Robert Goulding (2007), *Thomas Harriot's Artis Analyticae Praxis* (Dordrecht), 1.

law that he had carried out at Aylesbury's house with Harriot.[188] Named after a 1621 discovery by the Dutch astronomer and mathematician Willebrord Snell, Snell's law describes the relationship between the angles of the incidence and refraction of light.[189] As a great patron and collector of mathematical material, Aylesbury also apparently received concave glasses and some of the mathematical books that did not go to Digby when the Oxford mathematician Thomas Allen died in 1632.[190]

In addition to his mathematical pursuits, Aylesbury was politically active. When George Villiers, Marquess of Buckingham, was made Lord High Admiral of England in 1619, Aylesbury was kept on as his secretary. He was appointed surveyor-general in 1625, and baronet and warden of Cranborne Chase, adjoining Windsor Great Park, in April 1627. A few years later, in 1631, Aylesbury helped to develop new machinery for stamping (rather than hammering) coins, and he was granted a patent the following year for the making and issuing of weights for coinage. He was promoted to commissioner of the mint in 1635 and was made its master by July 1640. Starting in 1643, Aylesbury also served as Master of Requests at the University of Oxford, and he may well have partaken in John Wilkins's gatherings at Wadham College, which famously functioned as a precursor to the meetings of the Royal Society.[191] In December 1649, Aylesbury fled to Antwerp, where he would have undoubtedly interacted with the Cavendishes, before moving to Breda, from where he sent his receipt to William in February 1654 and where he died in June 1658.

Margaret Boswell (1595–1682)

The "Lady Boswell att the Hage, Our resident wife", who bestowed upon Margaret a recipe for "How to fatt Chickens", is Lady Margaret Boswell, wife of the long-term resident (or ambassador) in The Hague, Sir William Boswell (d. 1650).[192] Sir William was not only a diplomat but

[188] See Noel Malcolm and Jacqueline Stedall (2005), *John Pell (1611–1685) and His Correspondence with Sir Charles Cavendish: The Mental World of an Early Modern Mathematician* (Oxford).

[189] See John W. Shirley (1951), An Early Experimental Determination of Snell's Law. *American Journal of Physics* 19.9: 507–508.

[190] See Wood, *Athenae Oxonienses*, Vol. 1, 493.

[191] See Mordechai Feingold (1984), *The Mathematicians' Apprenticeship: Science, Universities and Society in England 1560–1640* (Cambridge), 107, 112, 137, and 152.

[192] Transcription, 225.

also a man of great learning, who corresponded with numerous contemporary philosophers, mathematicians, poets, and theologians, including James Ussher, John Pell, G.J. Vossius, Fabri de Peiresc, Samuel Hartlib, and Meric Casaubon.[193] Boswell was a noted Royalist and was tasked with the important role of negotiating the Netherlands' neutrality during the Civil Wars.

Having returned to England from a diplomatic trip to Savoy in February 1629, Boswell married his cousin, Margaret Boswell (or perhaps Bosvile), who was related to Sir Ralph Bosvile of Eynsford in Kent.[194] The couple do not appear to have had children, and little is known about Margaret's life in The Hague, though the recipe in Pw V90 suggests that she was in contact with Margaret Cavendish at this time. After her husband's death, Margaret Boswell was left with a collection containing his manuscripts and correspondences, which was apparently sent to King Charles II at the request of his majesty. At least some of these papers, however, seem to have been taken into the possession of Boswell's nephew, William Raymond, though he supposedly detested Margaret.[195] Margaret died in 1682, having endowed the "Sir William Boswell Scholarships" that funded pupils from Sevenoaks School to attend Jesus College, Cambridge, where her husband had been a student.

Edward Conway (bap. 1594–1655)

In Pw V90, a rather simple, entirely herbal "pouder for the Collick and stone" is offered by "lord conaway".[196] This is almost certainly a reference to Edward Conway, 2nd Viscount Conway and 2nd Viscount Killultagh, who was a notable bibliophile and politician. While Edward matriculated at Queen's College, Oxford, aged sixteen, it is unclear whether he finished his degree; all we know is that he was knighted in 1619.[197] He proceeded to marry Frances Popham (1596/7–1671), daughter of Sir Francis Popham (1572/3–1644) and his wife Anne Dudley (1575–1641), in 1621. Between 1624 and 1628, Conway served

[193] Alan Stewart (2009), Boswell, Sir William (d. 1650). *ODNB*.

[194] Gareth Davies (1917), Appendix II: Life of Sir William Boswell. *Autobiography of Thomas Raymond*, ed. Gareth Davies (London), 69–84 (79).

[195] Davies, Appendix II, 77–79.

[196] Transcription, 202.

[197] James Knowles (2008), Conway, Edward, second Viscount Conway and second Viscount Killultagh (bap. 1594, d. 1655), politician and book collector. *ODNB*.

as MP for several ridings including Warwick and the Isle of Wight, and, following his father's death in 1631, he succeeded him as 2nd Viscount Conway. During the Civil Wars, Conway's position was complicated by the fact that he had family members on both sides, but he ultimately allied himself with the Royalist cause, and was in fact briefly imprisoned in the spring of 1643 for his role in Edmund Waller's plot to seize London on behalf of the King. After his release from prison, Conway joined the King in Oxford, where he composed a book of Royalist poetry, much of which was published in 1662 as *Rump: Or an Exact Collection of the Choycest Poems & Songs*.[198]

According to the German poet Georg Rudolf Weckherlin, Conway was a "very learned man [...] curious for all manner of rarities".[199] He had a vast literary network that extended across Europe, and had some of the same associates as William Cavendish (not least Jonson). Conway also corresponded with a number of individuals in Pw V90, including Digby—who brought him romance novels from Paris and briefly lived with him during the 1630s—and Mayerne, who served as his physician.[200] Edward Hyde, 1st Earl of Clarendon, described Conway as "a voluptuous man in eating and drinking", and, sure enough, he exchanged satirical letters with Mayerne "for making fat men lean".[201] Conway's vast library matched his rather extensive network, with a sequestration list from 1643 indicating that it housed over 5000 volumes, while another list (from his Irish library) shows that it contained between 8000 and 9900 books.[202] In his *ONDB* article, James Knowles notes that

[198] Huntington Library, Hastings 16522.

[199] Timothy Raylor (1994), *Cavaliers, Clubs, and Literary Culture: Sir John Mennes, James Smith, and the Order of the Fancy* (Newark), 91–97.

[200] For Mayerne's and Digby's interactions with Conway, see Anne Conway (1992), *The Conway Letters: The Correspondence of Anne, Viscountess Conway, Henry More, and Their Friends, 1642–1684*, eds. Marjorie H. Nicolson and Sarah Hutton (Oxford), 20–28. On the romance books, see Daniel Starza Smith (2012), "La conquest du sang real": Edward, Second Viscount Conway's Quest for Books. *From Compositors to Collectors: Essays on the Book Trade*, eds. Matthew Day and John Hinks (London), 210–211.

[201] Hyde, *History of the Rebellion*, Vol. 1, 186. Daniel Starza Smith (2014), *John Donne and the Conway Papers: Patronage and Manuscript Circulation in the Early Seventeenth Century* (Oxford), 117.

[202] See Ian Roy (1968), The Libraries of Edward, 2nd Viscount Conway, and Others: An Inventory and Valuation of 1643. *Historical Research* 41.103: 35–46 and Thomas A. Birrell (1991), Reading as Pastime: The Place of Light Literature in Some Gentlemen's

> Conway had a particular interest in burlesque and satire and was an intelligent literary critic, neatly differentiating English plays from classical models, boasting later to his daughter-in-law Anne Conway that 'if the English language were understood by other nations, they would certainly imitate them'.[203]

As this suggests, Conway was also related to another important seventeenth-century female thinker, Anne Conway, who has herself often been associated with Margaret Cavendish, since they were both among seventeenth-century England's leading female philosophers.

Much like William, Conway was sequestered as a delinquent, but he had begun to compound for his estates by 1644. As a result of his diminishing rental income, he was heavily indebted by 1646–1647. In the early 1650s, he retired to Petworth, Sussex, to the house of Algernon Percy, 10th Earl of Northumberland, but he travelled abroad shortly after, at which point he likely passed his recipe onto the Cavendishes. Conway died in Lyons, France, on 26 June 1655.

Kenelm Digby (1603–1665)

Kenelm Digby was the son of Sir Everard Digby (c. 1578–1606), who was later executed for his role in the Gunpowder Plot, and his wife Mary (c. 1581–1653), heir of William Mulsho. The Digbys were an ancient aristocratic family, known in England for being steadfast Roman Catholics. As a young man, Kenelm was probably tutored by the Jesuit John Percy and the physician and astrologer Richard Napier, who may well have fostered his early medical interests. It was also during childhood that Digby came to know Venetia Stanley (1600–1633), daughter of Sir Edward Stanley and his wife, Lady Lucy Percy. The romance of Digby and Venetia would become one of the best-known tragic love stories in English history.

Following a trip to Spain from 1617 to 1618 with his cousin, Sir John Digby, Kenelm entered Gloucester Hall, Oxford, where he was taught by Thomas Allen, who was not only a mathematician and assiduous manuscript collector, but also a renowned Roman Catholic.[204]

Libraries of the Seventeenth Century. *Property of a Gentleman: The Formation, Organisation and Dispersal of the Private Library, 1620–1920*, eds. Robin Myers and Michael Harris (Winchester), 113–131.

203 Knowles, Conway, Edward; also see Conway, *The Conway Letters*, 30–32.

204 See Michael Foster (1982), Thomas Allen (1540–1632), Gloucester Hall and the Survival of Catholicism in Post-Reformation Oxford. *Oxoniensia* 46: 99–128.

Rather than taking his B.A., however, Digby embarked on a grand tour. According to his "Loose Fantasies", the already elderly Queen Marie de' Medici fell in love with him when he visited Paris, and, fearing the consequences of rejecting her, Digby feigned his death and fled to Florence in November 1620. From there, he penned letters to reassure Venetia of his safety, but his mother disapproved of the match and intercepted them; Venetia accordingly believed that Digby was dead. While in Florence, Digby obsessively collected books and manuscripts, and delivered three addresses to the Accademia dei Filomati in Siena. It was also here that an old Carmelite apparently bestowed the famous recipe for the "powder of sympathy" upon him, though we have reason to question the verisimilitude of this story.[205] Before returning to England, Kenelm assisted his cousin, John, with his effort to negotiate the marriage of Charles, Prince of Wales, to the Spanish infanta. Despite their negotiations ending in failure, Digby made enough of an impression on the prince to be knighted in October 1623 and he was also made a member of his privy chamber.

In December 1623, Digby assisted with the arrangement of Prince Charles's marriage to Henrietta Maria, and, probably in January 1625, Digby secretly wedded Venetia. After Charles's ascent to the throne, Digby departed on a privateering voyage in a bid to seize French ships in the Venetian port of İskenderun, and he kept a daily log of his voyage.[206] The booty that he brought home from France, Spain, and Italy in February 1629, and the stories that accompanied his trip, made him something of a legend at court. Based on his acumen at sea, he was also made a junior principal officer of the navy in December 1630, which necessitated a Protestant conversion. Tragically for Digby, however, Venetia died of a cerebral haemorrhage in April or May 1633, and some speculated that the viper wine that he recommended she take had brought it on. Mourning the death of his wife, Digby retreated to Gresham College in 1633, where he dressed in the garb of a hermit and pursued chymical experiments with his Hungarian lab-hand, Johannes Hunyades.[207] He migrated to Paris in September 1635 and took rooms in the Collège de Boncourt, located near the Sorbonne, the Jardin des Plantes de Paris, and the Royal laboratory. It was probably at this point

205 Digby, *A Late Discourse*, 11–12.

206 For an account of Digby's life in these years, see Moshenska. *A Stain in the Blood*.

207 Aubrey, *Brief Lives*, Vol. I, 226–227.

that Digby undertook chymical experiments with Davisson (who later advised William).[208] He also reconverted to Catholicism, which he made public in his *A Conference with a Lady about Choice of Religion* (1638). Around this time, Digby also commenced a lifelong friendship with the unorthodox English Catholic philosopher Thomas White, and cultivated relationships with luminaries such as Mersenne, Descartes, Hobbes, and Gassendi.

Digby returned to England in 1638 and was involved in Henrietta Maria's scheme to raise money from Catholics to aid her husband's campaign against the Scots. The Long Parliament ultimately called him to account for his role in this exploit, and he, along with all other recusants, was removed from court. Upon returning to France, Digby challenged Baron Mont le Ros to a duel for insulting Charles I and killed him; Louis XIII excused Digby on the understanding that he would leave the country immediately. Digby was, however, arrested in England in November 1642 when the Civil Wars broke out. In prison, he penned (apparently within twenty-four hours) his critique of Sir Thomas Browne's 1642 *Religio medici*, and, inspired by the aforementioned *novatores*, but also drawing heavily on chymical and Aristotelian ideas, it is here that he probably produced the first book of his 1644 magnum opus, *Two Treatises*.[209]

Upon his release from prison in July 1643, Digby returned to Paris yet again, and he was made chancellor to Henrietta Maria when she established court there the following year. At this point, he was in regular contact with Margaret, and reflected on their correspondence with delight in a 1657 epistle to her.[210] In February 1645, he left Paris for Rome—where he desperately pleaded the Pope for funds on Henrietta Maria's behalf—but he moved on from there too at the end of 1645 with an advance of 20,000 crowns. Once again in need of money, Digby swiftly returned to Rome in October 1646, but his luck was running out. Eventually, he appealed to Pope Innocent X (in 1647) for succour, but he became "high and hectored with His Holiness" and was ultimately unsuccessful.[211] He left Rome empty-handed, never to return. Digby was finally

[208] See Kenelm Digby (1661), *Discourse Concerning the Vegetation of Plants* (London), 82–83 and William Davisson (1641), *Philosophia pyrotechnica* (London), 73–74.

[209] Malcolm, *Thomas Hobbes, Vol. 6: The Correspondence, Vol I*, 830.

[210] William Cavendish and Margaret Cavendish (1678), *A Collection of Letters and Poems* (London), 65.

[211] Aubrey, *Brief Lives*, Vol. I, 225–226.

permitted to re-enter England in 1653, at which point he managed to free his estates from sequestration, and he spent the rest of the decade in England, settling in Covent Garden after the Restoration. Digby's efforts to promote his "powder of sympathy" had begun during his last stint in Paris, and were advanced in print with his 1658 *Late Discourse* [...] *Touching the Cure of Wounds by the Powder of Sympathy*. With the help of George Hartman, he proceeded to carry out chymical experiments in London, and Hartman published a posthumous record of these in two volumes: *Choice and Experimental Receipts in Physick and Chirurgery* (1668) and *Chymical Secrets and Rare Experiments in Physick and Philosophy* (1683). The latter contained the first published record of how to make the "Sympathetical Powder" that Digby had recommended to the Cavendishes many years prior.[212] Its placement at the end of the volume suggests that at least Hartman regarded it as Digby's crowning glory.

While Digby was absent from the first meeting of the Royal Society on 28 November 1660, he was made one of its original fellows owing to his natural philosophical and medical contributions. During the Society's early years, he was a relatively active member, reading a paper on plants in January 1661, which was turned into the Society's first formally authorised publication: *A Discourse Concerning the Vegetation of Plants* (1661). This was also Digby's last work. He did, however, prove an excellent target for the inveterate Society critic Henry Stubbes, who singled him out as "the Pliny of our age for lying".[213] Furthermore, according to John Henry, there is some evidence that, while Digby was a full member of the Society, some wanted him to be sidelined for his religious views, which had equally attracted the criticisms of Stubbes and others.[214] After

[212] Kenelm Digby and George Hartman (1683), *A Choice Collection of Rare Chymical Secrets and Experiments in Physick & Philosophy* (London), 270–272.

[213] Henry Stubbe (1670), *The Plus Ultra Reduced to a Non Plus* (London), 161. Also see Hedrick, Romancing the Salve.

[214] John Henry (2010), Sir Kenelm Digby, Recusant Philosopher. *Insiders and Outsiders in Seventeenth-Century Philosophy*, eds. Graham A.J. Rogers, Tom Sorell, and Jill Kraye (London), 43–75 (62–63). Henry cites "Cave Beck to Oldenburg, August 15, 1668" (see Henry Oldenburg (1968), *Henry Oldenburg, The Correspondence of Henry Oldenburg Vol. 5 1668–1669*, eds. and trans. Alfred Rupert Hall and Marie B. Hall (Madison), 16).

a long life full of adventure and controversy, Digby died on 11 June 1665, and was buried alongside Venetia in Christ Church, Newgate.

Robert Dudley (1574–1649)

One might naturally surmise that the "The Duke of Northumberland" who provides many "Receipts for my Lord Marquesse of Newcastell" is John Dudley, the legitimate 1st Duke of Northumberland.[215] But the evidence in Pw V90 indicates otherwise. In particular, Mayerne responded to William's 1649 enquiry about Dudley's cure by referring to "the Admirable Receipts of y^t^ Noble Man made Duke of Northum=berland by y^e^ Pope".[216] This is a confusing passage, not least because it is hard to see how the Pope could *make* anyone Duke of Northumberland. Instead, Mayerne likely confused the Pope (who would have been Paul V in 1620) with Ferdinand II, who, as we will see, recognised Dudley as the Duke of Northumberland in 1620. The fact that Dudley, and not his grandfather, is the "Duke of Northumberland" in Pw V90 is substantiated by the author of the recipe's assertion that "it hath ben tried often with Admirable Successe; By coman=dement of the great Duke of Tuscany", who (as we will see) was one of Dudley's chief patrons.

The Sir Robert Dudley in question, then, was the son of Robert Dudley, 1st Earl of Leicester (1532/3–1588), and Douglas Sheffield, Lady Sheffield (1542/3–1608). Whether or not the couple married was (and remains) a subject of some confusion. During the early 1580s, the young Dudley appears to have moved back and forth between his parents, but, after the death of Leicester's legitimate son, Robert (1581–1584), Lord of Denbigh, he became the father's favourite.[217]

Dudley matriculated at Christ Church, Oxford, in May 1588, where he spent two years studying under Thomas Chaloner. He then came into his inheritance in February 1590 upon the death of Leicester's brother, the 3rd Earl of Warwick.[218] Perhaps inspired by his tutor, but also by his father who had mathematical and maritime interests, Dudley became

[215] Transcription, 140.

[216] Transcription, 137.

[217] While it does not dwell on his medical preoccupations, the fullest biography of Dudley, from which the present account is partially derived, remains Arthur S.G. Lee (1964), *The Son of Leicester: The Story of Robert Dudley, Titular Earl of Warwick, Earl of Leicester, and Duke of Northumberland* (London).

[218] Simon Adams (2008), Dudley, Sir Robert (1574–1649). *ODNB*.

increasingly preoccupied with navigation after leaving Oxford, and he began to collect a wide range of naval instruments, many of which are now housed in the Museo Galileo in Florence. These included Thomas Gemini's astrolabe and Charles Whitwell's lunar computer, the latter of which was designed to calculate the place of the moon over a thirty-year period. Putting the theoretical knowledge that he had attained into practice, Dudley embarked upon a voyage to the West Indies in November 1594.[219] He also commanded one of the Queen's warships to Cádiz in 1596, and was knighted when he returned in August of that year. While he never went to sea for the sake of battle or exploration again, Dudley did fund a disastrous circumnavigation in 1601, from which only one man returned.

Having married Alice Leigh (1579–1669), daughter of Sir Thomas Leigh, in 1597, Dudley opened a protracted legal case at the Star Chamber in 1604/5 to prove himself Leicester's legitimate son. Despite his mother's testimony that she had married Leicester, Dudley lost the case; this was probably due to the potential knock-on effects of overturning the existing settlement of the estates.[220] A few months after the verdict was delivered, a despondent Dudley embarked on what was intended to be a three-year continental voyage, accompanied by his cousin and lover, Elizabeth Southwell; he soon invented a contract to invalidate his marriage to Alice. While in Lyons in the winter of 1605/6, Dudley and Elizabeth converted to Catholicism, and, in defiance of the result of his legal case, Dudley began to style himself the Earl of Warwick and Leicester. Around the same time, he also approached the grand Duke of Tuscany, Ferdinand I, with an offer to build warships for a campaign against the Turks.[221] The duke eventually accepted the offer, and Dudley and Elizabeth relocated to Florence, where they remained for the rest of their lives. After Ferdinand's death, Dudley was patronised by his son Cosimo II, who granted him a patent to produce an improved silk weaving machine. Cosimo's patronage also allowed Dudley to compose two manuscript volumes on navigation in English (in 1610),

[219] The manuscript account of his voyage can be found in British Library Sloane 358. For a published version with a still useful preface, see George F. Warner (1899), *The Voyage of Robert Dudley…to the West Indies, 1594–1595* (London).

[220] See Lee, *The Son of Leicester*, 109–116.

[221] On Dudley in this context, see Alberto Viviani (1961), Sir Robert Dudley, Duke of Northumberland, at the Grand Ducal Court. *Florence* 12.4: 32–33.

and a manuscript of instructions for the Tuscan navy in Italian titled "Direttorio marittimo" (1619/20).[222] As the maps in these works indicate, Dudley was a proficient cartographer, who produced early nautical maps of Northern Japan among other places.[223]

Since Dudley remained in Florence even after the official revocation of his licence to travel in 1607, James I sold off many of his properties, and eventually his earldoms, in 1618. It was partly in response that Dudley sought patronage from the Medicis, and became the great chamberlain to consecutive grand duchesses: Maria Maddelena and Christina. As noted, he began to style himself Duke of Northumberland in 1620 thanks to Maria's appeals to Emperor Ferdinand II, which led the latter to recognise his claim to the title. A few decades later, in 1644, Charles I made Dudley's wife, Lady Alice, Duchess of Dudley, perhaps out of a sense of guilt for neither paying Dudley the full value of his Kenilworth estate nor giving Alice her pension, but also out a conviction that Dudley was Leicester's legitimate son. For his part, Dudley retired in the early 1640s to Villa di Castello. It is here that he composed his naval encyclopaedia, *Dell'arcano del mare* ["Secrets of the Sea"], which was published in 1646/7. The book contained 130 original maps, and was dedicated to Ferdinand II. Dudley died at Villa di Castello in September of 1649.

Elizabeth Egerton [*née* Cavendish] (1626–1663)

The "Ladye Brackleye" in Pw V90 is almost certainly the middle daughter of William Cavendish and Elizabeth Bassett, also named Elizabeth. She married John Egerton (1623–1686) in 1641, who was then Lord Brackley but later became 2nd Earl of Bridgewater.[224] Given its placement in the manuscript, the relevant recipe for "Almond Butter" is probably from 1649 or 1650, meaning that Elizabeth would have already been married for nearly a decade by the time that the recipe was recorded.

Even after she married Egerton—and William migrated to the Continent—Elizabeth remained with her sisters at Welbeck. They twice entertained Charles I at the estate, before parliamentary forces took possession

[222] Much of this manuscript was included in Robert Dudley (1646), *Dell'Arcano del Mare* (Florence).

[223] See Josef F. Schütte (1969), Japanese Cartography at the Court of Florence: Robert Dudley's Maps of Japan, 1606–1636. *Imago Mundi* 23.1: 29–58. For Dudley's cartographic drawings, see Bayerische Staatsbibliothek, Icon. 138–140.

[224] Transcription, 185–186.

of it in 1645. That year, Elizabeth moved to the Egerton family estate in Ashridge, where she was largely insulated from the turmoil of war. Here she was in constant correspondence with family members, and, following in the footsteps of her father and stepmother (who was only three years her senior), Elizabeth penned various plays and poems, including her now well-known pious meditation, "Loose Papers".[225] During the early 1640s, Elizabeth also composed a veritable anthology of poems and plays with her sister, Jane. Their most famous composition was a play, written around 1645, titled "The Concealed Fansyes".[226] Elizabeth (and Jane) did not, however, follow Margaret's publishing path, and it has thus been left to contemporary scholars to track down and make accessible her manuscripts and those of her sister, which are scattered across England and America.[227] Elizabeth died giving birth to her tenth child in 1663, and Jane penned an epitaph "On the death of my Deare Sister", which reveals that she witnessed the traumatic event.[228]

Adrian Gilbert (1541–1628)

The Adrian Gilbert who provides a "Cordiall water" was Walter Raleigh's older half-brother, and the full brother of the famous explorer Humphrey Gilbert (1539–1583), who led a successful campaign in Ireland and travelled to North America where he claimed Newfoundland for Queen Elizabeth I.[229] Like Raleigh, Gilbert took a profound interest in gardening, chymistry, and mathematics, and he shared with Herbert a palpable enthusiasm for navigation. Suggesting that he also had spiritualist inclinations, Gilbert is said to have attended one of John Dee's famous conversations with angels in 1583.[230] He also enlisted Dee's technical expertise in 1577 after being granted, along with Herbert, a patent to search for a northwest passage from Europe to Asia, which resulted in

[225] See Betty Travitsky (1999), *Subordination and Authorship in Early Modern England: The Case of Elizabeth Cavendish Egerton and Her "Loose Papers"* (Tempe, AZ).

[226] See Elizabeth Brackley and Jane Cavendish (1996), The Concealed Fancies (c. 1645). *Renaissance Drama by Women: Texts and Documents*, eds. Susan P. Cerasano and Marion Wynne-Davies (New York), 127–155.

[227] Perhaps the richest resource is Bodleian Rawl. Poet 16, which is a presentation copy of the works of Elizabeth Egerton and Jane Cavendish.

[228] Huntington Library, Ellesmere 8353.

[229] Transcription, 171.

[230] See Peter French (1972), *John Dee: The World of an Elizabethan Magnus* (London), 62, 172, and 179–180.

a voyage that commenced in 1584. From 1585 to 1587, Gilbert also funded John Davis's three voyages.[231] Being born into wealth allowed Gilbert to finance such expeditions, but his fortunes dramatically shifted slightly later in life, not least because Raleigh owed him large sums for work on his estates, but, having been imprisoned in the Tower for nine years, was unable to pay his debts.

During the 1590s, Gilbert designed Raleigh's Sherborne lodging and gardens, and he also had an opportunity in around 1610 to put his chymical and medical knowledge to work when he settled at Wilton House, which Mary Herbert, Countess of Pembroke, had turned into a private college of sorts where many learned individuals famously gathered. Initially, Gilbert assisted Mary with her laboratory, but he was gradually tasked with designing the garden of her son William Herbert, 3rd Earl of Pembroke, who was Chancellor of the University of Oxford, and, with King James I, founder of Pembroke College, Oxford.[232] The garden, once finished, was widely praised, with the poet John Taylor, for instance, noting in 1623 that Gilbert had placed trees "in such admirable Artlike fashions, resembling both divine and morrall remembrances, as three Arbours standing in a Triangle, having each a recourse to a greater Arbour in the midst, resembleth three in one, and one in three". Taylor went on to style Gilbert "a true Adamist" and penned an anagram in praise of his ingenuity, which concluded with the lines: "So *Adryan Gilbert*, mendeth *Natures* features / By *Art*, that what *she* makes, doth seem his creatures".[233] In a less fawning tone, Aubrey referred to Gilbert "as a man of great parts", though he condemned him as "the greatest buffoon in England" for speaking his mind to all women regardless of their social standing.[234]

Speaking to Gilbert's wider interests, Aubrey also noted that he was vulgarly called "Dr Gilbert" (despite lacking a doctorate) and that, in addition to his acumen in gardening, he "did many admirable cures with

[231] See Richard Hakluyt (1599–1600), *The Principal Navigations, Voyages, Traffiques and Discoveries of the English Nation*, Vol. 2, Part 2 (London), 143–161.

[232] For a fuller picture of Gilbert's life at this time, see Cristina Malcolmson (2010), William Herbert's Gardener: Adrian Gilbert. *George Herbert's Pastoral*, ed. Christopher Hodgkins (Newark), 113–133.

[233] John Taylor (1623), *A New Discovery by Sea, with a wherry from London to Salisbury* (London), C3r–C3v.

[234] Aubrey, *Brief Lives*, Vol. I, 262.

his Chymicall Medicines. But those Secrets are lost, or died with him".[235] While this is for the most part true, a recipe for "Dr. Adrian Gilberts most Sovereign Cordial Water"—which is similar to the one in Pw V90—outlived him owing to its publication in *The Queens Closet Opened*.[236] This recipe includes a total of 42 plants, many of which were no doubt growing in the Wilton House garden. Having designed and tended to the Wilton House garden for nearly two decades, Gilbert appears to have passed away there in 1628.

Elizabeth Talbot Grey (1582–1651)

It has already been noted that the recipe for "The Countes of Kents Powder" refers to a recipe by Elizabeth Talbot Grey, Countess of Kent and heiress (along with her sister, Aletheia Talbot Howard (c. 1585–1654)) of Gilbert Talbot, 7th Earl of Shrewsbury (1552–1616).[237] As her title indicates, she married Henry Grey (c. 1583–1639), heir to the earldom of Kent, in 1601. While little is known about her early life, she was privately educated, probably by John Florio, who dedicated his 1603 translation of Michel de Montaigne's *Essays* to her. As a patron and skilled linguist in her own right, Grey later had various other translations and works in and about other languages dedicated to her, including Antimo Galli's *Rime all'illustrissima signora Elizabetta Talbot-Grey* (1609); Abraham Darcie's *Annales, the True and Royall History of* [...] *Elizabeth Queene of England* (1625), which was an English rendition of books 1–3 of William Camden's Latin *Annales*; and two works on Italian by Giovanni Torriano, *New and Easie Directions* (1639) and *The Italian Tutor* (1640).

Grey was a long-standing and active member of the court, serving as the first lady of the bedchamber to Queen Anne from 1617. For much of the 1630s, she lived with her husband in Wrest Park, Bedfordshire, where they surrounded themselves with renowned scholars, including the antiquary Robert Cotton, 1st Baronet of Connington; the cavalier poet

[235] John Aubrey (1685), Memoires of Naturall Remarques in the county of Wilts (Royal Society, MS/92), quoted in Malcomson, William Herbert's Gardner, 116.

[236] Henrietta Maria [and Walter Montagu] (1655), *Queens Closet Opened* (London), 11–13. W.M. was probably one of Queen Henrietta Maria's favourites, Walter Montagu (see Archer, The Queen's Arcanum).

[237] Transcription, 188. The Talbot family seems to have been a source of fascination for the Cavendishes. Charles Cavendish, for example, records in his notes an anagram "On the most Honorable Earle, Gilbert Earle of Shresbury" whom he proceeds to refer to as "Gilbertus Talbotius" (Harley 6083, f. 232r).

Thomas Carew; and the jurist and scholar John Selden. The last of these was the beneficiary of Grey's will, and Aubrey claims (without supporting evidence) that she married Selden following the death of her husband in 1639.[238] Following her husband's passing, Grey continued to patronise writers such as the poet and satirist Samuel Butler, while the poet John Suckling revealed that "in my Lady Kents well-being, much of ours consists".[239] As noted, she also became a close friend of Henrietta Maria, whom she swiftly followed into exile in 1642. This involved Grey with the Queen's court during a slightly earlier period than Margaret, and we have seen that the highly influential *Choice Manual of Rare and Select Secrets* (1653) that emerged from this setting was associated with Grey's name. Elizabeth died in London on 7 December 1651.

Aletheia Talbot Howard (c. 1585–1654)
The reference in "The manner how to make Gascons Powder which Mr Rust had from ye Countesse of Arendell" is to Aletheia Talbot Howard, Countess of Arundel, Surrey, and Norfolk, and 14th Baroness Talbot, 13th Baroness Furnivall, and 17th Baroness Strange of Blackmere.[240] Howard was William Cavendish's cousin, being the daughter of his aunt Mary Cavendish (1556–1632) and her husband Gilbert Talbot (1552–1616). She was also a devout Catholic, and her marriage to Thomas Howard, 4th Earl of Arundel and 4th Earl of Surrey (1585–1646), in 1606 made her a member of the foremost English Catholic family. Like her elder sister, Elizabeth, Aletheia was well educated in languages, art, and architecture. She was a prominent figure at the court of King James and Queen Anne, and performed in various court masques including Jonson's *Masque of Beauty* (1608) and *Masque of Queens* (1609). Howard was also a chief mourner at Queen Anne's funeral.

Along with her husband and the renowned architect Inigo Jones, Howard travelled to Heidelberg and then Siena (where she studied Italian) in 1613 and 1614. Upon returning to England, she supervised the latter's renovation of Arundel House, but a desire to travel led her back to Europe for another two years during the early 1620s, this time

[238] Aubrey, *Brief Lives*, Vol. II, 220–221.

[239] John Suckling (1659), *Letters to Several Persons of Honor* (London), 13.

[240] Transcription, 188. Much of this entry is derived from Michelle DiMeo, M. Howard (2008), Aletheia [née Talbot], countess of Arundel, of Surrey, and of Norfolk, and suo jure Baroness Furnivall, Baroness Talbot, and Baroness Strange of Blackmere. *ODNB*.

unaccompanied by her husband. She spent the preponderance of her time in Padua and Venice—where she patronised artists and collected their works—but she stopped in Antwerp on her way to Italy and had her portrait painted by Peter Paul Rubens; van Dyck was also probably part of her entourage.[241]

While the Howards were stalwart Royalists, they found themselves out of favour with Charles I when, in 1626, their son, Henry Frederick Howard, Lord Maltravers (1608–1652), secretly married Lady Elizabeth Stuart (1596–1662), whom the King had intended for Lord Lorne. In response, Charles imprisoned Thomas in the Tower, while confining Aletheia to her mother-in-law's house. Their punishments were not lifted until 1628, at which point Aletheia ingratiated herself with Henrietta Maria, and the King and Queen even toured Arundel House in the December of that year.[242] After 1633, Howard oversaw Nicholas Stone's renovation of Tart Hall, which she had purchased with Thomas, and it was here that she put her personal art collection on display.[243] In 1641, after their scheme to colonise Madagascar failed and with the outbreak of the Civil Wars, the Arundels escorted the Queen's mother, Marie de' Medici, to Italy. Aletheia subsequently established herself in Amsterdam, where she remained in exile for fourteen years and maintained a vast art collection and library. Her husband died in 1646 in Padua, and, with Aletheia absent, a fierce dispute over property inheritance afflicted the family since her son, Henry Frederick, then 15th Earl of Arundel, executed his father's will and usurped properties that should have been passed to Aletheia. Until her death in Amsterdam on 3 June 1654, Aletheia appears to have engaged in household experiments, though she probably contributed minimally to the famous collection of recipes attributed to her, *Natura Exenterata*, published in 1655.

John Mennes (1599–1671)

While the author of "A Remedie for a Tertian and Quarteran Ague" is listed in Pw V90 as "Sr John Mints", this is a reference to John

[241] David Howarth (1998), The Patronage and Collecting of Aletheia, Countess of Arundel, 1606–1654. *Journal of the History of Collections* 10: 125–137.

[242] DiMeo, Howard, Aletheia.

[243] See Jennifer Rabe (2016), Mediating Between Art and Nature: The Countess of Arundel at Tart Hall. *Sites of Mediation: Connected Histories of Places, Processes, and Objects in Europe and Beyond, 1450–1650*, eds. Susanna Burghartz, Lucas Burkart, and Christine Göttler (Leiden), 183–210.

Mennes.[244] John was born into a prominent Kent family, and was educated at the local grammar school.[245] While it is unclear whether he attended university, he spent his early twenties at sea. In November 1628, he convoyed a fleet to Glückstadt, and he frequently took distinguished travellers across the channel, including Rubens in May 1629. Around this time, Mennes also helped to form, and played a leading role in, a literary drinking club known as "The Order of Fancy", which was attended by the actor Thomas Pollard and the poets Philip Massinger and William Davenant, among others.[246] In 1640, Mennes was appointed to command a troop of Carabiniers in the war against the Scots, which resulted in a longer period onshore. He was knighted in 1642 for successfully escorting Henrietta Maria to the Netherlands, and, during these years, he would have certainly met William, and may well have first encountered Margaret.

Since the King no longer had a navy by 1643, Mennes transferred to the army full time. While he started as a general of artillery in Shrewsbury, he was soon appointed governor of North Wales (in 1644), and then became the King's nominal vice-admiral (in 1645). Heading to sea again in 1648, however, Mennes eventually attached himself to the exiled court in 1650, where he remained until the Restoration. During this period, he had a hand in publishing reams of (often profane) poetry including a 1654 volume, *Recreation for Ingenious Head-Peeces*, with his friend James Smith, a 1656 work titled *Musarum Deliciæ: Or, The Muses Recreation*, and a 1658 collection titled *Wit Restor'd*. More seriously, Mennes seems to have served as a secret agent, and may well have helped to negotiate Charles II's return to England.[247] For his loyalty to the King, Mennes was granted the prestigious roles of Commander in Chief in the Narrow Seas and Comptroller of the Navy in October 1661.[248] He commanded the fleet that delivered the Earl of Peterborough to Tangier and delivered Catherine of Braganza to England as Queen.

[244] Transcription, 206.

[245] Charles S. Knighton (2013), Mennes, Sir John (1599–1671). *ODNB*.

[246] See Raylor, *Cavaliers, Clubs, and Literary Culture*.

[247] Knighton, Mennes, Sir John.

[248] Joseph R. Tanner (1903), *A Descriptive Catalogue of the Naval Manuscripts in the Pepysian Library at Magdalene College, Cambridge*, Vol. 1 (London), 314 and 383.

Despite Mennes's successes, there was growing dissatisfaction with his service as the Navy's comptroller (or accounting supervisor), and, as his health deteriorated, attempts were made to have him removed.[249] While these were unsuccessful, William Brouncker, the first President of the Royal Society, was assigned to aid him with the treasurer's account, while William Penn, the English admiral and politician, assisted with his other tasks as victualler and purser. Notwithstanding Mennes's administrative shortcomings, he was widely respected as a virtuoso owing to his knowledge of art, poetry, and chymistry and it is also significant for the purposes of this volume that he occasionally acted as medical adviser to the exiled cavaliers. After the Restoration, in February 1663, an ill Samuel Pepys recorded that he was visited by Mennes who "would not have me take anything from the apothecary, but from him, his Venice Treakle being better than the others", and Pepys claimed that Mennes did indeed cure him after a night of "great sweat".[250] Mennes died in London in February 1671.

George Morley (1598?–1684)

Given the dates and context, the reference to "Doctor Morley" is almost certainly to the renowned divine George Morley. While he was born to an indebted father, and orphaned by the age of thirteen, Morley was evidently a bright child, for he was made a King's Scholar at Westminster School in 1611 and matriculated at Christ Church, Oxford, in 1618.[251] Morley graduated B.A. an impressive two days later, and proceeded to study for an M.A., which he obtained in June 1621. Part of the so-called "great Trew circle", centred around Lucius Cary, Viscount Falkland, during the 1620s and 1630s, Morley befriended some of the most significant English churchmen of his day including Henry Hammond, Gilbert Sheldon, Robert Payne, Robert Sanderson, and William Chillingworth. Morley was also closely associated with Thomas Tully and Barlow, yet

[249] Samuel Pepys (1995), *The Diary of Samuel Pepys*, Vol. 4, eds. Robert Latham and William Matthews (Berkeley), 98. Also see Robert Latham (1995), *Samuel Pepys and the Second Dutch War: Pepys's Navy White Book and Brooke House Papers* (Aldershot), 192.

[250] Pepys, *The Diary of Samuel Pepys*, Vol. 4, 39.

[251] The fullest biography of Morley is Andrew Thomson (2019), *Bishop Morley of Winchester 1598–1684: Politician, Benefactor, Pragmatist* (Winchester). Also see John Spurr (2004), Morley, George (1598?–1684). *ODNB*.

his general Calvinist bent (which the Cavendishes would have appreciated) left him on the fringes of the Laudian elite.[252] Morley was also a long-time friend of Clarendon—whose memoirs are an excellent source of information on Morley's life—and it may well have been Clarendon who helped Morley to secure his appointments as a Royal chaplain and Canon of Christ Church in 1641.[253]

Throughout the 1640s, Morley was a staunch supporter of the Royalist side, and he was stripped of his canonry in 1647 because of his role in the University of Oxford's committee to resist the parliamentary visitation. He was finally ejected from Parliament in spring 1648 and left the country the following year to join Prince Charles's court at St. Germain. For the next few years, he moved between Royalist communities before settling at Hyde's Antwerp home in November 1650. Morley would almost certainly have been in regular contact with the Cavendishes at this time, and there is evidence that he visited them at Rubenshuis in late 1653 before departing to Utrecht to stay with his friend, the fellow exiled churchman Robert Creighton.[254] Given that Morley's recipe—a "receite for the papall or vaticall pills"—comes relatively late in the manuscript, and that Pw V90 was abandoned by 1655, it is likely that he bestowed his recipe upon the Cavendishes between 1653 and 1655. Over the course of his travels, Morley frequently debated Catholic apologists (and those who claimed that Charles II was a Catholic), and it was probably during these travels that he was given the recipe for his "papal pills".

As the Protectorate crumbled in the late 1650s, Morley became an important mediator between the exiles and those Presbyterian ministers who were more sympathetic towards the Royalist cause. In the words of Richard Baxter, Morley sought to convince Presbyterians that "all

[252] On Morley as a Calvinist, see Wood, *Athenae Oxonienses*, Vol. 1, 344. For the religious allegiances of the Cavendishes, see Begley, *Margaret Cavendish, The Last Natural Philosopher*, 155–211.

[253] Edward Hyde (1843), *The Life of Edward, Earl of Clarendon*, Vol. 1 (Oxford), 55–57.

[254] See Cavendish and Cavendish, *A Collection of Letters and Poems*, 88.

Moderation was intended" upon the Restoration.[255] After the Restoration, Morley not only regained his canonry at Christ Church but was also made Dean of the College, and became the Bishop of Worcester. He was, furthermore, Bishop and Dean of the Chapel Royal and preached at Charles II's coronation in April 1661.[256] In the ensuing months, however, he became entangled in a controversy with Baxter who was outraged by his refusal to license him to preach in his erstwhile diocese of Worcester. Baxter was particularly vexed by the fact that Morley delivered "a long Invective" against him and Presbyterian theology in front of Baxter's former congregation at Kidderminster, which seemed to Baxter to be at variance with the moderation that Morley had promised in the late 1650s.[257] A paper war ensued, with Baxter publishing *The Mischiefs of Self-Ignorance and the Benefits of Self-Acquaintance* in 1662, and Morley swiftly responding with *The Bishop of Worcester's Letter to a Friend in Vindication of Himself from Mr. Baxter's Calumny*. Various other (mostly anonymous) tracts followed in defence of either side.

Notwithstanding, or indeed perhaps thanks to, these disputes, Morley was made Bishop of Winchester in May 1662, and, as his letters to Clarendon reveal, he exerted his influence to ensure that the Act of Uniformity was not diluted.[258] Morley was a regular at court until 1668, though he lost his position as Dean of the Chapel Royal with the fall of Clarendon. Beyond his political and institutional manoeuvrings, Morley was a serious thinker and his withdrawal from court afforded him more time to focus on writing. In addition to helping Clarendon with his *A Brief View* (1676), a reply to Hobbes's *Leviathan*, he famously supported Izaak Walton's endeavours, and Walton, in turn, dedicated his 1670 *Lives* to Morley.[259] In 1674, Morley also helped to spearhead a plan for general moral and religious reform, but nothing came of it, and he finally retired to Farnham Castle in 1679. Even during his retirement, he was unable to

[255] Richard Baxter (1696), *Vindiciæ anti-Baxterianæ: Or, Some Animadversions on a Book Intituled Reliquiæ Baxterianæ* (London), 208.

[256] See George Morley (1661), *A Sermon Preached at the Magnificent Coronation of the Most High and Mighty King Charles the IId King of Great Britain, France, and Ireland* (London).

[257] Baxter, *Vindiciæ anti-Baxterianæ*, 375.

[258] Spurr, Morley, George.

[259] See David Novarr (1951), Izaak Walton, Bishop Morley, and Love and Truth. *The Review of English Studies* 2.5: 30–39.

avoid controversies, and, in response to renewed allegations from Baxter, he penned *The Bishop of Winchester's Vindication* in 1683. Morley died at Farnham the following year and was buried in Winchester Cathedral.

Endymion Porter (1587–1649)

Endymion Porter, who provided advice to Glemham, was from a family of minor gentry.[260] His father, Edmund Porter (1562–1607), was a Spanish diplomat, and Endymion served as a member of Conde de Olivares's household in Spain as a young man. Returning to England in 1612, he began to work for Edward Villiers, and eventually married Olivia (or Olive) Boteler (1597–1663), one of Villiers's sisters, in 1619.[261] He was also Charles I's Groom of the Bedchamber, and was central to the artistic life of the court. Not only did Porter help Charles to form his art collection, but he was also instrumental in bringing van Dyck to England, and maintained a friendship with Rubens.[262] Porter was, moreover, an amateur poet, and wrote the final elegy to John Donne in the posthumously published *Poems by J.D. with Elegies on the Author's Death* (1633), as well as the prefatory poem to "Madagascar" (1638) by Davenant, who was poet in ordinary to the Porter family.[263] As the names of van Dyck and Davenant suggest, Porter patronised many of the same individuals as William Cavendish, and his efforts also guaranteed that various works were, in turn, dedicated to him, such as Thomas Dekker's *Dekker His Dreame* (1620) and Thomas May's *The Tragedy of Antigone* (1631).[264]

260 Transcription, 176.

261 For a thorough look at his early life, see Ronald G. Asch (2008), Porter, Endymion (1587–1649). *ODNB*. Older but still useful studies are Dorothea Townsend (1897), *Life and Letters of Mr. Endymion Porter* (London) and Gervas Huxley (1959), *Endymion Porter: The Life of a Courtier, 1587–1649* (London).

262 For Porter's purchase of van Dyck's "Rinaldo and Armida" in 1630 on Charles I's behalf, see British Library, Egerton 2553, f. 54v. Also see Lilian H. Zirpolo (2010), *Historical Dictionary of Baroque Art and Architecture* (Lanham), 218. For his relationship with Rubens, see William N. Sainsbury (1859), *Original Papers relating to Rubens* (London), 146, 203, 293, 324, and 353.

263 For Porter's poem appended to "Madagascar", see William Davenant (1673), *The Works of Sr. William Davenant* (London), 201–202, and, for Davenant's poems of thanks to Endymion and Olivia for patronage, see 165, 217–218, 223–224, 229, 233, 239–240, 243–244, and 320–321. For the elegy, see John Donne (1633), *Poems by J.D. with Elegies on the Author's Death* (London), 405–406.

264 See Thomas Dekker (1620), *Dekker His Dreame* (London), A2r–A3r and Thomas May (1631), *The Tragedy of Antigone* (London), A3r–A6r.

While Porter was generally well regarded at court, he found himself at the centre of a controversy in 1637 when Father Con, the papal agent in England, converted his wife, Olivia, a first lady of the court, to Catholicism.[265] An enthusiastic convert, Olivia attempted to persuade her friends and family to join the Church of Rome, and, indeed, she successfully converted her father, Lord Boteler, on his deathbed. All five of Olivia's and Endymion's sons received a Catholic education, though Endymion himself does not appear to have converted. It was presumed, however, that this was due to the political inexpediency of doing so, and he was therefore looked upon with suspicion in some quarters. William Prynne, for instance, publicly denounced Porter in 1643 as "addicted to the Popish religion" and "a bitter enemy of the King" who "reveales all his [Charles I's] greatest secrets to the Popes Legat".[266]

Porter was elected to the Long Parliament in November 1640, and, though there was an attempt to remove him from the Commons in February 1642, he retained his seat until March of the next year. He then became part of Charles I's personal entourage and a member of the Oxford Parliament, before migrating to Paris in July 1645. Shortly after arriving in France, Porter, who was William's old friend, apparently fomented a conflict between him and Margaret, having reported that he found her unapproachable at court, and that she had made a derogatory comment about the miniature that Porter had given her as a token of admiration. In a letter to William of that year, however, Margaret protested that "m^{rs} porter" was "a stranger to me" and that it was her custom to "never speek to ~~an~~ any man before they addres ~~the~~ them selves to me nor to look so much in ~~thes~~ ther face as to invit ther descours and i hop i never was ~~unsel~~ unsevell to any parson".[267] Despite this misunderstanding, William and Margaret stayed with Porter in Antwerp before they secured Rubenshuis (perhaps with his help).[268] In early 1649, Porter returned to London, and he died there in August of that year.

[265] Olivia's conversion is well documented in Asch, Porter, Endymion.

[266] William Prynne (1643), *Romes Master-Peece: Or, The Grand Conspiracy of the Pope* (London), 23.

[267] British Library, Add. 70499, f. 265r.

[268] Cavendish, *Life of William*, 63–64.

Elizabeth Throckmorton, Lady Raleigh (1565–1647)

The recipe for "Lady Throgmortons way to Stew a Carpe" was almost certainly by Walter Raleigh's wife, Elizabeth Throckmorton, Lady Raleigh, better known as Bess.[269] Bess was the daughter of the Parliamentarian Sir Nicholas Throckmorton (1515/15–1571) and his wife, Anne (c. 1520–1581). As a Gentlewoman of the Privy Chamber, Bess famously became pregnant with the child of Walter, whom she then secretly married in early 1591, to the great displeasure of Elizabeth I. When the unauthorised marriage was discovered in August 1591, both Raleigh and Bess were briefly imprisoned in the Tower of London. Bess was released on 22 December, and it has been suggested that the death of their son, Damerei, who disappears from the history books around this time, was a reason for this charity.[270] Bess and Raleigh had two other children in the ensuing years, Walter (1593–1618) and Carew (1605–1666).

In January 1592, when Walter and Elizabeth had partially reconciled, the Queen took a lease of ninety-nine years on the Sherborne estates of John Coldwell, Bishop of Salisbury, which Raleigh had long loved, and she immediately sublet them to him. Bess resided there in the ensuing years even as Walter went back and forth on voyages. But, in 1603, shortly after Elizabeth's death and in the early days of James I's reign, Walter was again imprisoned in the Tower for his presumed role in the "Main Plot", the aim of which was supposedly to foment rebellion and pave the way for a Spanish invasion. Bess famously begged for clemency, but Raleigh remained in prison for over a decade. While he was released in March 1616, his freedom was short-lived, as he was arrested in June 1618 and executed in October of that year for supposedly plotting to provoke a war between England and Spain. As is well known, Raleigh's head was placed in a "Red-leather bag" after his execution, which Bess apparently kept with her at all times.[271] In later life, Bess worked to restore Raleigh's lost properties and reputation, eventually earning a bill of restitution that allowed her eldest son to inherit many of the estates.

Bess was a woman of some celebrity, and her various medicinal and culinary recipes seem to have circulated widely in Royalist circles. Recipes

[269] Transcription, 197.

[270] See Mark Nicholls and Penry Williams (2015), Ralegh, Sir Walter (1554–1618). *ODNB*.

[271] Vincent T. Harlow, ed. (1932), *Ralegh's Last Voyage* (London), 309–310.

such as "The Lady Throgmortons Egg Pie" continued to crop up during the mid-seventeenth century in works such as the anonymous *The Ladies Companion* (1653), though no other examples of her carp recipe have yet come to light.[272]

Walter Raleigh (1554–1618)

Walter Raleigh, whose "Receipt for Stomack Pills" is in Pw V90, is of course the renowned courtier, author, and explorer.[273] He was the second son of Walter Raleigh (1504/5–1581), landowner of East Budleigh, and his third wife, Katherine (d. 1594).[274] It appears that, in 1569, Raleigh fought in the wars of religion with the Huguenot armies, but he returned to England after the peace of St. Germain in 1570. Raleigh then commenced studies at Oriel College, Oxford, in 1572, but left without a degree, and was admitted to Middle Temple in 1575. Perhaps introduced to the court by his half-brother, Sir Humphrey Gilbert (1537–1628), Raleigh soon developed close ties with Sir Francis Walsingham, the spy and secretary to Elizabeth I, and Robert Dudley, 1st Earl of Leicester (the father of the Dudley discussed above).

After immersing himself in the courtly environment, Raleigh went to sea, captaining the *Falcon* on Gilbert's voyage of discovery in 1578–1579. During the early 1580s, he was also a captain of the reinforcement fleet that countered the Desmond rebellion in Ireland. Upon returning to court in 1581, he began to attract the attention of Elizabeth I, and was made an esquire of the body. He penned her various poems, most famously "Farewell false love", and rumours about their relationship soon swirled around the court.[275] Indeed, Elizabeth granted Raleigh her handsome London dwelling, Durham Place on the Strand, in 1583, and he also received a patent for the sale of wine and licensing of vintners that May, which remained his principal source of income for years to come.[276]

[272] Anonymous (1653), *The Ladies Companion, or, A Table Furnished with Sundry Sorts of Pies and Tarts* (London), 59.

[273] Transcription, 199.

[274] The biographies of Raleigh are legion, but some classic overviews are Christopher Armitage (1987), *Sir Walter Ralegh: An Annotated Bibliography* (Chapel Hill) and Jerry L. Mills (1985), Recent Studies in Ralegh. *English Literary Renaissance* 15.2: 225–244.

[275] For a discussion of Raleigh's poetry in the context of his relationship to the Queen and Bess, see Alfred L. Rowse (1962), *Ralegh and the Throckmortons* (London), 150–169.

[276] See Aubrey, *Brief Lives*, Vol. 2, 183.

Raleigh was knighted on 6 January 1585 and made 2nd Earl of Bedford; he also became captain of the guard in 1587.

Having secured a patent for the purpose of colonisation in 1584, Raleigh sought the advice of Harriot about how best to proceed with this enterprise. He enlisted Richard Hakluyt (the elder) to compose his *Inducements to the Liking of the Voyage Intended towards Virginia* (1585), and had Hakluyt's more famous younger cousin write a *Discourse of Western Planting* (1584) in support of his plan.[277] Raleigh was part of several botched expeditions to New World in the subsequent years, first leading a fleet to Roanoke Island in 1585 with Sir Richard Grenville, and then serving under John White in 1587. A few years later, he published his own work of travel literature, *The Last Fight of the Revenge* (1591), in which he depicted the death of Grenville at the hands of a Spanish fleet.

In terms of his personal life, it has been noted that Raleigh courted, had a child with, and secretly married Elizabeth (Bess) Throckmorton in early 1591, which greatly displeased Queen Elizabeth, and briefly landed them in the Tower of London.[278] After Raleigh's release in late 1591, he was banished from the court, though, over the next few years, he became a successful Parliamentarian, sitting as MP for Dorset and Cornwall. In 1595 he embarked on another journey to the West Indies, which he described in his popular *The Discoverie of the Large, Rich and Beautiful Empire of Guiana* (1596). He subsequently took part in the sacking of Cádiz—which was recounted in the posthumously published *A Relation of the Action at Cádiz* (1700)—and his success here convinced Elizabeth to readmit him to court in 1597.

When Elizabeth died a few years later (in 1603), Raleigh's fortunes changed once again. Not much liked by James I, he was falsely implicated, thanks to the testimony of his courtly rival Henry Brooke, 11th Baron Cobham, in the so-called Main Plot. For this, Raleigh once again found himself in the Tower in July 1603. He kept himself busy in the Tower, especially with the composition of chymical cures, and it was here that he devised the famous "cordial" that he passed on to

[277] On Raleigh's relationship to Harriot, see John W. Shirley (1983), *Thomas Harriot: A Biography* (Oxford), 70–112.

[278] On Raleigh and the Throckmorton family, see Rowse, *Ralegh and the Throckmortons*.

Henry, Prince of Wales.[279] Moreover, in the process of informing Sir Ralph Winwood in 1612 about the death of Sidney's daughter, Elizabeth Manners, Chamberlain remarked that Raleigh had been "slaundered to have geven her certain pilles that dispatcht her", which were probably the same "stomach pills" found in Pw V90.[280] In addition to his medical concoctions, Raleigh also produced several works in the advice genre while in prison, including "A Discourse Touching a Marriage between Prince Henry of England and a Daughter of Savoy" (composed c. 1603–1612) and *Instructions to his Sonne and to Posterity* (composed in 1609 and published in 1632), along with his 1614 magnum opus, *The History of the World*, in which he sought to cover all major events from the creation to 146 BC.[281]

Raleigh was released from the Tower on 19 March 1616, and he immediately began to plan a fanciful, and ultimately ill-fated, expedition to the Orinoco river to discover El Dorado. Setting out in August 1617, he arrived in San Thomé in January 1618, and tried to take over the town. Eventually, it was burnt to the ground, and, since this violated their commission, Raleigh was once again arrested upon his return to England in June of that year. Charged with plotting with the French to provoke a war between England and Spain, he was sentenced to death on 28 October 1618, and executed the next morning.[282] He was buried at St. Margaret's, Westminster. In the ensuing years, Raleigh was gradually recreated as an English hero, especially after the appearance of Thomas Scott's often republished *Sir Walter Rawleighs Ghost, or, Englands Forewarner* (1626), and this salvaging of his reputation partly explains why he would have been granted a posthumous place in Pw V90.

Philip Stanhope (1583/4–1656)

The "Earle of Chesterfields Red Diaphorettick Powder" is the recipe of Philip Stanhope, 1st Earl of Chesterfield, a prominent Royalist

[279] See John Chamberlain (1939), *Letters of John Chamberlain*, Vol. 1, ed. Norman E. McClure (Philadelphia), 389.

[280] Chamberlain, *Letters of John Chamberlain*, Vol. 1, 374.

[281] For the context of the production of *History of the World*, see Nicholas Popper (2012), *Walter Ralegh's "History of the World" and the Historical Culture of the Late Renaissance* (Chicago).

[282] See Thomas N. Brushfield (1905–1906), Three State Documents Relating to the Arrest and Execution of Sir Walter Ralegh in 1618. *Report and Transactions of the Devonshire Association* 37: 284–324 and 38: 416–490.

nobleman.[283] Born to Sir John Stanhope (d. 1609) and his wife, Cordell (1574–1602?), Stanhope grew up in an established Nottinghamshire family, and, as a result, would have undoubtedly been acquainted with William. Stanhope was knighted in 1605, though, in November 1616, he paid £10,000 for the barony of Shelford, which enraged another leading Nottinghamshire man, Sir John Holles, who had paid a similar price for a barony. According to Holles, Stanhope was a disgrace to the crown due to his various run-ins with the law, including buggery at the assizes and even murder.[284]

Stanhope remained on the Nottinghamshire bench until July 1627, and Charles I made him the Earl of Chesterfield in 1628. He was also restored to the Nottinghamshire commission of the peace, from which he had been ousted following his legal controversy (though it seems that he preferred to send a proxy to the House of Lords over attending himself).[285] In March 1643, Stanhope was among the defenders of the close of Lichfield Cathedral when it was besieged by the Parliamentarian forces led by Lord Brook and Sir John Gell, at which point he was taken prisoner and sent to London. In June 1643, the Lords granted his request to live in a private house due to ill health, but he was faced with serious fines and forced to sell his Nottinghamshire properties in 1652. Three of his sons died fighting for the Royalist cause, and Stanhope himself passed away in September 1656.

III. Physicians and Apothecaries

Fabricius ab Acquapendente (1533–1619)

Hieronymus Fabricius ab Acquapendente (or in Italian Girolamo Fabrizi d'Acquapendente), who is alluded to in a note for "The manner of using Aquapen=dentis Pills", was a towering physician, anatomist, and professor at the University of Padua. Born in Acquapendente (as his name indicates), Fabricius was trained in humanist medicine and anatomy at the University of Padua under the direction of Fallopius (or Gabriele

[283] Transcription, 187.

[284] Much of what follows is based on Peter R. Seddon (2008), Stanhope, Philip, First Earl of Chesterfield (1583/4–1656). *ODNB*. Stanhope's questionable character is attested to in John Holles (1975), *Letters of John Holles, 1587–1637*, Vol. 1, ed. Peter R. Seddon (Nottingham), 142 and Chamberlain, *Letters of John Chamberlain*, Vol. 2, 33.

[285] Seddon, Stanhope, Philip.

Fallopio), and received his doctorate in 1559. He then practised medicine before succeeding Fallopius as Professor of Surgery and Anatomy. In this role, Acquapendente made important institutional contributions to the history of medicine and anatomy, most notably overseeing the creation of the anatomical theatre in Padua, the first of its kind, in 1594.[286] This theatre, in which Fabricius taught regularly, allowed up to 300 students to stand simultaneously no more than thirty feet from the cadaver on the dissecting table, and it was used until 1872.[287] More generally, Fabricius's thought, and especially his emphasis on dissection, contributed to the renaissance of an "empirical" Aristotelianism in sixteenth-century Padua.[288] In particular, his desire to learn not just about form but also function spurred him to make assiduous comparisons of the anatomies of various non-human animals.

Along with his teaching, Acquapendente was an acclaimed author who penned an influential book on the organs of sight, voice, and hearing, *De Visione, Voce, Auditu* (1600), and another on speech, *De locutione et ejus instrumentis tractatus* (1601).[289] He also composed a number of foundational works on reproduction and embryology, including *De formato foetu*

[286] Cynthia Klestinec (2004), A History of Anatomy Theaters in Sixteenth-Century Padua. *Journal of the History of Medicine and Allied Sciences* 59.3: 375–412.

[287] Sean B. Smith et al. (2004), Hieronymous Fabricius Ab Acquapendente (1533–1619). *Clinical Anatomy* 17.7: 540–543.

[288] Andrew Cunningham (1985), Fabricius and the 'Aristotle Project': Anatomical Teaching and Research at Padova. *The Medical Renaissance of the Sixteenth Century*, eds. Andrew Wear, Roger K. French, and Ian M. Lonie (Cambridge), 195–222. Still useful is John H. Randall (1940), The Development of Scientific Method in the School of Padua. *Journal of the History of Ideas* 1.2: 177–206. But some important correctives can be found in Charles B. Schmitt (1969), Experience and Experiment: A Comparison of Zabarella's View with Galileo's in *De motu*. *Studies in the Renaissance* 16: 80–138; William A. Wallace (1988), Randall Redivivus: Galileo and the Paduan Aristotelians. *Journal of the History of Ideas* 49.1: 133–149; and David M. Miller (2018), Regressus and Empiricism in the Controversy about Galileo's Lunar Observations. *Perspectives on Science* 26.3: 293–324.

[289] On Fabricius on the eye, see Tawrin Baker (2016), Why All This Jelly? Jacopo Zabarella and Hieronymus Fabricius ab Aquapendente on the Usefulness of the Vitreous Humor. *Early Modern Medicine and Natural Philosophy*, eds. Peter Distelzweig, Benjamin Goldberg, and Evan R. Ragland (Dordrecht), 59–88. Along with *De locutione et ejus instrumentis tractatus*, Fabricius wrote a treatise on the speech of animals, *De brutorum loquela* (Padua, 1603). On the latter, see Richard Serjeantson (2001), The Passions and Animal Language, 1540–1700. *Journal of the History of Ideas* 62.3: 425–444.

(1600) and *De formatione Ovi et Pulli* (1621).[290] The latter, a posthumous publication, contains the first description of the "bursa of Fabricius" (named after Acquapendente), which is the site of hematopoiesis in birds. Fabricius's works on embryology were also significant in their use of detailed engravings to illustrate aspects of the anatomical dissections being described.

Fabricius was also well known for his discovery of the valves of the veins (which he called "ostiola" in *De venarum ostiolis* (1603)), but, perhaps even more so, for being the teacher of William Harvey. Indeed, according to Boyle, Harvey's inquiry into the heart's motion was inspired by his puzzlement with Fabricius's account of the shape and function of the valves.[291] Acquapendente had a profound influence on Harvey and particularly his adaptation of Aristotelian insights, though Harvey ultimately sought to falsify Acquapendente's hypothesis that the valves of the veins are designed to stop the blood from falling downwards.[292] Fabricius also taught Julius Casserius, who succeeded him as Professor of Surgery and Anatomy at Padua, and Adriaan van den Spiegel, who succeeded Casserius. As a practising physician, he also treated Galileo Galilei, to whom he recommended the pills that appear in Pw V90.[293] Having had a long and successful career, Fabricius died on 21 May 1619, aged 86.

Mr Basincotes (active mid-seventeenth century)

The "Mr Basincotes" who supplied a recipe for "The true manner how to make Crocus-Metalorum and the use thereof" was, as Pw V90 reveals, "an Apothecary at Paris".[294] Having consulted numerous experts in seventeenth-century French medicine, however, we, unfortunately, cannot determine any further details about his life or practice.

Samuel Bispham (1612/13–1664)

The "Doctor Bispham" who provides both a "Receipt to make an excellent Balsum for any gre=ene wound" and "An Exellent water" is Samuel

[290] See Howard B. Adelmann (1967), *The Embryological Treatises of Hieronymus Fabricius of Aquapendente* (Ithaca).

[291] Robert Boyle (1688), *A Disquisition About the Final Causes of Natural Things* (London), 157–158.

[292] See Karin J. Ekholm (2010), Fabricius's and Harvey's Representations of Animal Generation. *Annals of Science* 67.3: 329–352 and James G. Lennox (2006), The Comparative Study of Animal Development: William Harvey's Aristotelianism. *The Problem of Animal Generation in Early Modern Philosophy*, Justin E.H. Smith (Cambridge), 21–46.

[293] Heilbron, *Galileo*, 415n.48.

[294] Transcription, 157.

Bispham.[295] Bispham first appears on the historical record during his student days, having begun his studies at the University of Leiden and incorporated at Oxford in May 1633.[296] As a physician in the early 1630s, he practised in Salford and Great Lever, treating John Bridgeman, Bishop of Chichester, in the latter location. From there, he moved to London, before fleeing the Civil Wars in 1643. At this point, he travelled through Genoa, Florence, Venice, and France, likely meeting William in Paris. Having served as a physician in ordinary to King Charles I, he asked that Charles II grant him the same position at the Restoration, but it is unclear whether he received this post.[297]

As with many other figures in Pw V90, Bispham was an established physician deeply indebted to chymical medicine. He penned a 46-page duodecimo of "Chymicall Notions", which dwells on distillation and is now catalogued as Harley 6940 in the British Library.[298] While the manuscript is simply addressed to "my very good Lord", it was no doubt intended for William, who scribbled a number of notes and recipes near the back of the manuscript.[299] For instance, before a recipe for "An Excellent Balsam for any greene wound inward bruse or hurt", which appears to be in Farr's hand, William remarked that "The dose for ann Inwarde brouse off this Exselente Balsumm, Is a Dramm or a Dramm & a Halfe, or towe Dramms to bee Takenn Inwardlye".[300] The existence of these recipes also suggests that this manuscript treatise was composed before the Cavendishes made Pw V90 (starting in 1647), and that it acted, much like Harley 6491, as a storehouse for their recipes, though many pages have been torn out of this manuscript, and only a few recipes in the hands of William and Farr remain. In addition to the balsam recipe and a recipe "For to make the Powder of Simp=athy" that is almost identical to the recipe in Pw V90 (Transcription, 172), there is also a recipe in Farr's hand from one "M$^{s.}$ Crass" titled "The most excellent rec=eipt in the world for Fresing a Pike" and a recipe in William's hand from "M$^{r.}$ Porter"

[295] Transcription, 171–172.

[296] Wood, *Athenae Oxonienses*, Vol. 1, 876.

[297] William Bispham (1890), *Memoranda Concerning the Family of Bispham in Great Britain and the United States of America* (New York), 101–103.

[298] See Harley 6940.

[299] Chapter 2, 19.

[300] Harley 6940, f. 47v.

(or Endymion Porter) for "howe to boyle a Pike".[301] There is also a set of prescriptions by "Bispham, Medicus Regius Juratus" in British Library, Sloane 933 titled "Selecta ex ejus præscriptis" from 1648. Taken together, these documents indicate that Bispham had close and long-standing ties to the Cavendishes.

In terms of his personal life, Bispham had a number of children who died in infancy, but the one who survived, Thomas Bispham (c. 1638–1677), matriculated as a gentleman commoner at Queen's College, Oxford, in 1638. In 1658, Thomas published a Latin poem titled *Iter Australe, a Reginesibus Oxon*, which provides an account of a journey made by the provost (Barlow) and fellows of Queen's college to visit some of the college properties, which Thomas dedicated to Barlow. Thomas was a barrister at Gray's Inn and married in 1663. Leaving behind a successful son, Samuel passed away the subsequent year.

Mathew Boucheret (d. 1671/2)

The "M. Boucheret" whose name is attached to a short note on "The Manner how to take the Steele in Syrope" is likely the apothecary Mathew Boucheret, who was born in Amiens, France, the son of Gedeon Boucheret of Lewes in Sussex, and his wife, Jane.[302] Boucheret appears to have been naturalised in England in 1662/3, and to have lived and worked at St. Giles, Cripplegate, in London, apparently serving as an apothecary to King Charles I.[303] Given that the nature of his contribution to Pw V90 is a piece of physic that served to supplement the salt of steel that is described as "Sr Theador Mayernes Invention", we might also surmise that he was one of Mayerne's apothecaries.[304] But unlike the other apothecary mentioned in this manuscript known to work with Mayerne, Calixt Rust, Boucheret's name only appears once in Pw V90.

Boucheret married Anne Tantaries of North Willingham at St. Ann Blackfriars, London, on 26 February 1643/4, with whom he had a

[301] Harley 6940, ff. 58r–63v.

[302] Transcription, 107.

[303] William A. Shaw (1911), *Letters of Denization and Acts of Naturalization for Aliens in England and Ireland, 1603–1800* (London), 81 and The National Archives, Parliamentary Archives, HL/PO/JO/10/1/316 (24 March 1663). There is some ambiguity regarding his naturalisation date, however, and it is recorded as 1644 in *Lincolnshire Pedigrees* (1902), Vol. I, ed. Rev. A.R. Maddison (London), 163. We surmise that the year of his marriage was confused in this case with the year of his naturalisation.

[304] Transcription, 102.

number of children.[305] According to Leslie G. Matthews, an apothecary named Mathew Boucheret married an heiress and became the proprietor of an estate.[306] While we cannot be sure that this is the same person, records exist showing that Boucheret owned property. Indeed, there was even a court case over one of his tenants: on 30 April 1667, the Fire Court heard a case between Katherine Cooling and Mathew Boucheret relating to Cooling's tenancy at a property of Boucheret's called the Green Dragon on Ludgate Hill in St. Bride's, London, which Boucheret had inherited from his wife Anne's parents.[307] Interestingly, two records indicate that Boucheret maintained contact with the Cavendish family years after the production of Pw V90. The first is a letter to Topp with advice for Frances Cavendish, the wife of Henry Cavendish, from August of 1666, in which Boucheret suggests that she avoid the usage of hot medicines, recommends a purging powder, and sends drugs along with a recipe for a "diette drink".[308] The second is a mortgage for Boucheret from Sir John Curzon related to various farms, which was witnessed by Henry Cavendish.[309] Boucheret apparently died of a "stop stomach" on 15 March 1671/2, and was buried at St. Ann Blackfriars.[310]

Thomas Cademan (1590?–1653)

Thomas Cademan, who advised the Cavendishes on several occasions, was born in Norfolk and matriculated as a pensioner at Trinity College, Cambridge, in around 1601.[311] Having received his M.A. in 1609, Cademan began working with Mayerne and others to petition James I

[305] Information on the marriages and baptisms of his children can be found in the parish registers of St. Ann Blackfriars, London, 1562–1726 (P69/ANN) and St. Martin's, Ludgate, London, 1558–1719 (P69/MTN1), respectively.

[306] Leslie G. Matthews (1974), London's Immigrant Apothecaries, 1600–1800. *Medical History* 18.3: 262–274 (269).

[307] Philip E. Jones, ed. (1966), *The Fire Court. Calendar to the Judgments and Decrees of the Court of Judicature Appointed to Determine Differences Between Landlords and Tenants as to Rebuilding After the Great Fire*, Vol. 1 (London), 37.

[308] Pw 1/31, Letter from Mathew Boucheret, London, to Mr. Francis Toppe, Welbeck; 16 Aug. 1666

[309] Nottinghamshire Archives, DPP83/13, 1666.

[310] Unclear, but perhaps a "stopped stomach": that is, gastroparesis. Parish registers of St Giles Cripplegate, London, 1667–1672 (P69/GIS).

[311] Venn, *Alumni Cantabrigienses*, Vol. 1, Part 2, 279. For Cademan's early life, see Brian Nance (2004), Cademan, Sir Thomas (c. 1590–1651). *ODNB*.

for a Royal charter for the Society of Apothecaries by 1617. He then proceeded (like many English physicians) to take his M.D. in Padua in 1620 where he remained for a year.[312] Shortly after his return to England (in 1623), he passed his examinations before the College of Physicians. But Cademan had a rocky early patch with the College. The College's Annals record the fellows' admonition of Cademan in 1623 for assuming that his Italian degree enabled him to practise without a license in the English capital, while there was some debate over whether Catholics such as Cademan should even be allowed to take the exam since they were not supposed to be admitted to the College.[313] Under the College microscope, Cademan was accused of malpractice for bleeding a patient with petechial fever in 1623, while a few years later (in 1625) he was formally charged, convicted, and fined for malpractice in a gonorrhoea case.[314]

Despite causing him some difficulties, Cademan's Catholicism had its upsides, for it brought him to the attention of Henrietta Maria, who made him a physician in ordinary in 1626. But there is no evidence that he later joined the Queen in exile, and the fact that he wrote to the Cavendishes (when they were stationed in Paris) indicates that he might have been advising them from afar. In any case, during the 1630s, Cademan and William Brouncker had a patent for distilling and brewing in a house at the back of St. James's Park. Co-opting Mayerne into the scheme, they incorporated as the Distillers of London in 1638. Here they approved of strong waters and vinegars, which is interesting in light of the fact that both Cademan and Mayerne dwell on the significance of various waters in the courses of treatment that they prescribed to the Cavendishes.[315] The Society of Apothecaries was not happy with this corporation, however, and petitioned against it in September 1638 as infringing upon its monopoly. But with such powerful individuals as Mayerne at the helm, nothing ultimately came of the complaint. Suggesting that the Society held a grudge against Cademan, his apothecary from 1641 onwards, Thomas Clarges, was classified as an "unlawful worker" in 1642, and

[312] See Edgar Morpurgo (1927), English Physicians—"Doctorati"—at the University of Padua in the "Collegio Veneto Artista" (1617–1771). *Proceedings of the Royal Society of Medicine* 90.2: 1369–1380.

[313] Munk, *The Roll of the Royal College of Physicians*, 200.

[314] Furdell, *The Royal Doctors, 1485–1714*, 105.

[315] See Thomas Cademan and Théodore de Mayerne (1639), *The Distiller of London* (London).

prohibited from setting up shop within the precincts of the company.[316] In due course, Clarges worked out the difficulties with the Society of Apothecaries, and, after paying a fine, was readmitted to the association.

Beyond his entrepreneurial exploits, Cademan was clearly admired in Royal circles. During the 1630s, he treated Davenant, who penned a glowing poem "To Doctor Cademan, Physitian to the Queen", which praises his "so small tyranny to so much Art" and thanks him "For setting now my condemn'd body free, / From that no God, but Devil Mercurie".[317] As this suggests, Cademan did not treat Davenant with quicksilver but rather more benign medicines, fit for "Eastern Queens that teeme" (or were Royal yet robust). Further suggesting their close proximity, Davenant married Cademan's widow, Anne, after his death.[318] In 1641, Cademan penned an account of the death of one of his patients, Francis Russell, 4th Earl of Bedford, titled *The Earle of Bedfords Passage to the Highest Court of Parliament*, in which he imputed his death to spending too much time in bed. More ambitiously, around the time that he was treating the Cavendishes, Cademan expended a great deal of energy producing a long (1318 folio page) and fascinating piece of medical semiology: "De signis morborum tractatus, cura Thamae Clargicii". This manuscript, which was dedicated to Henrietta Maria and is now kept in the Royal Society of Medicine, is a comprehensive attempt to catalogue symptoms and map them onto diseases. Cademan was made anatomy lecturer to the College of Physicians in 1649, but he died in May 1651, holding the role for less than two years.

William Davisson (1593–1669)

William Davisson—who wrote to William regarding Henry Cavendish's health, provided several recipes in Pw V90, and penned the treatise that makes up the bulk of Harley 6491—was born in Aberdeen to Duncan Davissone of Armakrone (d. 1600?) and Jonet Forbes.[319] After his father's death around 1600, relatives seized his mother's holdings (though, connected to these proceedings, Davisson managed to obtain

[316] See Thomas D. Whittet (1988), Sir Thomas Clarges, Apothecary and Envoy. *Journal of the Royal Society of Medicine* 81.8: 464–467.

[317] Davenant, *The Works of Sr. William Davenant*, 234–235.

[318] See Mary Edmont (2009), Davenant [D'Avenant], Sir William (1606–1668). *ODNB*.

[319] Transcription, 128–131 and 155.

a patent of nobility from Charles I in 1629).[320] He entered Marischal College in 1614, and graduated M.A. in 1617, at which point he relocated to France.[321] Davisson may have qualified for a doctorate in medicine at Montpellier, and began to lecture on chymistry in Paris. Around 1631 he produced a manuscript treatise titled "Elementa Physicae Practicae", while in 1641 he published his renowned textbook, *Philosophia pyrotechnica*, along with a work on salt, *Oblatio salis, sive, Gallia lege salis condita*.[322] Thanks to the intercession of Henrietta Maria and François Vautier, first physician to Louis XIII, Davisson became physician to the King in 1644. But when a professorship of chymistry was made available at the Jardin du Roi in 1647—the first of its kind in France—it was granted to Davisson, drawing his attention away from court.[323] His public lectures were well attended, with John Evelyn visiting in October 1649, and Hobbes going through a course of chymistry with Davisson at some point before 1648.[324] It has also been mentioned that Davisson dedicated "A Short Treatise of the Nature of Minerals" to William in 1645 during their time in Paris.[325]

In 1651, Davisson resigned his post in Paris, possibly because of a controversy sparked by his publication of two tracts that year: one titled *Observations sur l'antimoine*, which upheld the pharmaceutical use of antimony, and a more comprehensive defence of chymistry titled *Les élémens de la philosophie de l'art feu ou chemie*, which he published with the French industrial chymist Jean Hellot. These works put Davisson firmly on the side of the Montpellier Paracelsians (and against the Parisian Galenists). Following his resignation, and abetted by the Polish Queen Marie-Louise, Davisson obtained a post in Warsaw as first physician to the King of Poland, Jan II Kazimierz Waza, and Keeper of the Royal

320 For much of what follows, see Lawrence M. Principe (2004), Davisson [D'Avissone], William (c. 1593–1669). *ODNB*.

321 See William Davisson (1660), *Commentariorum in* [...] *Petri Severini Dani ideam medicinae, Philosophicae propediem proditurorum Prodromus* (The Hague), 407.

322 See St. John's College, Cambridge, L.2 (James 359).

323 See Ernest Hamy (1898), William Davisson, intendant du Jardin du Roi et professeur de chemie, 1647–1651. *Nouvelles Archives du Muséum d'Histoire Naturelle* (Paris) and John Read (1961), The First British Professor of Chemistry. *Ambix*, 9.2: 70–101.

324 Aubrey, *Brief Lives*, Vol. I, 336.

325 Harley 6491.

Garden.[326] While in Poland, Davisson completed a lengthy commentary on the *Idea medicinae philosophicae* of the great Dutch Paracelsian physician Petrus Severinus, which he published in 1660 (with a more condensed commentary following in 1663).[327] After the death of Marie-Louise in 1667, however, Davisson returned to Aberdeen, though his 1668 tract on scalp disease, *Plicomastix seu Plicae numero morborum*, was published in Danzig under the pseudonym Theophrastus Scotus. Davisson died in 1669, having returned to Paris earlier that year, where Louis XIV ratified his patent of nobility and thus guaranteed the nobility of his descendants.

Richard Farrar (active mid-seventeenth century)

The "Doctor Farrar" who provides a wide range of recipes—intended to treat issues from sterility to afflictions of the spleen—is Richard Farrar, best known for having penned an epitaph for Digby that praises him as "This age's wonder for his noble parts; / Skill'd in six tongues, and learn'd in all the arts".[328] It is also almost certainly via Digby that the Cavendishes became acquainted with Farrar, for his first letter refers to the fact that "Sr Kellam Digby (my Noble and best freind) hath written to me soe effectually, Concerning Sr Thomas Glemham as he could not possible in y^{t} kinde expresse his affection more to yor Lord.^{ip} then he did".[329] Not only did Farrar write an epitaph to Digby, who appears to have acted as a mediator between Farrar and courtiers, but it is also evident from Boyle Papers 29 that Farrar and Digby undertook chymical experiments together.[330] Farrar's profound interest in chymistry is further

[326] On his time in Poland, see John Small (1874), Notice of William Davidson, MD (Gulielmus Davissonus), First Professor of Chemistry, and Director of the Jardin des Plantes, Paris, Afterwards Physician to the King of Poland. *Proceedings of the Society of Antiquaries of Scotland* 10: 265–280.

[327] Davisson, *Commentariorum in* [...] *Petri Severini.*

[328] Transcription, 173–184 and Digby and Hartman, *A Choice Collection of Rare Chymical Secrets*, A5r.

[329] Transcription, 173–174.

[330] Royal Society, Boyle Papers RB/1/29, ff. 136r–139r. The same recipe can be found in Digby's experiment notes in Wellcome Library, 2124. Also see William R. Newman (1994), *Gehennical Fire: The Lives of George Starkey, an American Alchemist in the Scientific Revolution* (Cambridge, MA), 67.

evident from his astounding offer of five thousand pounds to George Starkey in exchange for his "metallurgical secrets".[331] Pw V90 does not, however, reflect Farrar's interest in metallurgy, since the recipes supplied here tend to be non-chymical concoctions involving herbal and animal ingredients, such as wormwood, mutton, and good English ale, a far cry from a recipe of his recorded in the Boyle Papers and reported by Newman on the use of a "Volatile salt of tartar, or fixed spiritus vini, to dissolve gold".[332]

It is also evident from one of Starkey's later letters to Boyle (from spring 1651) that Farrar offended Starkey with his offer to purchase his metallurgical secrets and "*luna fixa*", and Starkey refused thenceforth to show Farrar "ye least familiarity".[333] More generally, the recipes clustered around fols. 136r–139r of Boyle Papers 29 are very similar (and indeed often identical) to those in Digby's *Choice Collection of Rare Chymical Secrets*, which again suggests that Farrar and Digby regularly collaborated.[334] Perhaps most interesting for our purposes, however, is Farrar's cure for a dropsy in *Choice Collection*, which is very similar to his "Decoction for ye Dropsie" in Pw V90.[335] It is written here that

> a Famous Cure of a desperate Dropsie, done by Dr. Farrar upon an eminent Lord, who was over-grown with the Dropsie, his Belly and Stomach swelled to a Prodigious bigness, and was given over by the ablest Physicians as incurable. Sir K. D. made the bargain between the Lord and the Doctor, who was to have five hundred pounds for the Cure: But when the Lord was Cured, he would give the Doctor no more than three hundred pounds, saying, that five hundred pounds was too much

[331] William R. Newman, and Lawrence M. Principe (2005), *Alchemy Tried in the Fire: Starkey, Boyle, and the Fate of Helmontian Chymistry* (Chicago), 156.

[332] Newman, *Gehennical Fire*, 300 n.94. Newman here refers to Boyle Papers 29, ff. 136r–139r.

[333] Royal Society, Boyle Letters VI, ff. 99r–100v; transcribed in William R. Newman (1978), Newton's Clavis as Starkey's Key. *Isis* 78.4: 564–574 (572).

[334] William R. Newman (2002), George Starkey and the Selling of Secrets. *Samuel Hartlib and the Universal Reformation: Studies in Intellectual Communication*, eds. Mark Greengrass, Michael Leslie, and Timothy Raylor (Cambridge), 193–210 (209–210).

[335] Transcription, 176–179.

money, and that all the Ingredients be used could not stand him in twenty shillings.[336]

The fact that the recipe provided is virtually identical to the one sent to the Cavendishes, that the letter was sent via Digby (as were the ones in Pw V90), and that the "Lord" was in need of money strongly suggests that this is a reference to William Cavendish. Digby apparently relayed this story frequently, which suggests that William may have gained something of a reputation for stinginess after 1649.

Robert Fludd (bap. 1574–1637)

"Doctor Floids ways to Cure a loosenesse" is attributable to the famous physician Robert Fludd.[337] Born in Kent to Sir Thomas Fludd (d. 1607) and Elizabeth (d. 1592), he entered St. John's College, Oxford, as a commoner in November 1592.[338] Here Fludd was surrounded by figures who later became reputed physicians including Richard Andrewes and Sir William Paddy. Even as a student, he appears to have been attracted to the occult, for he later recorded that he used astrology to locate a stolen object for a tutor.[339] Upon graduating M.A. in July 1598, Fludd embarked upon a six-year European tour, over the course of which he sought out many thinkers who were immersed in chymical and occult learning.[340]

Returning to Oxford in 1605, Fludd supplicated for an M.D. on 14 May and was granted a license to practise medicine two days later. He immediately set up shop on Fenchurch Street, London. His initial bid to join the College of Physicians in November 1605 was rejected on the grounds that he was insufficiently versed in Galenic medicine, though another reason was probably his suspected Paracelsian leanings.[341]

[336] Digby and Hartman, *A Choice Collection of Rare Chymical Secrets*, 201–202.

[337] Transcription, 156.

[338] The fullest study of Fludd remains William H. Huffman (1988), *Robert Fludd and the End of the Renaissance* (London), but also see Ian Maclean (2008), Fludd, Robert (bap. 1574, d. 1637). *ODNB*.

[339] William H. Huffman, ed. (2001), *Robert Fludd: Essential Readings* (Berkeley), 16.

[340] See Robert Fludd (1618), *De naturæ simia, seu technica macrocosmi historia* (Oppenheim), 3. For a study of one of Fludd's encounters, see Conrad H. Josten (1964), Robert Fludd's Theory of Geomancy and His Experiences at Avignon in the Winter of 1601 to 1602. *Journal of the Warburg and Courtauld Institutes* 27: 327–335.

[341] See Geoffrey Keynes (1966), *The Life of William Harvey* (Oxford), 134–135.

In February 1606, however, he was begrudgingly admitted, and, with some reservations, he was made a full fellow in September 1609. Thereafter Fludd threw himself into College activities: he acted as censor and inspector of London apothecaries, ratified the *Pharmacopoeia Londinensis* (along with Mayerne) in 1618, and gave a College anatomy lecture in June 1620.[342] These facts, along with the recipe attributed to Fludd in Pw V90 and Peter Cole's statement that Fludd preferred prescribing Galenic remedies to chymical ones, suggest that his practice was more conventional than his personal views.[343] Within the college, Fludd was friends with William Harvey, and his *Pulsus* (written in 1629 and published in 1631) is one of the first texts to advocate for Harvey's notion of the circulation of blood.[344]

In terms of his broader philosophical and spiritual ideas, Fludd published several significant and controversial works of a Neo-Platonic bent. These include his two-volume masterwork *Utriusque cosmi* (1617–1621), which presents his account of the relationship between the macrocosm and microcosm, and his four-part *Medicina catholica* (1629–1631), an unfinished work in which all diseases are imputed, in the final account, to original sin and the divine will. It should also come as no surprise, then, that Fludd was a supporter of the Rosicrucians, whom he defended in his 1617 *Tractatus apologeticus*, which included an appeal for admission to the brotherhood. His major publications appear to have been more popular on the Continent than in England, with Johann Bayer (1572–1625), the German lawyer and celestial cartographer, writing in 1618 that his philosophy "now casts its rays on our Germany, as its light does through all the places of the most learned of Europe".[345] With that said, Fludd refined many of his ideas alongside influential scholars such as Selden and Cotton as part of Grey's English circles.[346] Being thoroughly

[342] See Munk, *The Roll of the Royal College of Physicians*, 142.

[343] Peter Cole (1656), *Mr. Culpeper's Ghost, Giving Seasonable Advice to the Lovers of his Writings* (London), 5–9.

[344] See Allen G. Debus (1691), Robert Fludd and the Circulation of the Blood. *Journal of the History of Medicine and Allied Sciences* 16.4: 374–393.

[345] See British Library, Royal 12 C II, f. 7v; this is transcribed in William Huffman and Robert A. Seelinger (1978), Robert Fludd's "Declaratio Brevis" to James I. *Ambix* 25.2: 69–92.

[346] On Selden's friendship with Fludd, see Mordechai Feingold (1991), John Selden and the Nature of Seventeenth-Century Science. *In the Presence of the Past: Essays in*

embedded within the Royalist networks of his day, he also enjoyed the graces of James I and Charles I alike.[347]

Though quite conservative in his practice, Fludd was a prominent defender of Paracelsian thought in England, and his *Anatomiae amphitheatrum* (1623), for instance, advocated at some length for the "weapon salve", which Digby would promote in an adapted form.[348] This work prompted a backlash, most notably from the theologian William Foster, who argued that the cure was not only ineffective but inspired by the devil. He went so far as to nail a copy of the title page of his tract (*Hoplocrisma-spongus, or, A Sponge to Wipe Away the Weapon-Salve*) to Fludd's door in 1631. Fludd was no stranger to controversy, and his broader philosophical views also led him into spats with figures no less eminent than Gassendi, who confuted him in *Epistolica Exercitatio, in qua principia philosophiae Roberti Fluddi medici reteguntur* (1630), and Johannes Kepler, who challenged his ideas on celestial correspondence as overly reliant on scripture rather than observational evidence in an appendix to his *Harmonices mundi* (1619).[349] The latter sparked a pamphlet war, the result of which was Fludd's 1622 *Monochordum mundi*, which emphasises the numerical proportions between the monochord, the parts of the human body, and the stars. After a long and distinguished career as a physician and defender of Platonic and hermetic thought, Fludd died at his London home in September 1637.

Alexander Fraizer (1607?–1681)

The "Doctor frayser" who provides a "receipt for The colick and the spleen" is the Scottish physician Alexander Fraizer.[350] The son of Adam Fraser, he was educated in Aberdeen, admitted to the University of Montpellier in October 1635, and incorporated at the University of Cambridge

Honor of Frank Manuel, eds. Richard T. Bienvenu and Mordechai Feingold (Springer), 55–78.

[347] See Huffman, *Robert Fludd and the End of the Renaissance*, 48.

[348] See Allen Debus (1964), Robert Fludd and the Use of Gilbert's *De Magnete* in the Weapon-Salve Controversy. *Journal of the History of Medicine and Applied Sciences* 19.4: 389–417.

[349] Robert S. Westman (1984), Nature, Art, and Psyche: Jung, Pauli, and the Kepler-Fludd Polemic. *Occult and Scientific Mentalities in the Renaissance*, ed. Brian Vickers (Cambridge), 177–229.

[350] Transcription, 201.

in March 1637.[351] A few years later (in 1640), Fraizer was admitted to the College of Physicians as a candidate, and he became a fellow in November 1641. As a staunch Royalist, Fraizer served as Prince Charles's physician in 1645, and he was threatened with arrest in 1647 for apparently seeking to carry out designs against Parliament.[352] He eventually assisted Charles II in negotiating with the Scots.[353] From 1651 to 1652, Fraizer attended the Royal family in France, and he subsequently followed Charles to Cologne. In the spring of 1656, he returned to Paris to arrange an interview between Henrietta Maria and Charles, which indicates the trust that Charles put in him.

At the Restoration, Fraizer was knighted and made physician in ordinary to Charles II. According to Pepys, he was so great with "all the ladies at Court, in helping to slip their calves when there is occasion, and with the great men in curing of their claps, that he can do what he please with the King in spite of any man".[354] The King admired Fraizer to such an extent that when the justice of the peace Sir Edmund Bury Godfrey sent bailiffs to collect his unpaid debts, the King demanded that the bailiffs were lashed. Despite the statute limiting election to the College of Physicians to Englishmen, Fraizer was admitted in July 1666 as an elect for his services to this institution. In August 1673, Fraizer was made a baronet, and, in March 1677, he was permitted to build apartments in Greencloth Yard, which was a part of Whitehall Palace. He died at Whitehall a few years later, in April 1681. Helen Dingwall has lamented that "Few records of Fraizer's medical practice remain", but Pw V90 offers a new and precious piece of the puzzle.[355]

Johannes Hartmann (1568–1631)

Oblique references to Johannes Hartmann crop up numerous times in a letter and receipt that Davisson sent to William Cavendish, in which

[351] Much of this entry comes from Munk, *The Roll*, 232–234 and from Helen M. Dingwall (2004), Fraizer, Sir Alexander, first baronet. *ODNB*.

[352] Hutchinson, *Memoirs*, 302.

[353] See John Evelyn (1879), *Diary of John Evelyn*, Vol. 4, ed. William Bray (London), 251.

[354] Pepys, *The Diary of Samuel Pepys*, Vol. 5, 275.

[355] Dingwall, Fraizer, Sir Alexander.

he recommended that William use the "narcoticke Sulpher of Hartmannus".[356] This was a chymical medicine used to treat pain, outlined in Hartmann's *Praxis Chymiatrica* under the title "Laudanum sine opio" ["laudanum without opium"].[357]

Hartmann was born in Amberg, Germany, on 14 January 1568, the son of a weaver. He graduated with an M.A. in 1591 from the University of Marburg, and studied mathematics and other subjects at several universities including Wittenberg and Jena before returning to Marburg to receive his doctorate in 1606.[358] His subsequent work focused on the intersections of mathematics, chymistry, and medicine, and in 1609 he began giving lectures on these subjects in the "laboratorium chymicum publicum" of the University of Marburg. In the same year, Hartmann became the first chair in *Chymiatrica*, focusing on *materia medica* and how aspects of Paracelsian chymistry and traditional Galenic herbalism could be combined.[359] After becoming entangled in a number of disputes with the university—particularly the university rector, deans, and some fellow professors who charged him with contributing to the death of one of his patients through the overuse of purgatives—Hartmann moved to Kassel in 1621. While he never formally renounced his professorship, he took up the post of court physician in Kassel until the abdication of Maurice the Learned, Landgrave of Hesse-Kassel. He then assumed a post in medicine at the University of Kassel, where he remained until his death in December 1631. His posthumous *Praxis Chymiatrica* (1633) was reprinted throughout the century, and was a common reference point for those using chymical substances in physic.[360]

John King (bap. 1614–1681)

Pinning down the "Doctor King" who offers a "way to make Meade" is not straightforward, since, as William Birken has shown, there were at

[356] Transcription, 130.

[357] Johannes Hartmann (1639), *Praxis Chymiatrica* (Geneva), 44–45.

[358] Richard S. Westfall (1995), Hartmann, Johannes. *The Galileo Project*. http://galileo.rice.edu/index.html.

[359] Lynn Thorndike (1958), *A History of Magic and Experimental Science*, Vol. 8 (New York), 116–117.

[360] The best source on Hartmann remains Bruce T. Moran (1991), *Chemical Pharmacy Enters the University: Johannes Hartmann and the Didactic Care of Chymiatria in the Early Seventeenth Century* (Madison).

least two "Dr Kings" on the London medical scene at this point.[361] One was Dr. John Regius (alias King) (bap. 1614–1681) and the other was Dr. John King (1604–1688). These figures are easily confused since their dates heavily overlap and both studied at Leiden for their medical degrees.

While we cannot be certain about which King appears in Pw V90, John Regius seems like the most plausible candidate given the general calibre of the physicians in Pw V90. The other Dr. John King was of French descent, his father coming from Rouen to England in 1572. After completing his degree at Leiden, King practised medicine first in Hertford, then in St. Albans, and, eventually, on Aldersgate Street, London.[362] There is no record of him becoming a member of the Royal College of Physicians, nor any evidence of his interacting with Royalists, though his son, Sir John King, was knighted by Charles II in 1674 and was a man of some standing at court.[363]

Of the two possible figures, Dr. King (*née* Regius) seems to have been the more prominent and well regarded.[364] This King was the son of Joannes Regius (1574–1627), a minister at Austin Friars.[365] He obtained his M.D. from Leiden in 1638 with a dissertation on fevers, and he was then incorporated at Oxford in 1641.[366] King married one Joanna Marolois in April 1643, and was nominated to a deaconship in the Dutch Church (but not elected) in December of that year. Most significantly, he was elected to the Royal College of Physicians in 1643, though he was not formally recognised as a fellow until 1648, which is the year in which the Cavendishes most likely received his instructions. Along with attending the College's yearly meetings, King served as its censor in 1651, 1659, and 1661, and as college librarian (along with George Ent) from 1651

[361] Transcription, 191. See William Birken (1976), Dr. John King (1614–1681) and Dr. Assuerus Regemorter (1615–1650). *Medical History* 20.3: 276–295.

[362] John King (1855), *A Memoir of the Life and Death of Sir John King, Knight, Written by His Father in 1677* (London), 62.

[363] King, *A Memoir*, 46.

[364] Robert W.I. Smith (1932), *English-Speaking Students of Medicine at the University of Leyden* (London), 133–134.

[365] William J.C. Moens, ed. (1884), *The Marriage, Baptismal and Burial Registers, 1571–1874, and Monumental Inscriptions of the Dutch Reformed Church, Austin Friars, London* (London), 61.

[366] Much of what follows is derived from William Birken (2008), King, John (bap. 1614, d. 1681). *ODNB*.

to 1654. Having had a successful career—amassing a reasonable wealth and becoming close friends with eminent English physicians including Ent and Francis Glisson—he died of jaundice in 1681. It is finally notable that a nearly identical recipe to the one in Pw V90, attributed to a Dr. King, is found in the *Queens Closet Opened*, though no further identifying information is provided.[367]

Mr. Phillips (active mid-seventeenth century)
While there is minimal available information about "Mr Phillips", he was behind the creation of a number of interesting distilled waters in Pw V90, or, as advertised in the manuscript, "Mr Phillips waters, with their severall virtues". There is also a reference to "A most excellent Aqua Coelestis taught by Mr. Philips Apothecary" in the *Queens Closet Opened* (where Phillips is spelt with both one and two "ls"). Given the similarities between the recipes, and the fact that these works have an overlapping network of contributors and approvers, it is very likely to be the same individual in both cases. Mr. Phillips, then, was in all probability an English apothecary of some standing among English Royalists, though we have not managed to find any traces of him in the records of The Worshipful Society of Apothecaries.

Théodore de Mayerne (1573–1655)
Born into an influential Huguenot family as the son of the historian Louis de Mayerne-Turquet (1533/4–1618) and godson of the great theological reformer Theodore Beza, Mayerne's early schooling took place in his native Geneva. He took his B.A. at the University of Heidelberg before studying medicine at the University of Montpellier, where he graduated M.B. in 1596.[368] Mayerne's doctoral thesis, which he submitted the subsequent year, defended chymical remedies and thus laid the groundwork for his lifelong interest in the subject. It was probably his father's close friend, the renowned Paracelsian Joseph Duchesne, who inspired him to pursue this course.[369] Upon graduation, Mayerne departed for Paris to join Duchesne, who was *médecin ordinaire* to Henry IV, and, through Duchesne's influence, Mayerne began to work part-time for the

367 Henrietta Maria (W.M.), *Queens Closet Opened*, 284.

368 As noted, the most significant and thorough study of Mayerne is Trevor-Roper, *Europe's Physician*; also see Nance, *Turquet de Mayerne as Baroque Physician* and Hugh Trevor-Roper (2004), Mayerne, Sir Theodore Turquet de (1573–1655). *ODNB*.

369 Trover-Roper, *Europe's Physician*, especially 21, 30, and 80–100.

King himself. Feeling threatened by the chymical principles of Duchesne and Mayerne, the conservative medical faculty at the University of Paris sought to ban them from practising in the city, which drove Mayerne to publish his *Apologia* for chymical remedies in 1603.

Mayerne became known to English royalty as early as 1605 when he treated an English nobleman, Lord Norreys of Rycote, during a stopover in Paris. Impressed by his abilities, Norrey took Mayerne with him to England in order to treat Queen Anne. While he returned to Paris shortly after, Mayerne resolved to move from France to England a few years later owing to troubles with the Paris medical establishment and the political turbulence that culminated in the assassination of Henry IV. Remembering that he had cured his wife, James I made him chief physician in 1609. Having been made a nobleman in France that year, Mayerne dropped the "Turquet" from his name as too plebeian, and assumed the title Théodore de Mayerne. In England, his use of chymical cures was not only tolerated but appreciated, and he gradually became the most sought-after physician in London. He was made a fellow of the College of Physicians in 1616, and, acquiring the papers of the famous Elizabethan physician Thomas Moffett, he quickly renewed the project of forming a pharmacopoeia. Mayerne published the *Pharmacopoeia Londinensis* with a protracted introduction in 1618; this monumental work was the first comprehensive, "official" collection of recipes that apothecaries could create and it notably included a section on chymical and metallic remedies. In addition, Mayerne played a leading role in obtaining a charter for London apothecaries in 1620, while also using his vast European networks to act as an agent on behalf of James I. He was knighted by the King in 1624.

When Charles I took the throne in March 1625, however, Mayerne's fortunes changed: the new monarch preferred William Harvey to Mayerne, who was relegated to the role of the Queen's doctor. Charles was also sceptical about Mayerne's Huguenot convictions and banned him from travelling, though he was given (along with Cademan) a patent for distilling in 1636 as partial compensation. At this time, Mayerne stayed away from court, though his London home on St. Martin's Lane became a hub for thinkers and artists in its own right. Mayerne had a longstanding interest in painting, and, between 1620 and 1646, he produced an important document titled "Pictoria, sculptoria et quae subalternarum artium" that elucidated the processes involved in Baroque oil-painting, including the making of pigments, the priming of surfaces, and the conservation of paintings.[370] It was also during these pre-Civil Wars years that Mayerne and Digby became close friends and exchanged recipes, with Digby even sending Mayerne his viper-based recipe.[371]

When Charles I fled London in 1642 in anticipation of the Civil Wars, the by that point elderly Mayerne did not follow him (as Harvey did), but instead retreated to a house that he had built in Chelsea around 1639: Lindsey House. The house on this site, which may have incorporated parts of Mayerne's original structure, remains one of the oldest homes in the borough, and is the location from which Mayerne sent most of his letters to the Cavendishes.[372] During the early 1640s, Mayerne received no Royal salary, and some Royalists even deemed him a traitor, though he was nonetheless called to Exeter to oversee the birth of Henrietta, later Duchess of Orléans, in June 1644.[373] While he was nominally made physician to Prince Charles after his father's execution in 1649, Mayerne maintained a low profile and good relations with Parliament.

Along with being the physician to three consecutive English Kings and a Lord Protector, Mayerne treated thinkers ranging from Donne to, in his later years, Anne Conway.[374] Less is known about his retired life after the Restoration, when he concentrated on experiments, cookery, and gardening (mostly medicinal herbs), and invented sundry medical recipes including a cordial that would later be recommended to Napoleon when he entered Geneva as conqueror.[375] The letters in Pw V90 are in fact some of the only ones from Mayerne that survive in English, and, indeed, his primary curative in Pw V90, the salt of steel, is, as far as we know, otherwise unremarked upon in historical accounts of Mayerne. Mayerne died on 22 March 1655, aged 82, and was eulogised in a published funeral sermon by Thomas Hodges as "a living library so replenisht with

[370] See Théodore de Mayerne, "Pictoria, sculptoria et quae subalternarum artium" (Sloane 2052).

[371] For Mayerne's letters of advice to Digby from the late 1630s, see Royal College of Physicians, MS-MAYET/444, and, for Digby later sending Mayerne his viper wine, see Sloane 2071, f. 44r–v.

[372] Walter H. Godfrey (1913), *Survey of London: Volume 4, Chelsea, Pt II* (London), 37–39.

[373] Trevor-Roper, Mayerne, Sir Theodore Turquet de.

[374] See Conway, *The Conway Letters*, 20–23.

[375] Trevor-Roper, Mayerne, Sir Theodore Turquet de.

variety; that we may truely say he had a large Soule and wisedome as the Sand upon the sea shore".[376]

Théodore Naudin (1616–1675)

The "Doctor Nodin at Paris" who features in Pw V90 is the Parisian physician Théodore Naudin, the son of Pierre Naudin (c. 1585–after 1649) and Louise Gilbert (c. 1590–after 1649), and the godson of Mayerne, after whom he was probably named.[377] Théodore's father, Pierre, was *apothécaire et valet de chamber du roi*, and his house on Rue de Seine was a notable Huguenot hub. Pierre was also Mayerne's apothecary in Paris during the early 1600s, and remained Mayerne's close friend throughout his life.[378] It was almost certainly through Mayerne that the Cavendishes were introduced to Théodore, who supplied the Cavendishes with recipes that were either from, or inspired by, Mayerne, including "The true Preparation of Crocus=Metalorum", "The true Preparation of the Spirit of Salt", and "The true Preparation of the Salt of Steele".

Although the Cavendishes would have known Naudin for his medical practices, he is perhaps better remembered for his Anabaptist views, with the fullest expression of his theological ideas appearing in a 1654 publication against infant baptism titled *Conference touchant le pedobaptesme*. A few years earlier, in 1651, Naudin had presented the thrust of this work in an oration for the Parisian Reformed congregation of the Geneva-born Huguenot theologian and anti-Catholic polemicist Jean Mestrezat.[379] While this indicates that Naudin was in Paris at the same time as the Cavendishes, he left for England shortly after and became an active member of the radical congregation of Peter Chamberlen, a Royal physician, which gathered in Lothbury Square, London. As the Lothbury Square church book reveals, Naudin engaged in many of the "dialogues" that occurred there in early 1654 on topics such as "whether women may speak" in church.[380] He soon fell out with Chamberlen, however, and left

[376] Thomas Hodges (1655), *Inaccessible Glory. Or The Impossibility of Seeing Gods Face, Whilst We Are in the Body* (London).

[377] Transcription, 164–165.

[378] Trevor-Roper, *Europe's Physician*, 52–53 and 380–381.

[379] Théodore Naudin (1654), *Conference touchant le pedobaptesme tenue a Paris entre le Sieur Jean Mestrezat, pasteur de l'Eglize reformée de Charenton lés Paris* (London).

[380] Bodleian, Rawl. D.828.

the congregation along with John More (who was famous for his radical interpretation of Daniel) later that year.

From a political perspective, Naudin's most significant speech was the one that he gave at Lothbury Square on 7 February 1654 in opposition to Cromwell taking sole authority over the Commonwealth. Along with Baron de Baas—whom Cardinal Mazarin had recently dispatched to England—Naudin became wrapped up later in the year in a botched attempt to turn the army against Cromwell and to convince Thomas Harrison to take the helm. This landed him in the Tower of London from June until December 1654.[381] While they would not have sympathised with his religious views, the Cavendishes might have applauded Naudin's efforts to oust Cromwell.

Francis Prujean (bap. 1597–1666)

The "Doctor Prigion" who offers a "receit for The Collick in the stomacke" is no doubt Sir Francis Prujean, the eldest son of the Reverend Francis Prujean (d. 1624).[382] Educated by his father at their Lincolnshire home, Prujean matriculated as a sizar at Gonville and Caius College, Cambridge, in 1610 where he gained his B.A. in 1614 and his M.A. in 1617.[383] He married Margaret Legatt (1596?–1661), probably in 1621, and, by December of that year, he was made a Licentiate of the College of Physicians and practiced in London, though he was not granted his M.D. until 1625. In the subsequent year, he was elected as a fellow to the College of Physicians, at which point he returned to Lincolnshire, where he ran a successful practice for thirteen years. But, in 1639, Prujean returned to London and was elected censor of the College of Physicians, which is a position that he held until 1647, along with serving as college registrar from 1641. Despite his high standing within the College, Prujean's home was ransacked in 1643 on the suspicion that he was a papist, and, in 1658, he was brought before a London court for supposedly housing popish sympathies; he resolutely denied such accusations.

[381] John Thurloe (1742), *A Collection of the State Papers of John Thurloe* (London), 351–352 and Oliver Cromwell (1974), *The Writings and Speeches of Oliver Cromwell*, Vol. 3, eds. Wilbur C. Abbott and Catherine Crane (Cambridge, MA), 522.

[382] Transcription, 208.

[383] Venn, *Alumni Cantabrigienses*, Vol. 1, Part 3, 404.

Rising in the College ranks, Prujean was elected president from 1650 to 1653, and, when the college elected William Harvey to the position in 1654, the latter declined and recommended that Prujean continue in this role. The college library was built in 1654, under Prujean's watch. Even when Edward Alston took over the presidency in 1655, Prujean remained Treasurer (until 1663) and served as consiliarius, which was a role that involved making key College decisions, from 1656 to his death in 1666.[384] He was knighted by Charles II in April 1661 for his efforts, and was also credited with curing Queen Catherine of Braganza of typhus with a cordial in 1663. As suggested by the many references to Prujean in the *Severall Chirurgicall Treatises* (1676) by Richard Wiseman—the renowned surgeon (and sergeant-surgeon to Charles II from 1646 to 1651)—Prujean also took Wiseman under his wing from at least the Restoration until his death.[385]

Beyond his medical expertise, John Evelyn noted that Prujean was profoundly interested in the mechanical arts, collected paintings, and was a talented musician.[386] Various works dedicated to him—including a 1642 satirical broadside by Henry Peacham titled "The World is Ruled & Governed by Opinion" and a 1661 *Discourse on the Invention of Navigation* by one Thomas Philipot—testify to his wide-ranging interests. While he did not publish himself, Prujean's recipes and advice seem to have circulated widely in manuscript.[387] He passed away in London on 23 June 1666 and was buried with his first wife and son at Hornchurch, Essex. As per Prujean's wishes, the physician and medical biographer Baldwin Hamey composed a Latin epitaph praising his accomplishments. Pepys also noted that he died a wealthy man, having married Margaret Fleming (c. 1620–1679), the widow of Sir Thomas Fleming and wealthy daughter of Edward Gorges, 1st Baron Gorges of Dundalk, in 1664.[388]

Calixt Rust (d. 1658/1659)

The "Mr Rust" who is a continual presence in Pw V90 is Calixt Rust. Originally from Norfolk (and born at Nacton), Rust seems to have set

[384] Munk, *The Roll*, 185–188.

[385] Michael McVaugh (2007), Richard Wiseman and the Medical Practitioners of Restoration London. *Journal of the History of Medicine and Allied Sciences* 62.2: 125–140.

[386] John Evelyn (1901), *The Diary of John Evelyn*, Vol. 1 (London), 349.

[387] See, for example, British Library, Add 61489 f. 8, 27, 36 and Sloane 2984, f. 3.

[388] Pepys, *The Diary of Samuel Pepys*, Vol. 5, 318–319.

up shop on the Strand, and he was made a freeman of the Society of Apothecaries on 6 March 1647.[389] As the letters in Pw V90 reveal, Rust also worked as Mayerne's apothecary from at least the late 1640s until the latter's death in 1655, not only compiling and sending replies to patients, but also composing letters on Mayerne's behalf and supplying his own medical advice. Rust married one Rebecca, who had passed away by the time that Rust drew up his will on 21 December 1658, and he had a daughter, Margaret, who was still alive at his death. His will was proved on 17 June 1659, which means that he must have died between 21 December 1658 and 17 June 1659.[390] In his will, he asked to be buried in the parish church of St. Martin in the Fields, London. The will reveals little about his wealth, but he left the totality of his estate to his only daughter. He named his "loving cousin" John Cremer of Setchey, Norfolk, and mother-in-law, Margaret Cremer of King's Lynn, as joint executors. Along with mentioning two brothers, Thomas and Edward, and a sister Dorothy, it refers to one Francis Meeres (originally from Uppingham in Rutland) who was apprenticed to Rust in 1655.

IV. Servants and Untitled Characters

Mrs. Alsberger

Unknown. She provides a "way to Make Metheglin".[391]

Mr. Foscues

Unknown. Our presumption is that he is a French apothecary, though we have not found any record of him in the archives. He provides a cure "For an ague".[392]

Mary

Unknown. The first-name entry suggests that it might have been someone close to the Cavendishes. There are at least two possible Marys who were related to Margaret Cavendish. The first, and most probable, is Margaret's sister Mary Lucas (1606/7?–1648?), who married Peter Killigrew (c. 1593–1668) in October 1625. Margaret would have known her

[389] For the will of Calixt Rust, see the Records of the Prerogative Court of Canterbury, 11/292/308. Many thanks to Peter Elmer for pointing us towards this source.

[390] The National Archives, Kew, PROB 11/292/308.

[391] Transcription, 190–191.

[392] Transcription, 206–207.

well, being from a close family. The second possible candidate is the wife of Margaret's brother Thomas (1597/8–1648/9), Lady Mary Lucas (née Byron) (b. 1603/4?). She was married to Thomas for much of Margaret's upbringing. The Mary in question provides a milk-based remedy for a flux that is similar to Margaret's.[393]

Thomas Michall
Unknown. There were various men named Thomas Michell around at the time, but it is impossible to settle which this one is. He offers a "way to boyle a Carpe".[394]

A nun at Antwerp
Unknown. A "Nun, att Antwerpe" provides "An approved medicine agaynst melancholy".[395] While it is impossible to identify this individual with any degree of certainty, it is notable that Margaret is referenced in a chronicle history of a Carmel convent—titled a "Short Colection of the Beginings of Our English Monastery of Teresians in Antwerp"—that J.P. Vander Motten and Katrien Daemen-de Gelder of Ghent University have uncovered in the Antwerp City Archives.[396] In the conclusion to the entry on the "Life of Mary Cotton" (1629–1694), it is written that "when she came to be Religious the Dutchess of New Castle being then here was much taken with her as being extream pretty, entertaind her at her own house dressed her with her own hands for her Entery like a Nimph and lead her in her self".[397] Given that the recipe in question appears immediately after a letter from January 1654, the recipe was probably given to the Cavendishes in this year, which is before Margaret conducted Cotton to the convent in 1656. For this reason, the recipe is probably not by Cotton, though it is notably recorded in the "Life of Mary Cotton" that, during her final sickness, "the Doctor ordered her Medicines, which she

[393] Transcription, 164.

[394] Transcription, 193.

[395] Transcription, 200–201.

[396] Jean Pierre Vander Motten and Katrien Daemen-de Gelder (2014), Margaret Cavendish, the Antwerp Carmel and *The Convent of Pleasure*. *Archiv für das Studium der Neueren Sprachen und Literaturen* 251.1: 134–145.

[397] Caroline Bowden, Laurence Lux-Sterritt, Nicky Hallett, and Katrien Daemen-de Gelder, eds. (2013), *English Convents in Exile, 1600–1800, Vol. 4: Life Writing II* (London), 515.

as having skill, believed quite contrary to her distemper".[398] Cotton, in other words, was known for her medical prowess, and may well have had an association with the convent before her official admittance.

In relation to the question of medicine and other women in the Antwerp Carmelite community, one Mary Harcourt, Sister Mary of the Angels, who was a nun at Antwerp in 1651, is said to have had "Skill to help" her poor neighbours "in their Sicknesses".[399] Whoever provided the cure, Rubenshuis and its adjacent properties bordered on the convent, and this recipe provides concrete proof that the Cavendishes had made inroads with the convent several years before Margaret conducted Cotton there, and suggests that Margaret's ties to the community may have been deeper than the evidence has hitherto suggested.[400] The recipe in Pw V90 also makes further sense of Cavendish's notion that those who live "Monastical lives" are able to provide great insights into "all Vegetables, Minerals, and Animals, their constitutions, their sympathies, and antipathies, their extractions, and applications which they apply, for health, and prolonging of life".[401]

Mr. Puet

Unknown. Puet sends a "Leter to His Exelency ye Lord Marquis of Newcastle" with a "mass of Pills" and a balsam. We presume that Puet was an apothecary, but we have not been able to find any information about him in the historical records.

398 Bowden, Lux-Sterrit, Hallett, and Daemen-de Gelder, *English Convents in Exile*, 234.

399 Bowden, Lux-Sterrit, Hallett, and Daemen-de Gelder, *English Convents in Exile*, 227.

400 The convent was in the Hopland area of the city (see Prosper Arents (2001), *De Bibliotheek van Pieter Pauwel Rubens: een reconstructive* (Antwerp), 72–75).

401 *Philosophical and Physical Opinions*, A2v. For a fuller discussion, see Justin Begley (2019), Confessional Disputes in the Republic of Letters: Susan Du Verger and Margaret Cavendish. *The Seventeenth Century* 34.2: 181–207 (195–197).

Bibliography

Manuscripts

Bayerische Staatsbibliothek, Cod.Icon., 138–140. Robert Dudley's "Arcano del mare", Tomes I-III, 1636.

Bodleian Library, Clarendon 109. A treatise of advice to King Charles II, 1660/1661.

Bodleian Library, Rawlinson D.828. Papers relating to French Protestants, including the Lothbury Square church book, 1654.

Bodleian Library, Rawlinson Poet 16. Presentation copy of the works of Elizabeth Egerton and Jane Cavendish, 17th century.

British Library, Additional 45865. William Cavendish's "Witts Triumvirate, or The Philosopher", 1635.

British Library, Additional 61489. Papers of Robert Spencer, 1663–1665.

British Library, Additional 70499. Cavendish Papers, 1604–1659.

British Library, Additional 78337. Evelyn Papers, 1651–1799.

British Library, Egerton 2553. Sir Anthony Van Dyck: Royal Warrant for £78 for Rinaldo and Armida, 1629.

British Library, Harley 4955. Verse compilation ("The Newcastle MS"), 1620–1634.

British Library, Harley 6083. Mathematical papers of Charles Cavendish, 1621–1654.

British Library, Harley 6491. William Davisson's "A Short Treatise of the Nature of Minerals", 1645.

British Library, Harley 6758. Kenelm Digby's "Loose Fantasies", 1628.

J. Begley and B. Goldberg, *The Medical World of Margaret Cavendish*, Palgrave Studies in Medieval and Early Modern Medicine, https://doi.org/10.1007/978-3-030-92927-5

British Library, Harley 6940. Samuel Bispham's "Chymicall Notions", 17th century.

British Library, Royal 12 C II. Robert Fludd's "Declaratio brevis to James I", c. 1618.

British Library, Sloane 358. Voyage of Robert Dudley to the West Indies, 1594.

British Library, Sloane 2052. Théodore de Mayerne's "Pictoria, sculptoria et quae subalternarum artium", 1620–1646.

British Library, Sloane 2071. Théodore de Mayerne's "Ephemerides", 17th century.

British Library, Sloane 2149. "Bustorum aliquot reliquiae ab anno 1628, qui mihi primus fuit conducti seorsim a parentibus, non inauspicato, hospitii. Authore Balduino Hamey, M.D., socio et benefactore collegii medicorum Londine", 17th century.

British Library, Sloane 2984. Material from the 10th to the 18th century, including "Sir Francis Prujean Medical receipt", n.d.

Huntington Library, Ellesmere 8353. Papers of the Earl of Bridgwater, Epitaph, verses, etc, 1660s.

Huntington Library, Hastings 16522. Collection of ballads, etc., 17th century.

Nottinghamshire Archives, DPP83/13. Assignment of a mortgage from Sir John Curzon to Matthew Boucheret, 1666.

Nottingham University Library, Pw 1/12-24. Correspondence from Mark Anthony Benoist to the Cavendishes, 1667–1673.

Nottingham University Library, Pw 1/31. Letter from Matthew Boucheret, London, to Mr. Francis Toppe, Welbeck, 1666.

Nottingham University Library, Pw 1/315-317. "Confession touching a conspiracy against Margaret Cavendish [née Lucas], Duchess of Newcastle", 1671.

Nottingham University Library, Pw 1/371-384. Letters from William Pierrepont to Henry Cavendish and Frances [Pierrepont], his wife, 1657–1674.

Nottingham University Library, Pw 1/592. "A perticular of such bills as was paid to Mr. Clayton in the Accounts of Mr. John Rolleston from tyme to tyme", 1661–1665.

Nottingham University Library, Pw V25. Poems and dramatic works mostly by William Cavendish, 1620–1644.

Nottingham University Library, Pw V26. Poems and dramatic works mostly by William Cavendish, 1620–1644.

Nottingham University Library, Pw V33. "The Humorists" by Thomas Shadwell, 1670–1671.

Nottingham University Library, Pw V34. "The Sullen Lovers: or, The Impertinents" by Thomas Shadwell, 1667–1668.

Nottingham University Library, Pw V90. Cavendish medical manuscript, c. 1647–1654.

Royal Collections, Archive Constantijn Huygens, G1-9.1. Correspondence between Margaret Cavendish and Constantijn Huygens, 20 March 1657; 27 October 1658; 11 January 1659.
Royal College of Physicians, MS-MAYET/444. An original record book of the cases and consultations of Sir Théodore de Mayerne, 1607–1651.
Royal Society, MS/92. John Aubrey's "Memoires of Naturall Remarques in the county of Wilts", 1685.
Royal Society, RB/3/6. Robert Boyle's miscellaneous correspondences, 17th century.
Royal Society, RB/1/29. Robert Boyle's scientific work, medical receipts, 17th century.
St Ann, Blackfriars, London, 1562–1726, P69/ANN. Parish registers.
St Giles Cripplegate, London, 1667–1672, P69/GIS. Parish registers.
St John's College, Cambridge, L.2 (James 359). William Davisson's "Elementa Physicae Practicae", c. 1631.
St Martin's, Ludgate, London, 1558–1719, P69/MTN1. Parish registers.
The Hartlib Papers, 26/72A-B. Memo on The Countess of Arundel's "Natura Exenterata", 17th century.
The Hartlib Papers, 29/4/21A. Ephemerides, including medical receipts, 1654.
The National Archives, PROB 11/292/308. Will of Calixt Rust, Apothecary of Saint Martin in the Fields, Middlesex, 1659.
Parliamentary Archives, HL/PO/JO/10/1/316. Records of naturalisation, including that of Matthew Boucheret, 1663.
The National Archives, PROB 11/309/148. Will of Edward Waldron of Mansfield, 1662.
Wellcome Library, 2124. Notes on Digby's experiments, 1668.

Printed Works (Before 1900)

ab Aquapendente, Fabricius. 1603. *De brutorum loquela*. Padua.
ab Aquapendente, Fabricius. 1624. *De locutione et ejus instrumentis tractatus*. Frankfurt.
Allestree, Richard. 1673. *The Ladies Calling*. Oxford.
Anonymous. 1653. *The Ladies Companion, or, A Table Furnished with Sundry Sorts of Pies and Tarts* [...] *By Persons of Quality Whose Names Are Mentioned*. London.
Anonymous. 1657. *A Physical Dictionary*. London.
Anonymous. 1670. *Bazilica Chymica, & Praxis Chymiatricae, or Royal and Practical Chymistry*. London.
Anonymous. 1676. *The Apothecaries Vindication: In Answer to an Abusive Book Entitled The Family Physician and House Apothecary*. London.
Ariosto, Ludovico. 1543. *Orlando Furioso*. Venice.

Aristotle. 1979. *De anima*, ed. and trans. William D. Ross. Oxford: Clarendon Press.

Aristotle. 2014. *Generation of Animals*, ed. and trans. Arthur L. Peck. Cambridge, MA: Harvard University Press.

Atkyns, Robert. 1689. *An Enquiry into the Power of Dispensing with Penal Statutes.* London.

Atkyns, Robert. 1734. *Parliamentary and Political Tracts*. London.

Aubrey, John. 1898. *Brief Lives, Chiefly of Contemporaries, Set Down by John Aubrey, Between the Years 1669 and 1696*, ed. Andrew Clark. Oxford: Clarendon Press.

Bacon, Francis. 1862. *The Works of Francis Bacon*, eds. James Spedding, Robert Ellis, and Douglas D. Heath. London: Longmans and Co.

Baldi, Baldo. 1646. *Del vero opobalsamo orientale discorso apologetico dell'illustrissimo.* Rome.

Bate, George. 1694. *Pharmacopœia Bateana: or, Bate's Dispensatory*. London.

Batman, Stephen. 1582. *Batman uppon Bartholome: His Book De proprietatibus rerum*. London.

Baxter, Richard. 1696. *Vindiciæ anti-Baxterianæ: or, Some Animadversions on a Book Intituled Reliquiæ Baxterianæ*. London.

Begbie, J. Warburton. 1875. Address in Medicine. *The British Medical Journal* 2.762: 164–171.

van Beverwyck, Jan. 1644. *Autarkeia Bataviae, sive introductio ad medicinam indigenam.* Leiden.

Bispham, William. 1890. *Memoranda Concerning the Family of Bispham in Great Britain and the United States of America*. New York.

Boulton, Samuel. 1656. *Medicina magica tamen physica*. London.

Boyle, Robert. 1661. *The Sceptical Chymist*. London.

Boyle, Robert. 1663. *Some Considerations Touching the Usefulness of Experimental Naturall Philosophy*. London.

Boyle, Robert. 1688. *A Disquisition About the Final Causes of Natural Things*. London.

Brackley, Elizabeth and Jane Cavendish. 1996. The Concealed Fancies (c. 1645). *Renaissance Drama by Women: Texts and Documents*, eds. Susan P. Cerasano and Marion Wynne-Davies, 127–155. New York: Routledge.

Browne, Thomas. 1643. *Religio Medici*. London.

Cademan, Thomas and Théodore de Mayerne. 1639. *The Distiller of London*. London.

Cavendish-Bentinck, William John Arthur Charles James, et al. 1891–1931. *The Manuscripts of His Grace the Duke of Portland, Preserved at Welbeck Abbey*, Vols. 1–10. London: Royal Commission on Historical Manuscripts.

Cavendish, Margaret. 1653. *Philosophicall Fancies*. London.

Cavendish, Margaret. 1655. *Philosophical and Physical Opinions*. London.

Cavendish, Margaret. 1655. *The World's Olio*. London.
Cavendish, Margaret. 1664. *Philosophical Letters*. London.
Cavendish, Margaret. 1664. *Sociable Letters*. London.
Cavendish, Margaret. 1667. *Life of William Cavendishe*. London.
Cavendish, Margaret. 2001. *Observations Upon Experimental Philosophy*, ed. Eileen O'Neill. Oxford: Oxford University Press.
Cavendish, William and Margaret Cavendish. 1678. *A Collection of Letters and Poems*. London.
Chalmers, Alexander. 1812. *The General Biographical Dictionary*, Vol. 3. London.
Chamberlain, John. 1939. *Letters of John Chamberlain*, ed. Norman E. McClure. Philadelphia: The American Philosophical Society.
Chambers, Ephraim. 1728. *Cyclopædia, or, An Universal Dictionary of Arts and Sciences*, Vol. 1. London.
Charleton, Walter and Jan Baptista van Helmont. 1650. *A Ternary of Paradoxes*. London.
Cobbett, William. 1809. *Cobbett's Parliamentary History of England*, Vol. 5. London.
Coelson, Lancelot. 1656. *The Poor-Mans Physician*. London.
Cogan, Thomas. 1636. *The Haven of Health*. London.
Cole, Peter. 1656. *Mr Culpeper's Ghost, Giving Seasonable Advice to the Lovers of his Writings*. London.
Conway, Anne. 1992. *The Conway Letters: The Correspondence of Anne, Viscountess Conway, Henry More, and Their Friends, 1642–1684*, eds. Marjorie H. Nicolson and Sarah Hutton. Oxford: Clarendon Press.
Cornacchini, Marco. 1619. *Methodus qua omnes humani corporis affectiones ab humoribus copia*. Florence.
Cotgrave, Randle. 1611. *A Dictionarie of the French and English Tongues*. London.
Cromwell, Oliver. 1974. *The Writings and Speeches of Oliver Cromwell*, Vol. 3, eds. Wilbur C. Abbott and Catherine Crane. Cambridge, MA: Harvard University Press.
Culpeper, Nicholas. 1649. *A Physicall Directory, or, A Translation of the London Dispensatory Made by the Colledge of Physicians in London*. London.
Culpeper, Nicholas. 1653. *Pharmacopeia Londinensis, or, The London Dispensatory*. London.
Culpeper, Nicholas. 1659. *School of Physick*. London.
Daniell, Francis H. Blackburne, ed. 1895 [1968]. *Calendar of State Papers, Domestic Series, January to November, 1671*. London.
Davenant, William. 1673. *The Works of Sr. William Davenant*. London.
Davisson, William. 1641. *Philosophia pyrotechnica*. London.

Davisson, William. 1660. *Commentariorum in* [...] *Petri Severini Dani ideam medicinae, Philosophicae propediem prodituororum Prodromus*. The Hague.
Dekker, Thomas. 1620. *Dekker His Dreame*. London.
Descartes, René. 1983. *Oeuvres de Descartes*, Vol. 1, eds. Charles Adam and Paul Tannery. Paris: Librairie Philosophique J. Vrin.
Digby, Kenelm. 1644. *Two Treatises*. London.
Digby, Kenelm. 1658. *A Late Discourse* [...] *Touching the Cure of Wounds by the Powder of Sympathy*. London.
Digby, Kenelm. 1661. *Discourse Concerning the Vegetation of Plants*. London.
Digby, Kenelm. 1669. *The Closet of the Eminently Learned Sir Kenelme Digbie*. London.
Digby, Kenelm and George Hartman. 1683. *A Choice Collection of Rare Chymical Secrets and Experiments in Physick & Philosophy*. London.
Dodoens, Rembert. 1578. *A Niewe Herball, or Historie of Plantes*, trans. Henry Lyte. London.
Donne, John. 1633. *Poems by J.D. with Elegies of the Author's Death*. London.
Duchesne, Joseph. 1590. *The Sclopetarie of J. Quercetanus, Phisition, Or His Book Containing the Cure of Wounds Received by Shot of Gunne Or Such Like Engines of Warre*, trans. John Hester. London.
Dudley, Robert. 1646. *Dell'Arcano del Mare*. Florence.
Dunglison, Robley. 1857. *Medical Lexicon: A Dictionary of Medical Science*. Philadelphia: Lea Brothers.
Etienne, Charles. 1616. *Maison Rustique, or, The Countrey Farme*, trans. Gervase Markham. London.
Evelyn, John. 1879. *The Diary of John Evelyn*, Vol. 4, ed. William Bray. London: Bickers and Son.
Evelyn, John. 1901. *The Diary of John Evelyn*, Vol. 1, ed. William Bray. London: M. Walter Dunn.
Evelyn, Mary and John Evelyn. 1690. *Mundus muliebris: or, The Ladies Dressing-Room Unlock'd, and Her Toilette Spread in Burlesque*. London.
Fernel, Jean. 1608. *Universa medicina, tribus et viginti libris absoluta*. Paris.
Fludd, Robert. 1603. *Uranometria*. Augsburg.
Fludd, Robert. 1618. *De naturæ simia, seu technica macrocosmi historia*. Oppenheim.
Gabrieli, Vittorio. 1968. *Loose Fantasies*. Rome: Edizioni di storia e letteratura.
Galen. 1916. *On the Natural Faculties*, ed. and trans. Arthur J. Brock. Cambridge, MA: Harvard University Press.
Galen. 2011. *Method of Medicine*, eds. and trans. Ian Johnston and Greg H.R. Horsley. Cambridge, MA: Harvard University Press.
Galen. 2020. *On Temperaments; On Non-Uniform Distemperment; The Soul's Traits Depend on Bodily Temperament*, ed. and trans. Ian Johnston. Cambridge, MA: Harvard University Press.

Galen. 2021. De compositione medicamentorum secundum locus. *Perseus Digital Library: Galen*. https://scaife.perseus.org/library/urn:cts:greekLit:tlg0057/.

Gerard, John. 1633. *The Herball or Generall Historie of Plantes*. London.

Glaser, Christophe. 1667. *Traité de la chymie*. Paris.

Glauber, Johann Rudolph. 1661. *Furni Novi Philosophici oder Beschreibung einer New-erfundenen Distillir-Kunst*. Amsterdam.

Glauber, Johann Rudolf. 1689. *The Works of the Highly Experienced and Famous Chymist, John Rudolph Glauber*, trans. Christopher Packe. London.

Green, Mary A. (ed.). 1889–1892. *Calendar of the Proceedings of the Committee for Compounding*. London.

Grey, Elizabeth Talbot. 1653. *A Choice Manual of Rare and Select Secrets*. London.

Hakluyt, Richard. 1599–1600. *The Principal Navigations, Voyages, Traffiques and Discoveries of the English Nation*. London.

Hamilton, William. 1866. *Discussions on Philosophy and Literature, Education and University Reform*. Edinburgh: Blackwood and Son.

Hamy, Ernest. 1898. William Davisson, intendant du Jardin du Roi et professeur de chemie, 1647–51. *Nouvelles Archives du Muséum d'Histoire Naturelle*. Paris: Masson et Cie.

Hartmann, Johannes. 1635. *Tractatus physico-medicus de opio*. Wittenberg.

Hartmann, Johannes. 1639. *Praxis chymiatrica*. Geneva.

Harvey, Gideon. 1676. *The Family Physician and the House Apothecary*. London.

Harvey, William. 1628. *De motu cordis*. Frankfurt.

van Helmont, Jan Baptiste. 1644. *Opuscula medica inaudita*. Cologne.

van Helmont, Jan Baptiste. 1664. *Works Containing His Most Excellent Philosophy, Physick, Chirurgery, Anatomy*, trans. John Chandler. London.

Hernández de Maceras, Domingo. 1607. *Libro del Arte de Cozina*. Salamanca.

Highmore, Nathaniel. 1651. *The History of Generation*. London.

Hill, John. 1751. *A History of the Materia Medica*. London.

Hippocrates. 1923. *Hippocrates, Volume I: Ancient Medicine; Airs, Waters, Places; Epidemics 1 and 3; The Oath; Precepts; Nutriments*, ed. and trans. William H.S. Jones. Cambridge, MA: Harvard University Press.

Hippocrates. 1923. *Hippocrates, Volume II: Prognostic; Regimen in Acute Diseases; The Sacred Disease; The Art; Breaths; Law; Decorum; Physician (Ch. 1); Dentition*, ed. and trans. William H.S. Jones. Cambridge, MA: Harvard University Press.

Hippocrates. 1923. *Hippocrates, Volume IV: Nature of Man; Regimen in Health; Humours; Aphorisms; Regimen 1–3; Dreams*, ed. and trans. William H.S. Jones. Cambridge, MA: Harvard University Press.

Hodges, Thomas. 1655. *Inaccessible Glory: Or The Impossibility of Seeing Gods Face, Whilst We Are in the Body*. London.

Holland, Philemon (trans). 1601. *The Historie of the World, Commonly called, The Naturall Historie of C. Plinius Secundus*. London.
Holles, John. 1975. *Letters of John Holles, 1587–1637*, ed. Peter R. Seddon. Nottingham: Thoroton Society Record Series.
Howard, Aletheia Talbot. 1655. *Natura Exenterata*. London.
Hutchinson, Lucy. 1846. *Memoirs of the Life of Colonel Hutchinson*. London.
Hyde, Edward. 1702–1704. *History of the Rebellion*. London.
Hyde, Edward. 1843. *The Life of Edward, Earl of Clarendon*, Vol. 1. Oxford.
Johnson, Samuel. 1818. *A Dictionary of the English Language*, Vol. 4. London.
King, John. 1855. *A Memoir of the Life and Death of Sir John King, Knight, Written by His Father in 1677*. London: Bell and Daldy.
Knutsson, Bengt. c. 1485. *A Litill Boke Necessarye [and] Behovefull A[g]enst the Pestilence*. London.
Lémery, Nicolas. 1685. *Recueil de curiosités rares et nouvelles*. London.
Lémery, Nicolas and Walter Harris. 1686. *A Course of Chemistry*. London.
Lloyd, David. 1668. *Memoires of* [...] *Excellent Personages*. London.
Llwyd, Humphrey. 1585. *Treasury of Healthe Conteynyng Many Profitable Medycines Gathered Out of Hypocrates, Galen and Auycen, by One Petrus Hyspanus*. London.
Machin, Lewis and Gervase Markham. 1608. *Dumbe Knight*. London.
Maria, Henrietta (W.M.). 1655. *Queens Closet Opened*. London.
Markham, Gervase. 1631. *The English House-Wife*. London.
May, Thomas. 1631. *The Tragedy of Antigone*. London.
de Mayerne, Théodore. 1676. *A Treatise of the Gout*. London.
de Mayerne, Théodore. 1703. *Opera Medica*. London.
Moens, William J.C. (ed.). 1884. *The Marriage, Baptismal and Burial Registers, 1571–1874, and Monumental Inscriptions of the Dutch Reformed Church, Austin Friars, London*. London.
Morley, George. 1661. *A Sermon Preached at the Magnificent Coronation of the Most High and Mighty King Charles the IId King of Great Britain, France, and Ireland*. London.
Munk, William. 1878. *The Roll of the Royal College of Physicians of London*. London: Royal College of Physicians.
Naudin, Théodore. 1654. *Conference touchant le pedobaptesme tenue a Paris entre le Sieur Jean Mestrezat, pasteur de l'Eglize reformée de Charenton lés Paris*. London.
O'Dowde, Thomas. 1664. *The Poor Man's Physician, Or the True Art of Medicine*. London.
Oldenburg, Henry. 1968. *Henry Oldenburg, The Correspondence of Henry Oldenburg 1668–1669*, Vol. 5, eds. and trans. Alfred Rupert Hall and Marie B. Hall. Madison: University of Wisconsin Press.

Ovid. 1929. *Art of Love; Cosmetics; Remedies for Love; Ibis; Walnut-tree; Sea Fishing; Consolation*, eds. and trans. John H. Mozley and Georges P. Goold. Cambridge, MA: Harvard University Press.
Paracelsus. 1573. *Philippi Aureoli Theophrasti Paracelsi Bombast eremitae, summi inter Germanos medici & philosophi. Chirurgia magna, in duos tomos digesta*. Strasbourg.
Partridge, John. 1591. *The Treasurie of Commodious Conceits*. London.
Pepys, Samuel. 1995. *The Diary of Samuel Pepys*, Vol. 4, eds. Robert Latham and William Matthews. Berkeley: University of California Press.
Plat, Hugh. 1653. *Jewel House of Art and Nature*. London.
Plater, Felix. 1664. *Platerus Golden Practice of Physick*. London.
Primrose, James. 1651. *Popular Errours of the People in the Matter of Physick*. London.
Prynne, William. 1643. *Romes Master-Peece: Or, The Grand Conspiracy of the Pope*. London.
R.B. 1689. *The Life of Oliver Cromwel*. London.
Read, Alexander. 1636. *The Chirurgicall Lectures of Tumors and Ulcers*. London.
Roux, François. 1809. *Nouveau Dictionnaire*. Halle.
Sainsbury, William N. 1859. *Original Papers Relating to Rubens*. London.
Sendivogius, Michael. 1650. *New Light of Alchymie Taken Out of the Fountaine of Nature, and Manuall Experience: To Which Is Added a Treatise of Sulphur*, trans. John French. London.
Severinus, Marcus Aurelius. 1650. *Vipera Pythia, id est, de viperae natura, veneno, medicina, demonstrationes*. Padua.
Shakespeare, William. 1608. *King Lear*. London.
Simmonds, Peter L. 1858. *Dictionary of Trade Products*. London.
Small, John. 1874. Notice of William Davidson, MD (Gulielmus Davissonus), First Professor of Chemistry, and Director of the Jardin des Plantes, Paris, Afterwards Physician to the King of Poland. *Proceedings of the Society of Antiquaries of Scotland* 10: 265–280.
Starkey, George. 1660. *George Starkey's Pill Vindicated from the Unlearned Alchymist*. London.
Stubbe, Henry. 1670. *The Plus Ultra Reduced to a Non Plus*. London.
Suckling, John. 1659. *Letters to Several Persons of Honor*. London.
Taylor, John. 1623. *A New Discovery by Sea, with a wherry from London to Salisbury*. London.
Temple, William. 1720. Of Health and Long Life. *The Works of William Temple*, 272–289. London.
Thibaut, Pierre. 1668. *The Art of Chymistry*. London.
Thorold Rogers, James E. 1872. *A History of Agriculture and Prices in England: 1401–1582*, Vol. 4. Oxford: Clarendon Press.
Thurloe, John. 1742. *A Collection of the State Papers of John Thurloe*. London.

Townsend, Dorothea. 1897. *Life and Letters of Mr. Endymion Porter*. London: T. F. Unwin.
Tozzetti, Giovanni Targioni. 1780. *Notizie degli aggrandimenti delle scienze fisiche accaduti in Toscana nel corso di anni LX del secolo XVII*, Vol. 3. Florence.
Virgil. 1999. *Eclogues, Georgics, Aeneid: Books 1–6*, eds. and trans. H. Rushton Fairclough and Georges P. Goold. Cambridge, MA: Harvard University Press.
Vives, Juan Luis. 2000. *The Education of a Christian Woman: A Sixteenth-Century Manual*, ed. Charles Fantazzi. Chicago: University of Chicago Press.
Warner, George F. 1899. *The Voyage of Robert Dudley…to the West Indies, 1594–1595*. London: Printed for the Hakluyt Society.
Watson, William H. 1827. *A Practical Treatise on the Law Relating to the Office and Duty of Sheriff*. London.
Wood, Anthony. 1691. *Athenae Oxonienses*, Vol. 1. London.
Woolley, Hannah. 1677. *The Compleat Servant-Maid; or, The Young Maidens Tutor*. London.

Printed Works (After 1900)

Adams, Simon. 2008. Dudley, Sir Robert (1574–1649). *Oxford Dictionary of National Biography*.
Adelmann, Howard B. 1967. *The Embryological Treatises of Hieronymus Fabricius of Aquapendente*. Ithaca: Cornell University Press.
Allen, Katherine. 2016. Hobby and Craft: Distilling Household Medicine in Eighteenth-Century England. *Early Modern Women* 11.1: 90–114.
Alonso-Almeida, Francisco. 2001. Punctuation Practice in a Late Medieval English Medical Remedybook. *Folia Linguistica Historica* 22.1–2: 207–232.
Alsbury, Colin. 2008. Aylesbury, Thomas (bap. 1597, d. 1660/61). *Oxford Dictionary of National Biography*.
Anstey, Peter. 2011. *John Locke and Natural Philosophy*. Oxford: Oxford University Press.
Archer, Jayne. 2002. The Queen's Arcanum: Authority and Authorship in *The Queens Closet Opened* (1655). *Renaissance Journal* 1.6: 14–26.
Archer, Jayne. 2016. Women and Chymistry in Early Modern England: The Manuscript Receipt Book (c. 1616) of Sarah Wigges. *Gender and Scientific Discourse in Early Modern Culture*, ed. Kathleen P. Long, 191–216. London: Routledge.
Arents, Prosper. 2001. *De Bibliotheek van Pieter Pauwel Rubens: een reconstructie*. Antwerp: Vereniging der Antwerpse Bibliofielen.
Armitage, Christopher. 1987. *Sir Walter Ralegh: An Annotated Bibliography*. Chapel Hill: University of North Carolina Press.

Asch, Ronald G. 2008. Porter, Endymion (1587–1649). *Oxford Dictionary of National Biography.*
Avramescu, Cătălin. 2009. *An Intellectual History of Cannibalism.* Princeton: Princeton University Press.
Baker, Tawrin. 2016. Why All This Jelly? Jacopo Zabarella and Hieronymus Fabricius ab Aquapendente on the Usefulness of the Vitreous Humor. *Early Modern Medicine and Natural Philosophy*, eds. Peter Distelzweig, Benjamin Goldberg, and Evan R. Ragland, 59–88. Dordrecht: Springer.
Baldwin, Martha. 1993. Alchemy and the Society of Jesus in the Seventeenth Century: Strange Bedfellows? *Ambix* 40: 41–64.
Barry, Jonathan. 2019. Educating Physicians in Seventeenth-Century England. *Science in Context* 32.2: 137–154.
Bassnett, Madeline. 2009. Restoring the Royal Household: Royalist Politics and the Commonwealth Recipe Book. *Early English Studies* 2: 1–32.
Bator, Magdalena and Marta Sylwanowicz. 2017. Measures in Medieval English Recipes—Culinary Vs. Medical. *Studia Anglica Posnaniensia* 52.1: 28–52.
Bayer, Penny. 2021. Women Alchemists and the Paracelsian Context in France and England, 1560–1616. *Early Modern Women* 15.2: 103–111.
de Beer, Esmond S. 1943. The House of Lords in the Parliament of 1680. *Historical Research* 20.59: 22–37.
Begley, Justin. 2017. "The minde is matter moved": Nehemiah Grew on Margaret Cavendish. *Intellectual History Review* 27.4: 493–514.
Begley, Justin. 2019. Confessional Disputes in the Republic of Letters: Susan Du Verger and Margaret Cavendish. *The Seventeenth Century* 34.2: 181–207.
Begley, Justin. 2021. Margaret Cavendish and Vegetable Life. *Vegetative Powers: The Roots of Life in Ancient, Medieval and Early Modern Natural Philosophy*, eds. Fabrizio Baldassarri and Andreas Blank, 305–324. Dordrecht: Springer.
van Beneden, Ben and Nora de Poorter (eds.). 2006. *Royalist Refugees: William and Margaret Cavendish in the Rubens House, 1648–1660.* Antwerp: Rubenshuis & Rubenianum.
Biagioli, Mario. 2006. *Galileo's Instruments of Credit.* Chicago: University of Chicago Press.
Birken, William. 1976. Dr. John King (1614–1681) and Dr. Assuerus Regemorter (1615–1650). *Medical History* 20.3: 276–295.
Birken, William. 2008. King, John (bap. 1614, d. 1681). *Oxford Dictionary of National Biography.*
Birrell, Thomas A. 1991. Reading as Pastime: The Place of Light Literature in Some Gentlemen's Libraries of the Seventeenth Century. *Property of a Gentleman: The Formation, Organisation and Dispersal of the Private Library, 1620–1920*, eds. Robin Myers and Michael Harris, 113–131. Winchester: St Paul's Bibliographies.

Blaen, Tom. 2012. *Medical Jewels, Magical Gems: Precious Stones in Early Modern Britain*. Crediton: The Medieval Press.

Blake, Liza. 2018. Pounced Corrections in Oxford Copies of Cavendish's *Philosophical and Physical Opinions;* or Margaret Cavendish's Glitter Pen. *New College Notes* 10.6: 1–11.

Boulton, Jeremy. 1996. Wage Labour in Seventeenth-Century London. *The Economic History Review* 49.2: 268–290.

Bowden, Caroline, Laurence Lux-Sterritt, Nicky Hallett, and Katrien Daemen-de Gelder (eds.). 2013. *English Convents in Exile, 1600–1800,* Vol. 4. London: Pickering & Chatto.

Boyle, Deborah. 2017. *The Well-Ordered Universe: The Philosophy of Margaret Cavendish*. Oxford: Oxford University Press.

Broad, Jacqueline. 2011. Cavendish, van Helmont, and the Mad Raging Womb. *The New Science and Women's Literary Discourse: Prefiguring Frankenstein*, ed. Judy A. Hayden, 47–46. New York: Palgrave Macmillan.

Brockbank, William. 1964. Sovereign Remedies: A Critical Depreciation of the 17th-Century London Pharmacopoeia. *Medical History* 8.1: 1–14.

Brooks, Christopher W. 2019. Religion and Law in Early Modern England. *Law, Lawyers and Litigants in Early Modern England: Essays in Memory of Christopher W. Brooks*, eds. Michael Lobban, Joanne Begiato, and Adrian Green, 327–365. Cambridge: Cambridge University Press.

Browning, Andrew. 1951. *Thomas Osborne, Earl of Danby and Duke of Leeds, 1632–1712*, Vol. 1. Glasgow: Jackson, Son & Co.

Brushfield, Thomas N. 1905–6. Three State Documents Relating to the Arrest and Execution of Sir Walter Ralegh in 1618. *Report and Transactions of the Devonshire Association* 37: 284–324 and 38: 416–490.

Caminietzki, Carlos Z. 2001. Jesuits and Alchemy in the Early Seventeenth Century: Father Johannes Roberti and the Weapon-Salve Controversy. *Ambix* 48.2: 83–101.

Campbell, Andrew, Lorenza Gianfrancesco, and Neil Tarrant. 2018. Alchemy and the Mendicant Orders of Late Medieval and Early Modern Europe. *Ambix* 65.3: 201–209.

Christensen, Ann C. 2014. Words About Women's Work: The Case of Housewifery in Early Modern England. *Early Modern Studies Journal* 6: 1–28.

Churchill, Wendy D. 2016. *Female Patients in Early Modern Britain: Gender, Diagnosis, and Treatment*. London: Routledge.

Clairhout, Isabelle. 2014. Erring from Good Huswifry? The Author as Witness in Margaret Cavendish and Mary Trye. *Renaissance and Reformation* 37.2: 81–114.

Clucas, Stephen (ed.). 2003. *A Princely Brave Woman: Essays on Margaret Cavendish, Duchess of Newcastle*. Aldershot: Ashgate.

Condren, Conal. 1993. Casuistry to Newcastle: *The Prince* in the World of the Book. *Political Discourse in Early Modern Britain*, eds. Nicholas Phillipson and Quentin Skinner, 164–186. Cambridge: Cambridge University Press.

Console, Renzo. 2013. Pharmaceutical Use of Gold from Antiquity to the Seventeenth Century. *A History of Geology and Medicine*, eds. Christopher Duffin, Richard T.J. Moody, and Christopher Gardner-Thorpe, 171–189. London: Geological Society of London.

Cook, Harold. 1986. *The Decline of the Old Medical Regime in Stuart London*. Ithaca: Cornell University Press.

Cook, Harold. 1994. Good Advice and Little Medicine: The Professional Authority of Early Modern English Physicians. *Journal of British Studies* 33.1: 1–31.

Cook, Harold. 2007. *Matters of Exchange: Commerce, Medicine, and Science in the Dutch Golden Age*. New Haven: Yale University Press.

Cooper, Alix. 2003. The Indigenous Versus the Exotic: Debating Natural Origins in Early Modern Europe. *Landscape Research* 28.1: 51–60.

Cooper, Alix. 2006. Homes and Households. *The Cambridge History of Science*, Vol. 3, eds. Katherine Park and Lorraine Daston, 224–237. Cambridge: Cambridge University Press.

Cooper, Alix. 2007. *Inventing the Indigenous: Local Knowledge and Natural History in Early Modern Europe*. Cambridge: Cambridge University Press.

Crawford, Patricia. 2005. Women's Published Writings 1600–1700. *Women in English Society, 1500–1800*, ed. Mary Prior, 158–209. London: Routledge.

Crellin, John K. 2019. Gascoigne's Powder: A British Prescription and Home Medicine, 1600s to early 1900s. *Pharmaceutical Historian* 49.1: 1–15.

Cunning, David. 2016. *Cavendish*. London: Routledge.

Cunningham, Andrew. 1985. Fabricius and the 'Aristotle Project': Anatomical Teaching and Research at Padova. *The Medical Renaissance of the Sixteenth Century*, eds. Andrew Wear, Roger K. French, and Ian M. Lonie, 195–222. Cambridge: Cambridge University Press.

Curth, Louise Hill. 2005. Astrological Medicine and the Popular Press in Early Modern England. *Culture and Cosmos* 9.1: 73–93.

Cust, Richard. 2013. Coventry, Thomas, First Baron Coventry (1578–1640). *Oxford Dictionary of National Biography*.

Davies, Gareth. 1917. Appendix II: Life of Sir William Boswell. *Autobiography of Thomas Raymond*, ed. Gareth Davies, 69–84. London: The Royal Historical Society.

Dear, Peter. 2006. The Meanings of Experience. *The Cambridge History of Science*, Vol. 3, eds. Katherine Park and Lorraine Daston, 106–131. Cambridge: Cambridge University Press.

Debus, Allen G. 1961. Robert Fludd and the Circulation of the Blood. *Journal of the History of Medicine and Allied Sciences* 16.4: 374–393.

Debus, Allen G. 1964. Robert Fludd and the Use of Gilbert's *De Magnete* in the Weapon-Salve Controversy. *Journal of the History of Medicine and Applied Sciences* 19.4: 389–417.

Debus, Allen G. 1965. *The English Paracelsians*. London: Oldbourne.

Debus, Allen G. 2002. *The Chemical Philosophy: Paracelsian Science and Medicine in the Sixteenth and Seventeenth Centuries*. Mineola: Dover Publications.

DiMeo, Michelle. 2008. Howard, Aletheia [née Talbot], Countess of Arundel, of Surrey, and of Norfolk, and suo jure Baroness Furnivall, Baroness Talbot, and Baroness Strange of Blackmere. *Oxford Dictionary of National Biography*.

DiMeo, Michelle. 2016. The Rhetoric of Medical Authority in Lady Katherine Ranelagh's Letters. *Women and Epistolary Agency in Early Modern Culture, 1450–1690*, eds. James Daybell and Andrew Gordon, 112–125. London: Routledge.

DiMeo, Michelle and Sara Pennell. 2013. *Reading and Writing Recipe Books, 1550–1800*. Manchester: Manchester University Press.

Dingwall, Helen M. 2004. Fraizer, Sir Alexander, First Baronet. *Oxford Dictionary of National Biography*.

Distelzweig, Peter, Benjamin Goldberg, and Evan R. Ragland (eds.). 2016. *Early Modern Medicine and Natural Philosophy*. Dordrecht: Springer.

Dixon, Laurinda S. 1993. Some Penetrating Insights: The Imagery of Enemas in Art. *Art Journal* 52.3: 28–35.

Doyle, Derek. 2005. Per Rectum: A History of Enemata. *The Journal of the Royal College of Physicians of Edinburgh* 35.4: 367–370.

Driver, Christopher (ed.). 1997. *John Evelyn, Cook: The Manuscript Receipt Book of John Evelyn*. London: Prospect Books.

Dutton, Danielle. 2016. *Margaret the First*. Melbourne: Scribe.

Eamon, William. 1994. *Science and the Secrets of Nature: Books of Secrets in Medieval and Early Modern Culture*. Princeton: Princeton University Press.

Edmont, Mary. 2009. Davenant [D'Avenant], Sir William (1606–1668). *Oxford Dictionary of National Biography*.

Ekholm, Karin J. 2010. Fabricius's and Harvey's Representations of Animal Generation. *Annals of Science* 67.3: 329–352.

Elmer, Peter. 1989. Medicine, Religion, and the Puritan Revolution. *The Medical Revolution of the Seventeenth Century*, eds. Andrew Wear and Roger French, 10–45. Cambridge: Cambridge University Press.

Estes, J. Worth. 1995. The European Reception of the First Drugs from the New World. *Pharmacy in History* 37.1: 3–23.

Eyre, Hermione. 2015. *Viper Wine*. London: Vintage Books.

Fabbri, Christiane N. 2007. Treating Medieval Plague: The Wonderful Virtue of Theriac. *Early Science and Medicine* 12.3: 247–283.

Feingold, Mordechai. 1984. *The Mathematicians' Apprenticeship: Science, Universities and Society in England 1560–1640*. Cambridge: Cambridge University Press.

Feingold, Mordechai. 1991. John Selden and the Nature of Seventeenth-Century Science. *In the Presence of the Past: Essays in Honor of Frank Manuel*, eds. Richard T. Bienvenu and Mordechai Feingold, 55–78. Springer: Dordrecht.

Feinstein, Sandy. 2021. Experience, Authority, and the Alchemy of Language: Margaret Cavendish and Marie Meurdrac Respond to the Art. *Early Modern Women* 15: 133–142.

Fissell, Mary. 1992. Readers, Texts, and Contexts: Vernacular Medical Works in Early Modern England. *The Popularization of Medicine, 1650–1850*, ed. Roy Porter, 72–96. New York: Routledge.

Fissell, Mary. 2004. *Vernacular Bodies: The Politics of Reproduction in Early Modern England*. Oxford: Oxford University Press.

Fissell, Mary. 2007. The Marketplace of Print. *Medicine and the Market in England and Its Colonies, c.1450-c.1850*, eds. Mark Jenner and Patrick Wallis, 108–132. Basingstoke: Palgrave Macmillan.

Fitzmaurice, James. 2000. Margaret Cavendish in Antwerp: The Actual and the Imaginary. *In-Between* 9.1–2: 29–40.

Foster, Michael. 1982. Thomas Allen (1540–1632), Gloucester Hall and the Survival of Catholicism in Post-Reformation Oxford. *Oxoniensia* 46: 99–128.

Fowler, Claire J. 2018. *Pharmacopoeia Londinensis and Its Descendants*. London: Royal College of Physicians.

Frank, Robert. 1980. *Harvey and the Oxford Physiologists: A Study of Scientific Ideas*. Berkeley: University of California Press.

French, Peter. 1972. *John Dee: The World of an Elizabethan Magnus*. London: Routledge.

French, Roger. 1994. *William Harvey's Natural Philosophy*. Cambridge: Cambridge University Press.

French, Roger and Andrew Cunningham. 1996. *Before Science: The Invention of the Friars' Natural Philosophy*. Aldershot: Ashgate Press.

Frye, Susan and Karen Robertson (eds.). 1999. *Maids and Mistresses, Cousins and Queens: Women's Alliances in Early Modern England*. Oxford: Oxford University Press.

Furdell, Elizabeth L. 2001. *The Royal Doctors, 1485–1714: Medical Personnel at the Tudor and Stuart Courts*. Rochester: University of Rochester Press.

Gilman, Ernest. 1999. The Arts of Sympathy: Dr. Harvey, Sir Kenelm Digby, and the Arundel Circle. *Opening the Borders: Inclusivity in Early Modern Studies*, ed. Peter Herman, 265–297. Newark: Associated University Press.

Godfrey, Walter H. 1913. *Survey of London: Chelsea, Pt II*, Vol. 4. London.

Goldberg, Benjamin. 2017. Epigenesis and the Rationality of Nature in William Harvey and Margaret Cavendish. *History and Philosophy of the Life Sciences* 39.8: 1–23.

Goldwater, Leonard J. 1935. A Short History of Iron Therapy. *Annals of Medical History* 7.3: 261–267.

Gordon-Roth, Jessica and Nancy Kendrick. 2019. Recovering Early Modern Women Writers: Some Tensions. *Metaphilosophy* 50.3: 268–285.

Göttler, Christine, Bart A.M. Ramakers, and Joanna Woodall (eds.). 2014. *Trading Values in Early Modern Antwerp*. Leiden: Brill.

Gowland, Angus. 2006. *The Worlds of Renaissance Melancholy: Robert Burton in Context*. Cambridge: Cambridge University Press.

Grafton, Anthony and William R. Newman. 2001. *Secrets of Nature: Astrology and Alchemy in Early Modern Europe*. Cambridge, MA: MIT Press.

Graham, Elspeth. 2016. 'An After-Game of Reputation': Systems of Representation, William Cavendish and the Battle of Marston Moor. *Authority, Authorship and Aristocratic Identity*, eds. Peter Edwards and Elspeth Graham, 83–110. Leiden: Brill.

Greenberg, Janelle. 2004. Atkyns, Sir Robert (bap. 1621, d. 1710). *Oxford Dictionary of National Biography*.

Hallock, William and Herbert T. Wade. 1906. *Outlines of the Evolution of Weights and Measures and the Metric System*. London: Macmillan.

Hareven, Tamara. 1991. The History of the Family and the Complexity of Social Change. *The American Historical Review* 96.1: 95–124.

Harkness, Deborah. 2007. *The Jewel House: Elizabethan London and the Scientific Revolution*. New Haven: Yale University Press.

Harkness, Deborah. 2008. A View from the Streets: Women and Medical Work in Elizabethan London. *Bulletin of the History of Medicine* 82.1: 52–85.

Harlow, Vincent T. (ed.). 1932. *Ralegh's Last Voyage*. London: Argonaut Press.

Harris, Frances. 1997. Living in the Neighbourhood of Science: Mary Evelyn, Margaret Cavendish and the Greshamites. *Women, Science and Medicine 1500–1700*, eds. Lynette Hunter and Sarah Hutton, 198–217. Thrupp: Sutton Publishing.

Härting, Ursula. 2002. Lord William Cavendish und Duchess Margaret Cavendish im Rubensgarten in Antwerpen. *Jahrbuch der Berliner Museen* 44: 15–28.

Haycock, David Boyd and Patrick Wallis. 2005. Quackery and Commerce in Seventeenth-Century London: The Proprietary Medicine Business of Anthony Daffy. *Medical History, Supplement* 25: 1–216.

Hebbert, Francis J. 1950. Some Historical Aspects of Iron Therapy. *Glasgow Medical Journal* 31.11: 385–396.

Hedesan, Georgiana. 2016. *An Alchemical Quest for Universal Knowledge: The 'Christian Philosophy' of Jan Baptist van Helmont (1579–1644)*. London: Routledge.

Hedrick, Elizabeth. 2008. Romancing the Salve: Sir Kenelm Digby and the Powder of Sympathy. *British Journal for the History of Science* 41.2: 161–185.

Heilbron, John L. 2012. *Galileo*. Oxford: Oxford University Press.

Hendriksen, Marieke. 2018. The Repudiation and Persistence of Lapidary Medicine in Eighteenth-Century Dutch Medicine and Pharmacy. *Gems in the Early Modern World: Materials, Knowledge & Global Trade, 1450–1800*, eds. Michael Bycroft and Sven Dupré, 197–220. Basingstoke: Palgrave Macmillan.

te Hennepe, Mieneke. 2012. Of the Fisherman's Net and Skin Pores: Reframing Conceptions of the Skin in Medicine 1572–1714. *Blood, Sweat and Tears: The Changing Concepts of Physiology from Antiquity into Early Modern Europe*, eds. Herman F.J. Horstmanshoff, Helen King, and Claus Zittle, 523–548. Leiden: Brill.

Henry, John. 2009. Sir Kenelm Digby, Recusant Philosopher. *Insiders and Outsiders in Seventeenth-Century Philosophy*, eds. Graham A.J. Rogers, Tom Sorell, and Jill Kraye, 43–75. London: Routledge.

Hill, Christopher. 1972. *The World Turned Upside Down: Radical Ideas During the English Revolution*. London: Temple Smith.

Hill, Christopher. 1974. The Medical Profession and Its Radical Critics. *Change and Continuity in Seventeenth-Century England*, ed. Christopher Hill, 157–178. London: Weidenfeld & Nicolson.

Himmelman, Kenneth P. 1997. The Medicinal Body: An Analysis of Medicinal Cannibalism in Europe, 1300–1700. *Dialectical Anthropology* 22.2: 183–203.

Hirai, Hiro. 2005. *Le concept de semence dans les théories de la matière à la Renaissance: De Marsile Ficin à Pierre Gassendi*. Turnhout: Brepols Publishers.

Holland, Bart K. 2000. Treatments of the Bubonic Plague: Reports from Seventeenth Century British Epidemics. *Journal of the Royal Society of Medicine* 93.6: 322–324.

Holt, James C. 2008. The Ancient Constitution of Medieval England. *The Roots of Liberty: Magna Carta, Ancient Constitution and the Anglo-American Tradition of the Rule of Law*, ed. Ellis Sandoz. Indianapolis: Liberty Fund.

Hopper, Andrew J. 2008. Glemham, Sir Thomas (1595–1649). *Oxford Dictionary of National Biography*.

Hosford, David H. 1976. *Nottingham, Nobles, and the North: Aspects of the Revolution of 1688*. Hamden: Archon Books.

Howarth, David. 1998. The Patronage and Collecting of Aletheia, Countess of Arundel, 1606–54. *Journal of the History of Collections* 10: 125–137.

Huffman, William H. 1988. *Robert Fludd and the End of the Renaissance*. London: Routledge.

Huffman, William H. (ed.). 2001. *Robert Fludd*. Berkeley: North Atlantic Books.

Huffman, William H. and Robert A. Seelinger. 1978. Robert Fludd's "Declaratio Brevis to James I". *Ambix* 25.2: 69–92.

Hulse, Lynn. 1995. "The King's Entertainment" by the Duke of Newcastle. *Viator* 26: 355–405.

Hunter, Lynette. 1997. Women and Domestic Medicine: Lady Experimenters, 1570–1620. *Women, Science and Medicine 1500–1700*, eds. Lynette Hunter and Sarah Hutton, 89–107. Thrupp: Sutton Publishing.

Hunter, Lynette. 1999. English Cookery Books of the Period 1500–1700. *The Oxford Companion to Food*, eds. Alan Davidson, Soun Vannithone, and Tom Jaine, 276–277. Oxford: Oxford University Press.

Hunter, Lynette. 2005. Women and Science in the Sixteenth and Seventeenth Centuries—Different Social Practices, Different Textualities, and Different Kinds of Science. *Men, Women, and the Birthing of Modern Science*, ed. Judith P. Zinsser, 123–140. DeKalb: Northern Illinois University Press.

Hunter, Michael. 1989. *Science and the Shape of Orthodoxy: Intellectual Change in Late Seventeenth-Century Britain*. Woodbridge: Boydell Press.

Hunter, Michael. 2007. *Editing Early Modern Texts: An Introduction to Principles and Practice*. Basingstoke: Palgrave Macmillan.

Hutton, Sarah. 1997. Anne Conway, Margaret Cavendish and Seventeenth-Century Scientific Thought. *Women, Science and Medicine 1500–1700*, eds. Lynette Hunter and Sarah Hutton, 218–234. Thrupp: Sutton Publishing.

Huxley, Gervas. 1959. *Endymion Porter: The Life of a Courtier, 1587–1649*. London: Chatto & Windus.

James, Susan. 1999. The Philosophical Innovations of Margaret Cavendish. *British Journal for the History of Philosophy* 7.2: 219–244.

Jenner, Mark and Patrick Wallis. 2007. The Medical Marketplace. *Medicine and the Market in England and Its Colonies, c. 1450–c. 1850*, eds. Mark Jenner and Patrick Wallis, 1–23. Basingstoke: Palgrave Macmillan.

Jones, Philip E. (ed.). 1966. *The Fire Court: Calendar to the Judgments and Decrees of the Court of Judicature Appointed to Determine Differences Between Landlords and Tenants as to Rebuilding After the Great Fire*, 2 Vols. London: W. Clowes & Sons Ltd.

Josten, Conrad H. 1964. Robert Fludd's Theory of Geomancy and His Experiences at Avignon in the Winter of 1601 to 1602. *Journal of the Warburg and Courtauld Institutes* 27: 327–335.

Kahn, Didier. 2007. *Alchimie et Paracelsisme en France à la fin de la Renaissance*. Geneva: Librairie Droz.

Kassell, Lauren. 2007. Magic, Alchemy and the Medical Economy in Early Modern England: The Case of Robert Fludd's Magnetical Medicine. *Medicine and the Market in England and Its Colonies, c. 1450–c. 1850*, eds. Mark Jenner and Patrick Wallis, 88–107. Basingstoke: Palgrave Macmillan.

Katritzky, Margaret A. 2007. *Women, Medicine and Theatre 1500-1750: Literary Mountebanks and Performing Quacks*. London: Routledge.

Keynes, Geoffrey. 1966. *The Life of William Harvey*. Oxford: Clarendon Press.

King, Helen. 2004. *The Disease of Virgins: Green Sickness, Chlorosis, and the Problems of Puberty*. New York: Routledge.

Klestinec, Cynthia. 2004. A History of Anatomy Theaters in Sixteenth-Century Padua. *Journal of the History of Medicine and Allied Sciences* 59.3: 375–412.

Knighton, Charles S. 2013. Mennes, Sir John (1599–1671). *Oxford Dictionary of National Biography*.

Knoppers, Laura L. 2007. Opening the Queen's Closet: Henrietta Maria, Elizabeth Cromwell, and the Politics of Cookery. *Renaissance Quarterly* 60.2: 464–499.

Knoppers, Laura L. 2019. Gender, Knowledge, and the Medical Marketplace: The Case of Margaret Cavendish. *The Routledge Research Companion to Women, Sex, and Gender in the Early British Colonial World*, eds. Kimberly A. Coles and Eve Keller, 270–285. New York: Routledge.

Knowles, James. 2008. Conway, Edward, second Viscount Conway and second Viscount Killultagh (bap. 1594, d. 1655), Politician and Book Collector. *Oxford Dictionary of National Biography*.

Kraye, Jill. 1988. Moral Philosophy. *The Cambridge History of Renaissance Philosophy*, eds. Charles Schmitt, Quentin Skinner, Eckhard Kessler, and Jill Kraye, 303–386. Cambridge: Cambridge University Press.

Landon, Michael. 1970. *The Triumph of the Lawyers: Their Role in English Politics, 1678–1689*. University: University of Alabama Press.

Laroche, Rebecca. 2016. *Medical Authority and Englishwomen's Herbal Texts, 1550–1650*. London: Routledge.

Latham, Robert. 1995. *Samuel Pepys and the Second Dutch War: Pepys's Navy White Book and Brooke House Papers*. Aldershot: Scholar Press for the Navy Records Society.

Lee, Arthur S.G. 1964. *The Son of Leicester: The Story of Robert Dudley, Titular Earl of Warwick, Earl of Leicester, and Duke of Northumberland*. London: Gollancz.

Leigh, Robert. 2015. *On Theriac to Piso, Attributed to Galen: A Critical Edition with Translation and Commentary*. Leiden: Brill.

Lennox, James G. 2006. The Comparative Study of Animal Development: William Harvey's Aristotelianism. *The Problem of Animal Generation Early Modern Philosophy*, ed. Justin E.H. Smith, 21–46. Cambridge: Cambridge University Press.

Leong, Elaine. 2014. "Herbals she peruseth": Reading Medicine in Early Modern England. *Renaissance Studies* 28.4: 556–578.

Leong, Elaine. 2018. *Recipes and Everyday Knowledge: Medicine, Science, and the Household in Early Modern England*. Chicago: University of Chicago Press.

Leong, Elaine and Sara Pennell. 2007. Recipe Collections and the Currency of Medical Knowledge in the Early Modern 'Medical Marketplace'. *Medicine and the Market in England and Its Colonies, c. 1450–c. 1850*, eds. Mark Jenner and Patrick Wallis, 133–152. Basingstoke: Palgrave Macmillan.

Levitin, Dmitri. 2015. *Ancient Wisdom in the Age of the New Science: Histories of Philosophy in England, c. 1640–1700*. Cambridge: Cambridge University Press.

Levitin, Dmitri. 2015. "Made Up from Many Experimentall Notions": The Society of Apothecaries, Medical Humanism, and the Rhetoric of Experience in 1630s London. *Journal of the History of Medicine and Allied Sciences* 70.4: 549–587.

Lobis, Seth. 2011. Sir Kenelm Digby and the Powder of Sympathy. *Huntington Library Quarterly* 74.2: 243–260.

MacInnes, Ian. 2012. The Political Worm: Invertebrate Life in the Early Modern English Body. *The Indistinct Human in Renaissance Literature*, eds. Jean E. Feerick and Vin Nardizzi, 253–273. Basingstoke: Palgrave Macmillan.

Maclean, Ian. 1990. *The Renaissance Notion of Woman: A Study in the Fortunes of Scholasticism and Medical Science in European Intellectual Life*. Cambridge: Cambridge University Press.

Maclean, Ian. 2002. *Logic, Signs and Nature in the Renaissance: The Case of Learned Medicine*. Cambridge: Cambridge University Press.

Maclean, Ian. 2008. Fludd, Robert (bap. 1574, d. 1637). *Oxford Dictionary of National Biography*.

Maclean, Ian. 2008. The Medical Republic of Letters Before the Thirty Years War. *Intellectual History Review* 18.1: 15–30.

Malcolm, Noel (ed.). 1994. *The Clarendon Edition of the Works of Thomas Hobbes, The Correspondence: 1622–1659*, Vol. I. Oxford: Clarendon Press.

Malcolm, Noel and Jacqueline Stedall. 2005. *John Pell (1611–1685) and His Correspondence with Sir Charles Cavendish: The Mental World of an Early Modern Mathematician*. Oxford: Oxford University Press.

Malcolmson, Cristina. 2010. William Herbert's Gardener: Adrian Gilbert. *George Herbert's Pastoral*, ed. Christopher Hodgkins, 113–133. Newark: University of Delaware Press.

Matthews, Leslie G. 1974. London's Immigrant Apothecaries, 1600–1800. *Medical History* 18.3: 262–274.

McCallum, R. Ian. 1999. *Antimony in Medical History: An Account of the Medical Uses of Antimony and Its Compounds Since Early Times to the Present*. Edinburgh: Pentland Press.

McKenzie, Donald F. 2002. The London Book Trade in 1644. *Making Meaning: "Printers of the Mind" and Other Essays*, eds. Peter D. McDonald and Michael F. Suarez, 126–143. Amherst: University of Massachusetts Press.

McVaugh, Michael. 2007. Richard Wiseman and the Medical Practitioners of Restoration London. *Journal of the History of Medicine and Allied Sciences* 62.2: 125–140.

Merton, Robert K. 1938. Science, Technology and Society in Seventeenth Century England. *Osiris* 4: 360–632.

Mikkeli, Heikki. 1999. *Hygiene in the Early Modern Tradition*. Helsinki: Academia Scientiarum Fennica.

Millen, Ron. 2002. The Manifestation of Occult Qualities in the Scientific Revolution. *Religion, Science, and Worldview: Essays in Honour of Richard S. Westfall*, eds. Margaret J. Osler and Paul L. Farber, 185–216. Cambridge: Cambridge University Press.

Miller, David M. 2018. Regressus and Empiricism in the Controversy About Galileo's Lunar Observations. *Perspectives on Science* 26.3: 293–324.

Mills, Jerry L. 1985. Recent Studies in Ralegh. *English Literary Renaissance* 15.2: 225–244.

Mintz, Samuel I. 1952. The Duchess of Newcastle's Visit to the Royal Society. *The Journal of English and Germanic Philology* 51.2: 168–176.

Moran, Bruce. 1991. *Chemical Pharmacy Enters the University: Johannes Hartmann and the Didactic Care of Chymiatria in the Early Seventeenth Century*. Madison: American Institute of the History of Pharmacy.

Moran, Bruce. 2019. *Paracelsus: An Alchemical Life*. London: Reaktion Books.

Morpurgo, Edgar. 1927. English Physicians—"Doctorati"—at the University of Padua in the "Collegio Veneto Artista" (1617–1771). *Proceedings of the Royal Society of Medicine* 90.2: 1369–1380.

Moshenska, Joe. 2016. *A Stain in the Blood: The Remarkable Voyage of Sir Kenelm Digby, Pirate and Poet, Courtier and Cook, King's Servant and Traitor's Son*. London: William Heinemann.

Munro, John. 1988. *The Consumption of Spices and Their Costs in Late-Medieval and Early-Modern Europe: Luxuries or Necessities?* www.economics.utoronto.ca/munro5/SPICES1.htm.

Nance, Brian. 2001. *Turquet de Mayerne as Baroque Physician: The Art of Medical Portraiture*. Amsterdam: Rodopi.

Nance, Brian. 2004. Cademan, Sir Thomas (c. 1590–1651). *Oxford Dictionary of National Biography*.

Nash, Richard. 2017. William Cavendish: Riding School and Race Track. *Authority, Authorship and Aristocratic Identity in Seventeenth-Century England*, eds. Peter Edwards and Elspeth Graham, 317–330. Leiden: Brill.

Newman, Peter R. 1981. *Royalist Officers in England and Wales, 1642–1660: A Biographical Dictionary*. New York: Garland.

Newman, William R. 1978. Newton's Clavis as Starkey's Key. *Isis* 78.4: 564–574.

Newman, William R. 1994. *Gehennical Fire: The Lives of George Starkey, an American Alchemist in the Scientific Revolution*. Cambridge, MA: Harvard University Press.

Newman, William R. 1995. The Philosophers' Egg: Theory and Practice in the Alchemy of Roger Bacon. *Micrologus* 3: 75–101.

Newman, William R. 2001. *Promethean Ambitions: Alchemy and the Quest to Perfect Nature*. Chicago: University of Chicago Press.

Newman, William R. 2002. George Starkey and the Selling of Secrets. *Samuel Hartlib and the Universal Reformation: Studies in Intellectual Communication*, eds. Mark Greengrass, Michael Leslie, and Timothy Raylor, 193–210. Cambridge: Cambridge University Press.

Newman, William R. 2006. *Atoms and Alchemy: Chymistry and the Experimental Origins of the Scientific Revolution*. Chicago: University of Chicago Press.

Newman, William R. and Lawrence M. Principe. 2005. *Alchemy Tried in the Fire: Starkey, Boyle, and the Fate of Helmontian Chymistry*. Chicago: University of Chicago Press.

Nicholls, Mark and Penry Williams. 2015. Ralegh, Sir Walter (1554–1618). *Oxford Dictionary of National Biography*.

Noble, Louise. 2011. *Medicinal Cannibalism in Early Modern English Literature and Culture*. New York: Palgrave Macmillan.

Novarr, David. 1951. Izaak Walton, Bishop Morley, and Love and Truth. *The Review of English Studies* 2.5: 30–39.

O'Neill, Eileen. 1997. Disappearing Ink: Early Modern Women Philosophers and Their Fate in History. *Philosophy in a Feminist Voice: Critiques and Reconstructions*, ed. Janet A. Kourany, 17–62. Princeton: Princeton University Press.

Pagel, Walter. 1958. *Paracelsus: An Introduction to Philosophical Medicine in the Era of the Renaissance*. Basel: S. Karger.

Pagel, Walter. 1982. *Joan Baptista van Helmont: Reformer of Science and Medicine*. Cambridge: Cambridge University Press.

Pagel, Walter. 1984. *Smiling Spleen: Paracelsianism in Storm and Stress*. London: Karger Publishers.

Palmer, Richard. 1991. Health, Hygiene and Longevity in Medieval and Renaissance Europe. *History of Hygiene*, eds. Yosio Kawakita, Shizu Sakai, and Yasuo Otsuka, 75–98. Tokyo: Ishiyaku.

Parsons, Thomas S. 1986. *The Vertebrate Body*, Vol. 7. Philadelphia: Saunders College Publishing.

Pelling, Margaret. 1997. Thoroughly Resented? Older Women and the Medical Role in Early Modern London. *Women, Science and Medicine 1500–1700*, eds. Lynette Hunter and Sarah Hutton, 63–88. Thrupp: Sutton Publishing.

Pelling, Margaret. 1998. *The Common Lot: Sickness, Medical Occupations and the Urban Poor in Early Modern England*. London: Longman.

Pelling, Margaret and Frances White. 2003. *Medical Conflicts in Early Modern London: Patronage, Physicians, and Irregular Practitioners, 1550–1640*. Oxford: Oxford University Press.

Pennell, Sara. 2016. *The Birth of the English Kitchen, 1600–1850*. London: Bloomsbury Academic.

Peña, Miriam Criado. 2020. Punctuation Practice in Early Modern English Scientific Writing: The Case of MS 3009 at the Wellcome Library, London. *Miscelánea: A Journal of English and American Studies* 61: 81–97.

Pérez-Ramos, Antonio. 1988. *Francis Bacon's Idea of Science and the Maker's Knowledge Tradition*. Oxford: Oxford University Press.

Peterman, Alison. 2018. Canonizing Cavendish. *HOPOS: The Journal of the International Society for the History of Philosophy of Science* 8.1: 191–197.

Pomata, Gianna. 2018. Innate Heat, Radical Moisture and Generation. *Reproduction: Antiquity to the Present Day*, eds. Nick Hopwood, Rebecca Flemming, and Lauren Kassell, 195–208. Cambridge: Cambridge University Press.

Poole, William. 2015. Margaret Cavendish's Books in New College, and Around Oxford. *New College Notes* 6: 1–7.

Popper, Nicholas. 2012. *Walter Ralegh's "History of the World" and the Historical Culture of the Late Renaissance*. Chicago: University of Chicago Press.

Porter, Roy. 1995. *Disease, Medicine, and Society in England, 1550–1860*. Cambridge: Cambridge University Press.

Principe, Lawrence. 1995. Virtuous Romance and Romantic Virtuoso: The Shaping of Robert Boyle's Literary Style. *Journal of the History of Ideas* 56.3: 377–397.

Principe, Lawrence. 2004. Davisson [D'Avissone], William (c. 1593–1669). *Oxford Dictionary of National Biography*.

Rabe, Jennifer. 2016. Mediating Between Art and Nature: The Countess of Arundel at Tart Hall. *Sites of Mediation: Connected Histories of Places, Processes, and Objects in Europe and Beyond, 1450–1650*, eds. Susanna Burghartz, Lucas Burkart, and Christine Göttler, 183–210. Leiden: Brill.

Radden, Jennifer. 2000. *The Nature of Melancholy: From Aristotle to Kristeva*. Oxford: Oxford University Press.

Ragland, Evan R. 2017. "Making Trials" in Sixteenth- and Early Seventeenth-Century European Academic Medicine. *Isis* 108.3: 503–528.

Randall, John H. 1940. The Development of Scientific Method in the School of Padua. *Journal of the History of Ideas* 1.2: 177–206.

Rankin, Alisha M. 2013. *Panaceia's Daughters: Noblewomen as Healers in Early Modern Germany*. Chicago: University of Chicago Press.

Rattansi, Piyo M. 1963. Paracelsus and the Puritan Revolution. *Ambix* 11.1: 24–32.

Rattansi, Piyo M. 1964. The Helmontian-Galenist Controversy in Restoration England. *Ambix* 12.1: 1–23.

Ray, Meredith. 2015. *Daughters of Alchemy: Women and Scientific Culture in Early Modern Italy*. Cambridge, MA: Harvard University Press.

Raylor, Timothy. 1994. *Cavaliers, Clubs, and Literary Culture: Sir John Mennes, James Smith, and the Order of the Fancy*. Newark: University of Delaware Press.

Raylor, Timothy. 2000. Newcastle's Ghosts: Robert Payne, Ben Jonson, and the "Cavendish Circle". *Literary Circles and Cultural Communities in Renaissance England*, eds. Claude J. Summers and Ted-Larry Pebworth, 92–114. Columbia: University of Missouri Press.

Raylor, Timothy. 2001. Hobbes, Payne, and *A Short Tract on First Principles*. *The Historical Journal* 44.1: 29–58.

Raymond, Joad. 2006. *Pamphlets and Pamphleteering in Early Modern Britain*. Cambridge: Cambridge University Press.

Read, John. 1961. The First British Professor of Chemistry. *Ambix* 9.2: 70–101.

Read, John. 1995. *From Alchemy to Chemistry*. New York: Dover Publications.

Reresby, John. 1936. *Memoirs of Sir John Reresby*, ed. Andrew Browning. Glasgow: Jackson.

Rijks, Marlise. 2019. 'Unusual Excrescences of Nature': Collected Coral and the Study of Petrified Luxury in Early Modern Antwerp. *Dutch Crossing: Journal of Low Country Studies* 43.2: 127–156.

Roling, Bernd. 2014. Der Wal als Schauobjekt: Thomas Bartholin (1616–1680), die dänische Nation und das Ende der Einhörner. *Zoology in Early Modern Culture: Intersections of Science, Theology, Philology, and Political and Religious Education*, eds. Karl A.E. Enenkel and Paul J. Smith, 172–196. Leiden: Brill.

Roos, Anna Marie. 2000. Luminaries in Medicine: Richard Mead, James Gibbs, and Solar and Lunar Effects on the Human Body in Early Modern England. *Bulletin of the History of Medicine* 74.3: 433–457.

Roos, Anna Marie. 2007. *The Salt of the Earth: Natural Philosophy, Medicine, and Chymistry in England, 1650–1750*. Leiden: Brill.

Rowe, Nick. 1994. 'My best patron': William Cavendish and Jonson's Caroline Drama. *The Seventeenth Century* 9.2: 197–212.

Rowse, Alfred L. 1962. *Ralegh and the Throckmortons*. London: Palgrave Macmillan.

Roy, Ian. 1968. The Libraries of Edward, 2nd Viscount Conway, and Others: An Inventory and Valuation of 1643. *Historical Research* 41.103: 35–46.

Sarasohn, Lisa. 1984. A Science Turned Upside Down: Feminism and the Natural Philosophy of Margaret Cavendish. *The Huntington Library Quarterly* 47.4: 289–307.

Sarasohn, Lisa. 2010. *The Natural Philosophy of Margaret Cavendish: Reason and Fancy During the Scientific Revolution*. Baltimore: Johns Hopkins University Press.

Scarborough, John. 2011. Ancient Medicinal Use of Aristolochia: Birthwort's Tradition and Toxicity. *Pharmacy in History* 53.1: 3–21.

Schiebinger, Londa. 1991. Margaret Cavendish, Duchess of Newcastle. *A History of Women Philosophers: Modern Women Philosophers, 1600–1900*, ed. Mary Ellen Waithe, 1–20. Boston: Kluwer Academic.

Schmitt, Charles B. 1969. Experience and Experiment: A Comparison of Zabarella's View with Galileo's in *De motu*. *Studies in the Renaissance* 16: 80–138.

Schütte, Josef F. 1969. Japanese Cartography at the Court of Florence: Robert Dudley's Maps of Japan, 1606–1636. *Imago Mundi* 23.1: 29–58.

Seddon, Peter R. 2004. Cavendish, Henry, Second Duke of Newcastle upon Tyne (1630–1691). *Oxford Dictionary of National Biography*.

Seddon, Peter R. 2008. Stanhope, Philip, First Earl of Chesterfield (1583/4–1656). *Oxford Dictionary of National Biography*.

Seltman, Muriel and Robert Goulding. 2007. *Thomas Harriot's Artis Analyticae Praxis*. Dordrecht: Springer.

Semler, Liam. 2011. The Magnetic Attraction of Margaret Cavendish and Walter Charleton. *Early Modern Englishwomen Testing Ideas*, eds. Jo Wallwork and Paul Salzman, 55–72. Farnham: Ashgate.

Serjeantson, Richard. 1999. Testimony and Proof in Early-Modern England. *Studies in History and Philosophy of Science* 30.2: 195–236.

Serjeantson, Richard. 2001. The Passions and Animal Language, 1540–1700. *Journal of the History of Ideas* 62.3: 425–444.

Shaw, William A. 1911. *Letters of Denization and Acts of Naturalization for Aliens in England and Ireland, 1603–1800*. London: Huguenot Society of London.

Shirley, John W. 1951. An Early Experimental Determination of Snell's Law. *American Journal of Physics* 19.9: 507–508.

Shirley, John W. 1983. *Thomas Harriot: A Biography*. Oxford: Clarendon Press.

Siegfried, Brandie. 2007. *Bonum Theatrale*: The Matter of Elizabeth I in Francis Bacon's *Of Tribute* and Margaret Cavendish's *The Blazing World*. *Resurrecting Elizabeth I in Seventeenth-Century England*, eds. Elizabeth Hageman and Katherine Conway, 185–204. Madison: Fairleigh Dickinson University Press.

Siegfried, Brandie and Lisa Sarasohn (eds.). 2014. *God and Nature in the Thought of Margaret Cavendish*. Farnham: Ashgate.

Siraisi, Nancy. 2012. *Communities of Learned Experience: Epistolary Medicine in the Renaissance*. Baltimore: Johns Hopkins University Press.

Skuse, Alanna. 2014. Wombs, Worms and Wolves: Constructing Cancer in Early Modern England. *Social History of Medicine* 27.4: 632–648.

Slack, Paul. 1979. Mirrors of Health and Treasures of Poor Men: The Uses of the Vernacular Medical Literature of Tudor England. *Health, Medicine and Mortality in the Sixteenth Century*, ed. Charles Webster, 237–273. Cambridge: Cambridge University Press.

Smith, Daniel Starza. 2012. "La conquest du sang real": Edward, Second Viscount Conway's Quest for Books. *From Compositors to Collectors: Essays on the Book Trade*, eds. Matthew Day and John Hinks. London: The British Library.

Smith, Daniel Starza. 2014. *John Donne and the Conway Papers: Patronage and Manuscript Circulation in the Early Seventeenth Century*. Oxford: Oxford University Press.

Smith, Hilda L. 2014. Claims to Orthodoxy: How Far Can We Trust Margaret Cavendish's Autobiography? *God and Nature in the Thought of Margaret Cavendish*, eds. Brandie R. Siegfried and Lisa T. Sarasohn, 14–26. London: Routledge.

Smith, Lisa. 2006. The Relative Duties of a Man: Domestic Medicine in England and France, ca. 1685–1740. *Journal of Family History* 31.3: 237–256.

Smith, Pamela H. 2004. *The Body of the Artisan: Art and Experience in the Scientific Revolution*. Chicago: University of Chicago Press.

Smith, Robert W.I. 1932. *English-Speaking Students of Medicine at the University of Leyden*. London: Oliver and Boyd.

Smith, Sean B. et al. 2004. Hieronymous Fabricius Ab Acquapendente (1533–1619). *Clinical Anatomy* 17.7: 540–543.

Snively, Samantha. 2015. "The Quintessence of Wit": Domestic Labor, Science, and Margaret Cavendish's Kitchen Fancies. *Newberry Essays in Medieval and Early Modern Studies* 9: 21–28.

Snook, Edith. 2007. "Soveraigne Receipts" and the Politics of Beauty in *The Queens Closet Opened*. *Early Modern Literary Studies* 15: 1–19.

Spiller, Elizabeth. 2008. *Seventeenth-Century English Recipe Books: Cooking, Physic and Chirurgery in the Works of Elizabeth Talbot Grey and Aletheia Talbot Howard*. Aldershot: Ashgate.

Spiller, Elizabeth. 2008. *Seventeenth-Century English Recipe Books: Cooking, Physic and Chirurgery in the Works of W.M. and Queen Henrietta Maria, and of Mary Tillinghast*. Aldershot: Ashgate.

Spiller, Elizabeth. 2010. Recipes for Knowledge: Maker's Knowledge Traditions, Paracelsian Recipes, and the Invention of the Cookbook, 1600–1660. *Renaissance Food from Rabelais to Shakespeare*, ed. Joan Fitzpatrick, 55–72. Farnham: Ashgate.

Spurr, John. 2004. Morley, George (1598?–1684). *Oxford Dictionary of National Biography*.

Stewart, Alan. 2009. Boswell, Sir William (*d.*1650). *Oxford Dictionary of National Biography.*

Stillman, John M. 1960. *The Story of Alchemy and Early Chemistry*. New York: Dover Publications.

Stobart, Anne. 2016. *Household Medicine in Seventeenth-Century England.* London: Bloomsbury Academic.

Sugg, Richard. 2006. 'Good Physic but Bad Food': Early Modern Attitudes to Medicinal Cannibalism and Its Suppliers. *Social History of Medicine* 19.2: 225–240.

Sugg, Richard. 2011. *Mummies, Cannibals, and Vampires: The History of Corpse Medicine from the Renaissance to the Victorians*. London: Routledge.

Tadmor, Naomi. 1996. The Concept of the Household-Family in Eighteenth-Century England. *Past and Present* 151: 111–140.

Tadmor, Naomi. 2010. Early Modern English Kinship in the Long Run: Reflections on Continuity and Change. *Continuity and Change* 25.1: 15–48.

Tanner, Joseph R. 1903. *A Descriptive Catalogue of the Naval Manuscripts in the Pepysian Library at Magdalene College, Cambridge*, Vol. 1. London.

Tebeaux, Elizabeth. 1997. Women and Technical Writing, 1475–1700. *Women, Science and Medicine, 1500–1700*, eds. Lynette Hunter and Sarah Hutton, 29–62. Thrupp: Sutton Publishing.

Thomas, Keith. 1987. Numeracy in Early Modern England. *Transactions of the Royal Historical Society* 37: 103–132.

Thomson, Andrew. 2019. *Bishop Morley of Winchester 1598–1684: Politician, Benefactor, Pragmatist*. Winchester: Winchester University Press.

Thorndike, Lynn. 1958. *A History of Magic and Experimental Science*. New York: Columbia University Press.

Thulesius, Olav. 1992. *Nicholas Culpeper: English Physician and Astrologer*. New York: St. Martin's Press.

Totelin, Laurence. 2004. Mithradates' Antidote—A Pharmacological Ghost. *Early Science and Medicine* 9.1: 1–19.

Travitsky, Betty. 1999. *Subordination and Authorship in Early Modern England: The Case of Elizabeth Cavendish Egerton and Her "Loose Papers"*. Temple: Arizona Center for Medieval and Renaissance Studies.

Trease, Geoffrey. 1979. *Portrait of a Cavalier: William Cavendish, First Duke of Newcastle*. London: Macmillan.

Trevor-Roper, Hugh. 2004. Mayerne, Sir Théodore Turquet de (1573–1655). *Oxford Dictionary of National Biography.*

Trevor-Roper, Hugh. 2006. *Europe's Physician: The Various Life of Sir Théodore de Mayerne*. New Haven: Yale University Press.

Vander Motten, Jean Pierre and Katrien Daemen-de Gelder. 2014. Margaret Cavendish, the Antwerp Carmel and *The Convent of Pleasure*. *Archiv für das Studium der Neueren Sprachen und Literaturen* 251.1: 134–145.

Venn, John. 2011. *Alumni Cantabrigienses: A Biographical List of All Known Students, Graduates and Holders of Office at the University of Cambridge, From the Earliest Times to 1900*. Cambridge: Cambridge University Press.

Viviani, Alberto. 1961. Sir Robert Dudley, Duke of Northumberland, at the Grand Ducal Court. *Florence* 12.4: 32–33.

Waddell, Mark A. 2003. The Perversion of Nature: Johannes Baptista Van Helmont, the Society of Jesus, and the Magnetic Cure of Wounds. *Canadian Journal of History* 38.2: 180–197.

Wall, Wendy. 2002. *Staging Domesticity: Household Work and English Identity in Early Modern Drama*. Cambridge: Cambridge University Press.

Wall, Wendy. 2015. *Recipes for Thought: Knowledge and Taste in the Early Modern English Kitchen*. Philadelphia: University of Pennsylvania Press.

Wallace, William A. 1988. Randall Redivivus: Galileo and the Paduan Aristotelians. *Journal of the History of Ideas* 49.1: 133–149.

Wallis, Patrick. 2008. Consumption, Retailing, and Medicine in Early-Modern London. *Economic History Review* 61.1: 26–53.

Wallis, Patrick. 2012. Exotic Drugs and English Medicine: England's Drug Trade, c.1550–c.1800. *Social History of Medicine* 25.1: 20–46

Walters, Lisa. 2014. *Margaret Cavendish: Gender, Science and Politics*. Cambridge: Cambridge University Press.

Wear, Andrew. 1992a. Making Sense of Health and Environment in Early Modern England. *Medicine in Society: Historical Essays*, ed. Andrew Wear, 119–148. Cambridge: Cambridge University Press.

Wear, Andrew. 1992b. The Popularization of Medicine in Early Modern England. *The Popularization of Medicine, 1650–1850*, ed. Roy Porter, 17–41. New York: Routledge.

Wear, Andrew. 1996. Religious Beliefs and Medicine in Early Modern England. *The Task of Healing: Medicine, Religion and Gender in England and the Netherlands 1450–1800*, eds. Hilary Marland and Margaret Pelling, 145–171. Rotterdam: Erasmus.

Wear, Andrew. 2000. *Knowledge and Practice in English Medicine, 1500–1680*. Cambridge: Cambridge University Press.

Webster, Charles. 1967. English Medical Reformers of the Puritan Revolution: A Background to the "Society of Chymical Physitians". *Ambix* 14: 16–41.

Webster, Charles. 1975. *The Great Instauration: Science, Medicine, and Reform, 1626–1660*. London: Duckworth.

Werrett, Simon. 2019. *Thrifty Science: Making the Most of Materials in the History of Experiment*. Chicago: University of Chicago Press.

Westfall, Richard S. (compiler). 1995. Hartmann, Johannes. *The Galileo Project*: http://galileo.rice.edu/index.html.

Westman, Robert S. 1984. Nature, Art, and Psyche: Jung, Pauli, and the Kepler-Fludd Polemic. *Occult and Scientific Mentalities in the Renaissance*, ed. Brian Vickers, 177–229. Cambridge: Cambridge University Press.

Weston, Corinne C. and Janelle R. Greenberg. 1981. *Subjects and Sovereigns: The Grand Controversy Over Legal Sovereignty in Stuart England*. Cambridge: Cambridge University Press.

Whitaker, Katie. 2004. *Mad Madge: Margaret Cavendish, Duchess of Newcastle, Royalist, Writer and Romantic*. London: Vintage.

Whittet, Thomas D. 1988. Sir Thomas Clarges, Apothecary and Envoy. *Journal of the Royal Society of Medicine* 81.8: 464–467.

Whittle, Jane and Mark Hailwood. 2020. The Gender Division of Labour in Early Modern England. *Economic History Review* 73.1: 3–32.

Woolley, Benjamin. 2004. *The Herbalist: Nicholas Culpeper and the Fight for Medical Freedom Freedom*. Toronto: Harper Collin.

Worsley, Lucy and Tom Addyman. 2002. Riding Houses and Horses: William Cavendish's Architecture for the Art of Horsemanship. *Architectural History* 45: 194–229.

Worsley, Lucy. 2004. Building a Family: William Cavendish, First Duke of Newcastle, and the Construction of Bolsover and Nottingham Castles. *The Seventeenth Century* 19.2: 233–259.

Worsley, Lucy. 2007. *Cavalier: A Tale of Chivalry, Passion, and Great Houses*. New York: Bloomsbury Publishing.

Zirpolo, Lilian H. 2010. *Historical Dictionary of Baroque Art and Architecture*. Lanham: Rowman & Littlefield.

Zupko, Ronald E. 1977. *British Weights & Measures: A History from Antiquity to the Seventeenth Century*. Madison: University of Wisconsin Press.

Zupko, Ronald E. 1981. *Italian Weights and Measures from the Middle Ages to the Nineteenth Century*. Boston: American Philosophical Society.

Zupko, Ronald E. 1990. Medieval Apothecary Weights and Measures: The Principal Units of England and France. *Pharmacy in History* 32.2: 57–62.

Unpublished Secondary Works

Begley, Justin. 2016. *Margaret Cavendish, The Last Natural Philosopher*. Dissertation: University of Oxford.

Connolly, Elizabeth. 2017. *A Dynamic Equilibrium: Doctors and Patients in Seventeenth-Century England*. Dissertation: University of Adelaide.

Grissom, Julie. 2014. *Parasitic Worms in Early Modern Science and Medicine, 1650–1810*. Dissertation: University of Oklahoma.

Stine, Jennifer K. 1996. *Opening Closets: The Discovery of Household Medicine in Early Modern England*. Dissertation: Stanford University.

Index

J. Begley and B. Goldberg, *The Medical World of Margaret Cavendish*, Palgrave Studies in Medieval and Early Modern Medicine,
https://doi.org/10.1007/978-3-030-92927-5

D

GPSR Compliance
The European Union's (EU) General Product Safety Regulation (GPSR) is a set of rules that requires consumer products to be safe and our obligations to ensure this.

If you have any concerns about our products, you can contact us on

ProductSafety@springernature.com

In case Publisher is established outside the EU, the EU authorized representative is:

Springer Nature Customer Service Center GmbH
Europaplatz 3
69115 Heidelberg, Germany

www.ingramcontent.com/pod-product-compliance
Ingram Content Group UK Ltd.
Pitfield, Milton Keynes, MK11 3LW, UK
UKHW022220051026
15624UKWH00002B/266
9783030929268